Land Drainage

Land Drainage

Planning and design of agricultural drainage systems

Lambert K. Smedema
PhD, MSc

David W. Rycroft
PhD, Dip.App.Sc, ARCATS, CEng, MICE

Cornell University Press
Ithaca, New York

For information address Cornell University Press,
124 Roberts Place, Ithaca, New York 14850.

First published 1983 by Cornell University Press

International Standard Book Number 0-8014-1629-9
Library of Congress Catalogue Card Number 83-45150

Printed in Great Britain

Contents

Preface

With the advances in soil physics and in hydrology during the last 20 to 25 years, knowledge and understanding of the causes, nature, and harmful effects of excess water in agricultural land, as well as of the remedial measures to be taken, have increased greatly. Land drainage has now attained a firm scientific basis. A large number of publications is evidence of this progress. Special mention should be made of the American Society of Agronomy monographs *Drainage of Agricultural Lands* (1957) and *Drainage for Agriculture* (1974), the publications of the International Institute of Land Reclamation and Improvement at Wageningen *Drainage Principles and Applications* (4 volumes 1972–74) and *Proceedings of the International Drainage Workshop* (1979), and the FAO Irrigation and Drainage Papers *Drainage Testing* (1976) and *Drainage Design Factors* (1981). These are high standard compilations to which leading drainage researchers and engineers from many countries contributed.

This book has been prepared as a modern textbook for a university level Land Drainage course and as such, it should, in principle, be read in chapter order. It is especially suitable for use in a one- to two-years Land Drainage course within either a BSc or an MSc programme for Agricultural or Civil Engineers specialising in 'soil and water' or 'irrigation engineering'.

Land drainage is treated as being a field of applied soil physics and applied hydrology. The approach is always thoroughly analytical, directed at giving students a good understanding of what is physically (or chemically) going on in the soil and on the land during different stages of the various drainage processes. This understanding will help students to appreciate the many factors and interrelationships to be taken into account in the diagnosis and solution of drainage problems, as well as to be aware of the reliability and limitations of existing theories and formulae. Universal relationships between the main design variables and soil, climatological and other relevant environmental conditions are stressed which gives the text a broad geographical scope.

The book is divided into six parts:

- I Basic soil and water; introduction to agricultural drainage
- II Drainage systems: description of components and functions, construction aspects, operation and maintenance
- III Planning and design of drainage systems: design variables, criteria and procedures
- IV Drainage for soil salinity control
- V Miscellaneous drainage problems (seepage; reclamations and drainage of unripened soils)
- VI Investigations (surveys; economic evaluation).

Classic drainage topics are covered as well as some of the less common. Considerable attention has been paid to the drainage of heavy land and to drainage for salinity control, both in the forefront of current interest. The entire drainage path from field to outlet is linked by a thorough discussion of the philosophy behind the 'design discharge', using both steady and recently developed non-steady approaches. Up-to-date factual information on the important technological aspects of drainage (drainage materials, machinery, construction, maintenance, etc) is also presented.

Much of the current knowledge of drainage has been accumulated steadily over many years and the authors have borrowed heavily, and built on, what others have developed. It may appear that justice has not been given to all who deserve it but references have been selected mainly with the needs of the students in mind.

The completion of this book was made possible largely by the faculty positions currently held by the authors at the Departments of Civil Engineering at Delft and Southampton University, and by the cooperation of staff colleagues at these institutions. Much of the authors knowledge and experience on drainage, however, derives from their earlier work with Euroconsult/Ilaco consultants (Smedema) and with the Field Drainage Experimental Unit of the UK Ministry of Agriculture and Sir M MacDonald and partners (Rycroft). Special acknowledgement is due to Messrs J H Boumans, J W van Hoorn and B D Trafford, all highly regarded drainage specialists, who read and commented on the manuscript. Thanks are especially due to a former student, Mr H van Drunen, who weeded out many inconsistencies and errors.

This book arose from a need for such a text and has been based on lecture notes prepared by Lambert Smedema for a Land Drainage course at the University of Nairobi. Under joint authorship, these notes have been completely revised and expanded. The authors gratefully acknowledge the pleasant coaching on publishing matters received from Thelma M Nye, senior editor at Batsfords.

Drainage is still largely confined to the developed countries, but undoubtedly has an important role to play in increasing the worlds agricultural production. It is particularly hoped that this book will help to promote the teaching and application of drainage to the rest of the world.

Arnhem/Winchester 1983 LKS/DWR

PART I
INTRODUCTION

Chapter 1 Water in the soil

The design of an agricultural drainage system requires a good understanding of the occurrence, nature and movement of water in the soil as well as of the drainage related hydrological processes. These subjects are discussed briefly in this chapter.

1.1 Hydrological cycle

Figure 1.1 illustrates the main components of the hydrological cycle insofar as they are related to drainage.

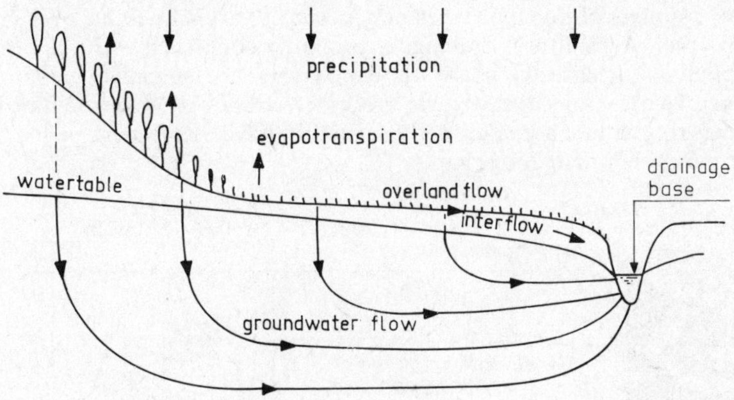

FIGURE 1.1 *Main components of the hydrological cycle*

The precipitation reaching the soil surface will partly enter the soil (*infiltration*) where it may be retained in the upper layers (*soil moisture storage*) or percolate through to the deeper layers (*deep percolation*). The deep percolation eventually reaching the groundwater is termed the *groundwater recharge*.

The rainfall which initially fails to infiltrate remains ponded on the land, mainly filling up the various (micro) depressions. Once the storage potential on the soil surface (*surface retention*) has been occupied, the water will start to move down the slope as *overland flow*. This water will eventually collect in the (natural or man-made) field drains and then discharge through the main drains to the outlet. The water in transit in overland flow and in the field drains constitutes the *surface detention* but once this water reaches the main drainage system it is referred to as *surface runoff*.

The groundwater recharge will cause the watertable to rise. When the watertable rises above the local *groundwater drainage base*, as formed by the waterlevel in the local drainage system, a hydraulic gradient is established which causes groundwater flow to this system. Part of the water that has infiltrated may also find its way towards the drains as lateral flow above the watertable (*inter-flow*), especially in the case of impeded percolation.

Transpiration by the vegetation and evaporation directly from the soil surface (jointly referred to as *evapotranspiration*) cause a water loss from the soil into the atmosphere. The evapotranspiration and continued deep percolation deplete the soil moisture storage; rainfall replenishes this storage.

1.2 Forms and nature of occurrence of water in the soil

Figure 1.2 illustrates the different forms of occurrence of water in the soil.

Groundwater

Groundwater fills the pores of the permeable strata in the earth's upper crust (*aquifers*). The groundwater in these strata may be free water under normal static or dynamic pressure or it may be subjected to an over-pressure. The latter may occur when an aquifer is overlain by a poorly permeable layer. Such aquifers are termed *confined aquifers*. Aquifers of different types may occur at varying depths, over or underlying each other. Agricultural drainage is primarily concerned with the free groundwater found in or directly below the soil layers, the so-called *phreatic groundwater*. Its surface is termed the *phreatic level* or *watertable*. However, confined groundwater occurring at much greater depths can also have a considerable local influence on agricultural drainage (see chapter 13).

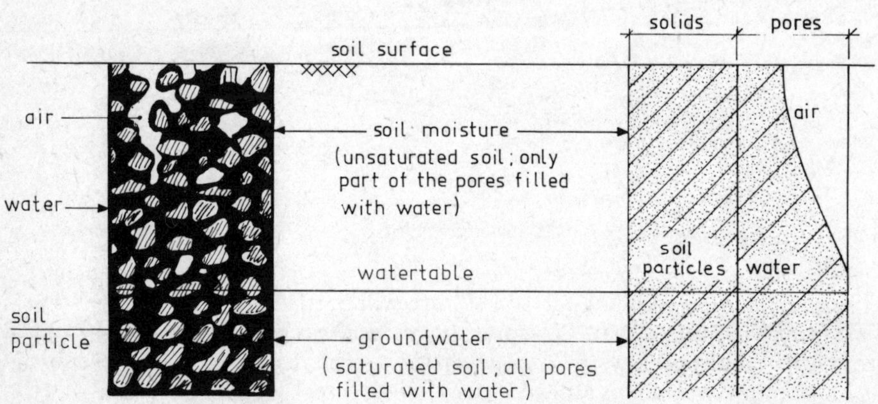

FIGURE 1.2 *Forms of occurrence of soil water*

Most land is underlain at some depth by phreatic groundwater. When it occurs deep below the soil surface (say at 25–50 m) it is of little direct concern to agricultural drainage. This condition exists across much of the land in arid/semi-arid climates and also in fairly humid climates when the land is underlain by permeable, freely draining strata, located high above the local groundwater drainage base. Groundwater at shallow depths (watertables < 5–10 m below the soil surface) has a much greater relevance to agricultural drainage. This situation is to be expected where the groundwater recharge is high (high rainfall or high irrigation losses, seepage, etc) or the groundwater drainage is slow (poorly permeable substrata, poorly developed drainage system, high drainage base, etc).

Watertable (phreatic level)

The watertable marks the division between the groundwater and the moisture zones in the soil. Its location is found by sinking a borehole into the groundwater body. Water from the surrounding soil will flow into the hole and fill it to the watertable level (figure 1.3a). For regular watertable depth measurements the borehole may be fitted with a perforated pipe in order to give the hole a degree of permanence.

When the deep percolation is impeded by a poorly permeable layer, a so-called *perched watertable* may develop (see figure 1.3b). Its occurrence is temporary as the impeded water will continue to seep through to deeper layers or drain laterally. It may, however, exist long enough to cause a serious problem of excess water in the soil. Perched watertables may be detected by drilling a borehole into (but not through) the impeding layer.

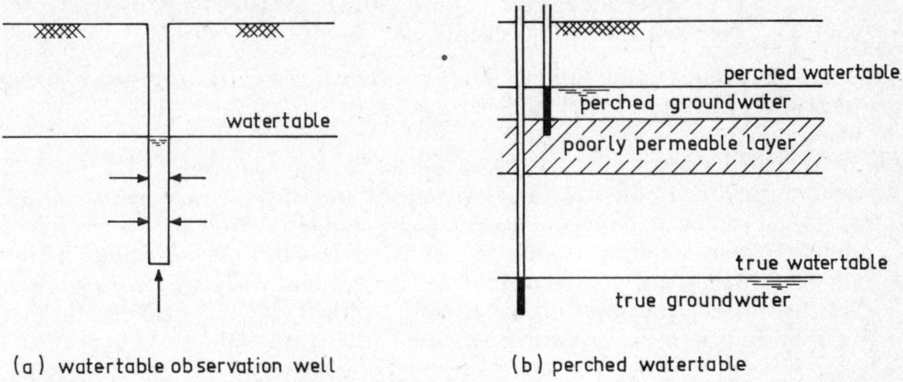

FIGURE 1.3 *Watertables in the soil*

Soil moisture

In the unsaturated soil above the watertable the pores are partly occupied by water and partly by air. The soil water in this zone is commonly referred to as *soil moisture*. The amount of soil moisture varies greatly with depth and in time. The soil moisture profile as depicted in figure 1.2 would be typical of a soil with a fairly high watertable (within 3 m from the soil surface) shortly after a prolonged period of rainfall. In a narrow zone above the watertable, pores fill by capillary rise from the groundwater. In the lower part of this capillary zone all pores are filled with water, making the soil in this so-called *capillary fringe* as saturated as in the groundwater below the watertable. Nevertheless, the water in the capillary fringe is under negative pressure while the pressure in the groundwater is positive, (see section 1.3). Above the capillary zone, pores fill with water mostly by retaining part of the percolating rain (or irrigation) water.

When the groundwater is very deep (> 10–20 m below the soil surface) the soil may be moist in the upper layers (retained rain or irrigation water) and also in a zone immediately above the watertable (capillary zone) while, in between, the soil may be much drier. The soil moisture content in the upper soil layer (down to 0·5–1·0 m) is particularly variable, mainly due to variations in daily weather conditions (especially rainfall variations). Deeper down, variations occur over a longer term, in parallel with seasonal weather variations.

1.3 Pressures in the soil water

Pressures, which are forces per unit area, will be designated by the symbol 'P'. In the SI system (*Système International*) the unit of force is the Newton (= force needed to give 1 kg mass an acceleration of 1 $m.sec^{-2}$); the unit of pressure is the Pascal = Newton per m^2. In a (soil) water system, pressures are often more conveniently expressed as an equivalent water gauge reading (*hydraulic head* or just *head*) where:

$$P\,(\text{in Pascal}) = P\,(\text{in m head}) \times \rho_w \times g \qquad \text{......(eq. 1.1)}$$

with: ρ_w = density of water ($\sim 10^3$ $kg.m^{-3}$ at normal soil temperatures)
g = gravitational acceleration ($\sim 9{\cdot}8$ $m.sec^{-2}$)

For large pressures, the bar and the atmosphere are useful units:

1 bar = 10^5 Pascal and 1 atm = 1·0132 bar (or roughly
1 bar = 1 atm = 10 m head).

Negative pressures in the soil moisture are also conveniently expressed by their *pF-value* where $pF = \log_{10}|P|$, with P in cm head.

Pressures below the watertable

At the watertable level, the soil water is always at atmospheric pressure ($P = P_{atm}$). The forces prevailing in the groundwater below the watertable are the normal forces encountered in a standing or moving body of water. Flow velocities in the groundwater are always too low to generate any significant kinetic forces. Normal positive hydrostatic pressures prevail ($P > P_{atm}$ with $P_{atm} = 0$ as reference) which may conveniently be measured with a *piezometer* (see figure 1.4).

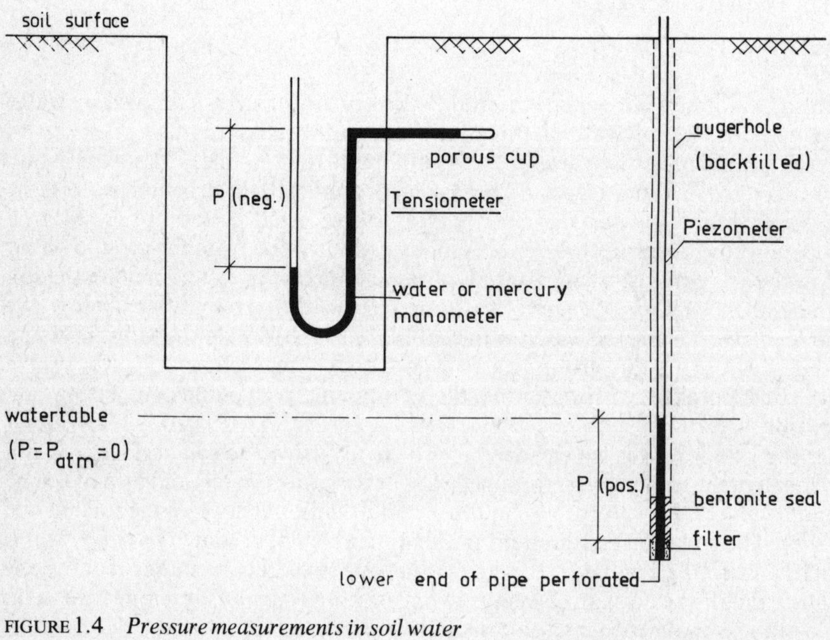

FIGURE 1.4 *Pressure measurements in soil water*

Pressures above the watertable

In the soil moisture above the watertable two types of forces prevail *viz*:

(a) *Capillary forces*: these are essentially surface tension forces, in this case activated by the adhesion between water and the soil and by the fineness of the pores. Capillary forces manifest themselves by the formation of curved air-water interfaces (*menisci*). The water pressure beneath the meniscus is negative, becoming increasingly so with increasing curvature of the meniscus (decreasing radius). The radius of curvature of the meniscus and the soil water pressure are related as:

$$P_{cap} = \frac{-2\tau}{R\rho_w g} \qquad \text{......(eq. 1.2)}$$

where P_{cap} = capillary pressure (m head)
τ = surface tension (0·073 Newton.m^{-1} for water at 15°C)
R = radius of curvature of the meniscus (m)
ρ_w = density of the water ($\simeq 10^3$ kg.m^{-3})
g = gravitational acceleration ($\simeq$ 9·8 m.sec^{-2})

In a pore the curvature of a meniscus attains a maximum value when its radius equals the radius of the pore, in which case $P_{cap} = -4\tau/D\,\rho_w g$ where D = diameter of the pore (or roughly $\boxed{P_{cap} = -3000/D}$ with P_{cap} in cm head and D in microns, μ). Such a meniscus can support a water column of length equal to P_{cap} (see figure 1.5).

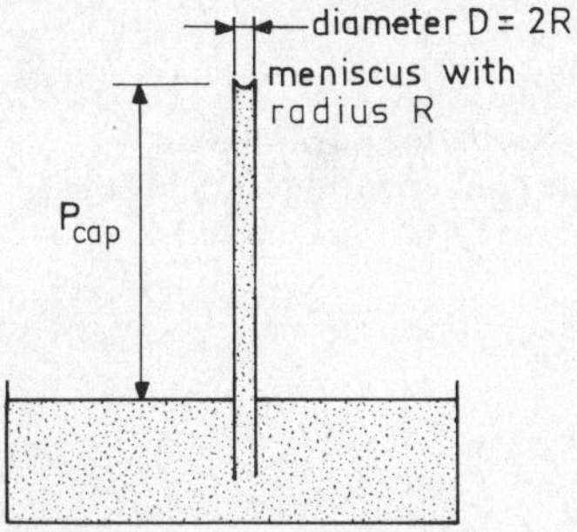

FIGURE 1.5 *Surface tension effects in a capillary tube*

(b) *Adsorption forces*: these include van der Waals and electrostatic forces exerted on the water by the charged colloidal surfaces of the soil particles. These forces exercise increasing influence with increasing colloidal surface area within the soil.

Both the capillary and the adsorption forces bind the soil moisture to the soil particles (see figure 1.6) making up the soil skeleton (*soil matrix*), thereby retaining it above the watertable against the gravitational pull. Pressures in the soil moisture are negative ($P < P_{atm}$). These negative pressures, commonly referred to as *tension* or *suction*, may be measured with a *tensiometer* (see figure 1.4). This instrument measures the so-called *matric suction* generated by the combined capillary and adsorption forces.*

* Positive pressures in the soil water below the watertable as measured with a piezometer, are also referred to as *piezometric pressures*; similarly negative pressures in the soil moisture above the watertable may be termed *tensiometric pressures*.

In saline soils *osmotic forces* generated by the salts in the water also need to be taken into account. These forces also retain the water (suction forces). They are of greater significance in the soil moisture than in the groundwater since salt concentrations in the unsaturated soil will normally be higher. As such they add to the matric suction of the soil moisture. Osmotic forces in the soil generally only need to be considered in the presence of a semi-permeable membrane such as the cell walls of a plant (see section 10.2.1).

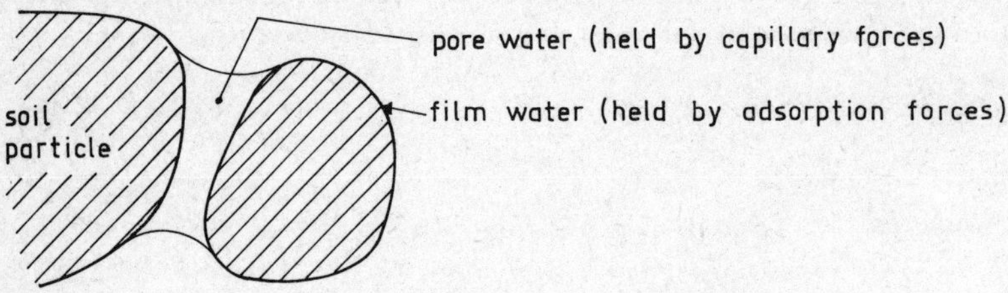

FIGURE 1.6 *Soil moisture held by capillary forces and adsorption forces*

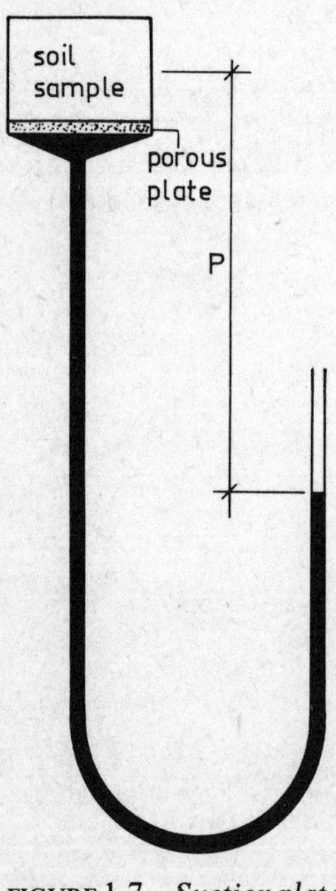

NB: suction plates are applicable for P > – 500 cm.
For lower values a pressure plate apparatus should be used.

FIGURE 1.7 *Suction plate apparatus for the determination of soil moisture retention curves*

1.4 Soil moisture characteristics

The relationship between the soil moisture content (θ) and the matric suction (P) is known as the *soil moisture characteristic*, or, when in graphical form, as the *soil moisture retention curve*. When a pF-scale instead of a P-scale is used, retention curves are also referred to as *pF-curves*. Examples are shown in figure 1.8.

Soil moisture retention curves may be determined in the laboratory by subjecting a soil sample to a stepped increase in suction (see figure 1.7). Initially the sample is saturated and is held at zero suction ($\theta_o = \theta_{sat}$; $P_o = 0$ cm). Subsequently the suction is increased to P_1 which forces all water to drain out of the pores above a certain size (for example when $P_1 = -100$ cm according to eq. 1.2 all menisci in pores with a diameter greater than 30 μ will break and these pores will drain completely). When the outflow has ceased, the moisture content of the sample is determined (θ_1). This value represents the moisture held in the still finer pores and as film water. As the suction increases, increasingly finer pores will drain until eventually even the film water (held mostly at strong suctions) is removed. The corresponding suction and moisture content values are plotted enabling the soil moisture retention curve to be drawn.

NB: A P-θ curve may also be obtained by starting with a dry soil sample and letting the soil absorb water. The resulting wetting curve will be different from the drying curve (at the same suction value a soil will normally hold more water on the drying than on the wetting curve). For drainage purposes the drying type retention curve is more relevant.

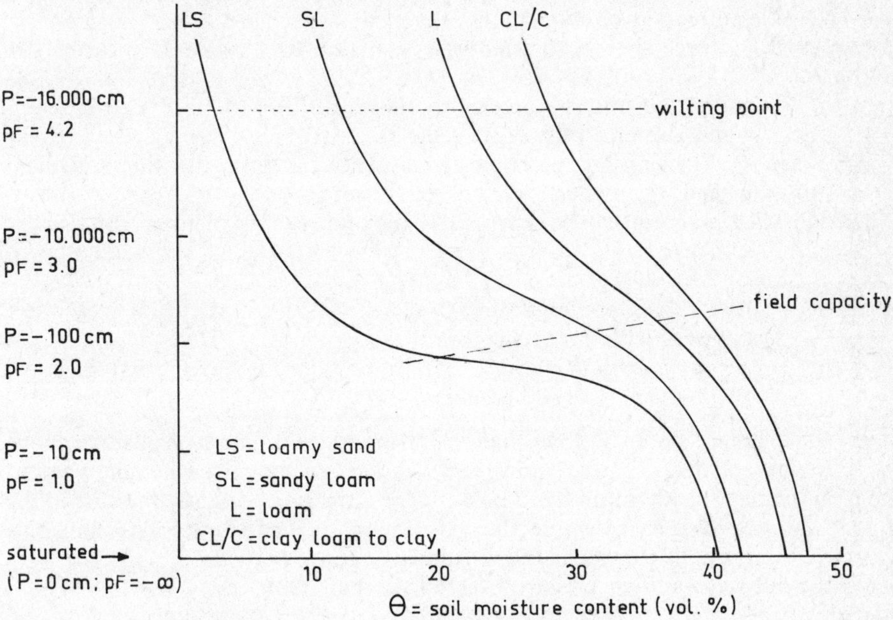

FIGURE 1.8 *Soil moisture retention curves*

Figure 1.8 shows that quite different curves apply to different soils. In the loamy sand, pores are predominantly medium fine which yield to suctions >70–80 cm (pores with D > 40 μ according to eq. 1.2). The colloid surface area, adsorbing film water, is relatively small so that the soil is only able to retain a small amount of moisture at high suctions (most water held at pF > 3 is film water). A high proportion of the soil moisture in clay loam/clay soils is film water, explaining the high soil moisture content at high suction in these soils. The curves for sandy loam and loam occupy intermediate positions. The moisture in these soils is proportionally well distributed, held as pore water and as film water, the loam having somewhat more film water and tightly held pore water than the sandy loam.

Retention curves not only vary with soil texture but also depend on the soil structure. In a well structured soil, a fair number of pores of all sizes exist while in a poorly structured soil there are few macro pores. In the former case, the soil moisture is retained over a wide distribution of suction forces while in the latter case the distribution is centred more towards the stronger suction forces.

1.5 SOIL WATER POTENTIAL

The soil water potential (φ) at a point in a soil water system is 'the work (energy) required to move a unit quantity of water from the reference state to that point'. In this case potential will be expressed per unit weight of water (this gives convenient dimensional units).

In a soil water system it is often convenient to take the energy state of the water at the watertable level as the reference state ($\varphi = 0$ when also taking $P_{atm} = 0$ as reference). Where there is no watertable, an imaginary free water surface at a convenient elevation may be taken as the reference state.

Energy (E) = force (F) × path length (L) = mass (M) × acceleration (a) × path length (L).

In the SI system the unit of force is the Newton ($F = M \times a = kg \times m.sec^{-2} = kg.m.sec^{-2}$) and the unit of energy is the Joule ($E = F \times L = kg.m.sec^{-2} \times m = kg.m^2.sec^{-2}$). The unit for potential, Joule per kg weight, has the equivalent dimension of m head [$(kg.m^2.sec^{-2})/(kg.m.sec^{-2}) = m$].

The soil water potential can be expressed as the sum of component potentials:

$$\varphi_{(total)} = \varphi_g{}'\varphi_p \qquad \text{......(eq. 1.3)}$$

where: φ_g = *gravitational potential*: energy expended or gained in moving water against or with the gravity force.

φ_p = *pressure potential*: energy expended or gained in moving water against or with the pressure difference.

For the reference state it follows that $\varphi_g = 0$ and $\varphi_p = 0$. The φ_g is positive above the reference level and negative below this level. The φ_p is positive when the water at the point under consideration has a positive pressure as is normally the case in the groundwater below the watertable; the φ_p is negative when the water has a negative pressure as is normally the case in the soil moisture above the watertable.

For the soil-water system in figure 1.9 it follows that:

– point A: $\varphi_A = z_A + (-P_A)$
– point B: $\varphi_B = -z_B + P_B$.

Figure 1.9 also illustrates the values of the component potentials at different depths in a static equilibrium soil water profile with a watertable. This represents the so-called field capacity situation some one to two days after a good soaking when all deep percolation has ceased but significant evapotranspiration losses have not yet occurred (see also section 1.9). The soil moisture will move until at all points the pressure potential (φ_p) has become equal but opposite in sign to the gravitational potential (φ_g). So at point A in figure 1.9 the water will percolate until the suction P_A equals Z_A. The total potential at point A is then:

$$\varphi_A = \varphi_g + \varphi_p = z_A + (-z_A) = 0.$$

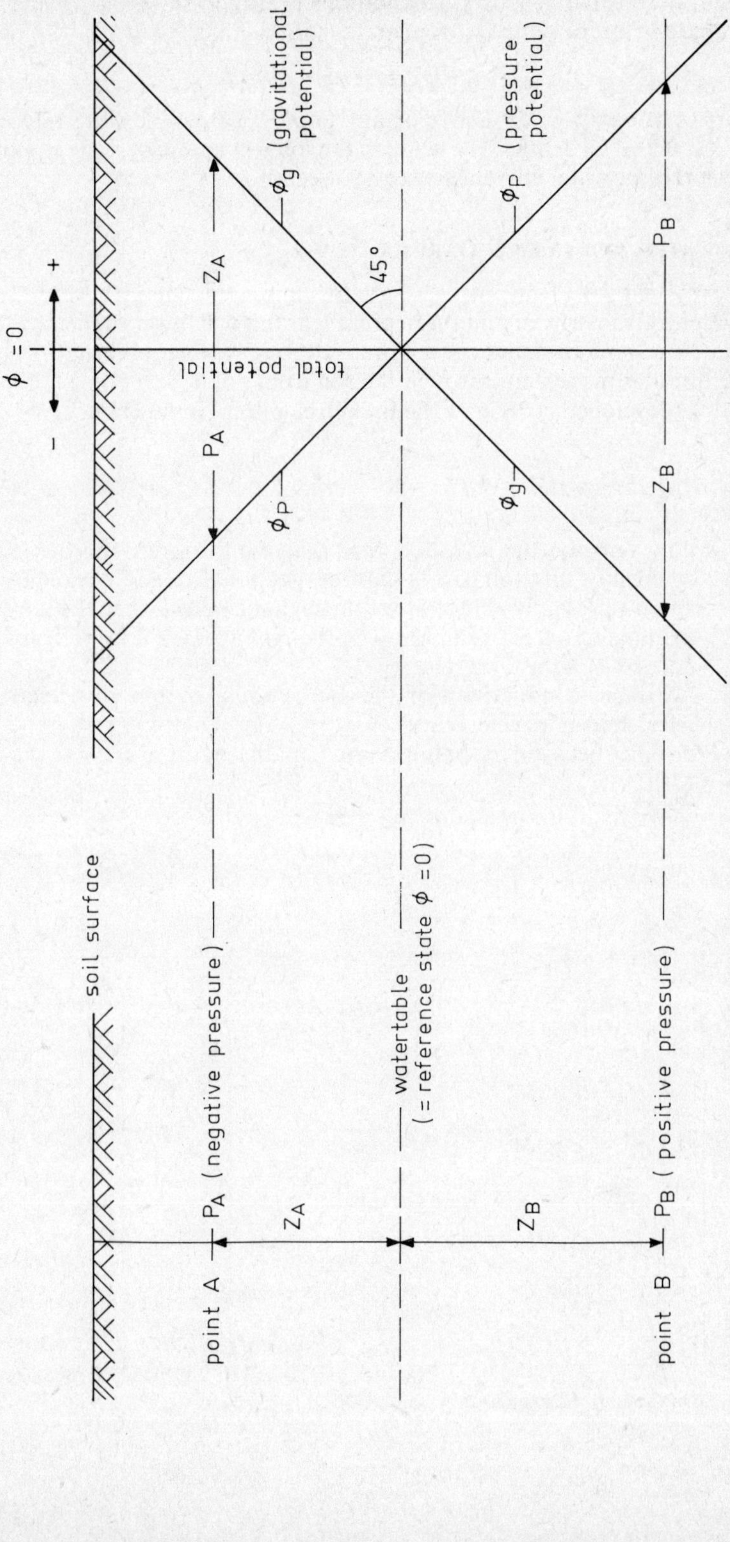

FIGURE 1.9 *Component and total soil water potentials in the static equilibrium soil water profile with a watertable*

For zero flow, normal hydrostatic conditions prevail in the groundwater below the watertable and for the potential at point B it follows that:

$$\varphi_B = \varphi_g + \varphi_p = (-z_B) + z_B = 0.$$

This applies at any point A above or any point B below the watertable and it follows that $\varphi = 0$ at all depths. The water in the soil is in a state of static equilibrium (no driving forces for water movement, see next section).

1.6 Soil water movement: Darcy's Law

A soil-water system will always strive to attain a minimum energy level. As long as $\varphi_A > \varphi_B$, water will move from point A to point B as this will lower the energy level of the system. The water movement will continue until $\varphi_A = \varphi_B$ (equilibrium).

The rate of water movement through the soil from point A to point B obeys the Law of Darcy (developed in 1856 by the French engineer Henri Darcy):

$$v = -K\frac{\varphi_B - \varphi_A}{L} \qquad \text{......(eq. 1.4)}$$

where: v = flow velocity (the so-called *filter velocity*) through a porous medium where flow through part of the cross-sectional area is blocked by solids and water is only able to move through the pores; when the porosity of the medium is for example 40% the real average flow velocity in the pores is $2\frac{1}{2} \times$ the filter velocity

K = hydraulic conductivity of the soil (popularly, but incorrectly, also referred to as 'permeability').

L = distance between the points A and B (path length)

$$\frac{\varphi_B - \varphi_A}{L} = \frac{\Delta\varphi}{L} = \text{hydraulic gradient } (i).$$

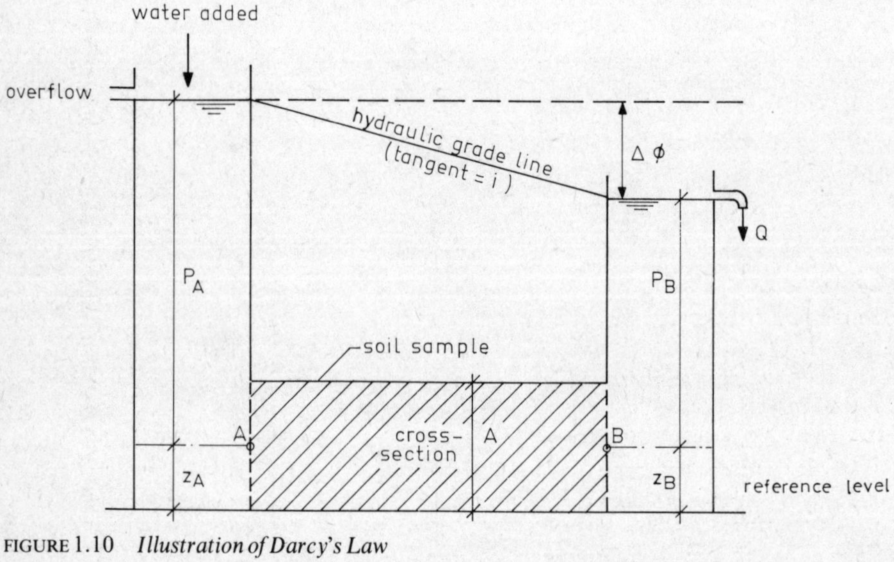

FIGURE 1.10 *Illustration of Darcy's Law*

DARCY's Law states that the rate of water movement through a soil is proportional to the gradient of the soil water potential (which is the 'driving force' behind the movement). The hydraulic conductivity is the constant of proportionality in this relationship, the negative sign accounting for the fact that the potential decreases in the flow direction.

In figure 1.10: $\varphi_A = z_A + P_A$.
$\varphi_B = z_B + P_B$

This shows that in this case $(\varphi_B - \varphi_A)$ equals the negative values $-\Delta\varphi$, so DARCY's Law becomes:

$$v = -K\frac{-\Delta\varphi}{L} = K\frac{\Delta\varphi}{L} = Ki$$

$$Q = vA = KiA.$$

Units: Velocity (v) in $cm.sec^{-1}$ or in $m.day^{-1}$
Gradient (i), dimensionless ($\Delta\varphi$ in cm or m head and L in cm or m length)
Area (A) in cm^2 or in m^2
Flow (Q) in $cm^3.sec^{-1}$ or in $m^3.day^{-1}$
Hydraulic conductivity (K) in $cm.sec^{-1}$ or in $m.day^{-1}$.

In the horizontal flow through a layer in the soil, the cross-section (A) may be replaced by the thickness of the layer D (ie the cross-sectional area of a unit width of the layer, equal to D m^2). Darcy's Law then becomes: Q = KiD where Q represents the discharge per m width (Q/m). In combination, K and D determine the opportunity for horizontal flow through a soil layer and as such they are often combined into one parameter, the KD-value (*transmissivity*) of the soil.

1.7 HYDRAULIC CONDUCTIVITY

The hydraulic conductivity (K-value) of the soil depends mainly on the geometry and distribution of the water-filled pores.* Values are low when the water has to follow a tortuose path through fine pores. This will generally be the case when the soil moisture content is low since under these conditions the water will be mainly present in the finer pores and as filmwater, forming an irregular hydraulic continuity with many 'bottle-necks'. Differences in the hydraulic conductivity of soils under saturated conditions reflect differences in the geometry of the total pore space of the

	K_{sat} (*order of magnitude*)
Coarse gravelly sand	10 – 50 $m.day^{-1}$
Medium sand	1 – 5 $m.day^{-1}$
Sandy loam/fine sand	1 – 3 $m.day^{-1}$
Loam/clay loam/clay, well structured	0·5 – 2.0 $m.day^{-1}$
Very fine sandy loam	0·2 – 0·5 $m.day^{-1}$
Clay loam/clay, poorly structured	0·02– 0·2 $m.day^{-1}$
Dense clay, not cracked and no bio-pores	<0·002 $m.day^{-1}$

* Temperature also has an influence on K. With rising temperature, water becomes less viscous and the K-value increases. In the deeper soil layers through which most drainage flow takes place, temperatures are rather uniform and steady, and its influence on K may generally be neglected.

soil. As such K_{sat} values are often well correlated with soil texture and structure. The presence and characteristics of bio-pores (root channels, worm holes and other small conduits left by biological processes in the soil), however, also greatly influences the hydraulic conductivity.

The hydraulic conductivity of a field soil may vary considerably across an area as well in depth due to variations in soil texture and soil structure. In a layered soil, K-values will generally differ between layers. Some soils also show a marked difference between hydraulic conductivity in horizontal and vertical directions (*anisotropic soil*, see chapter 16).

Figure 1.11 presents two examples of the decrease in K-value with decreasing soil moisture content. The general shape of the K–θ curves reflects the proportional distribution of the soil water held in macro pores, in micro pores and as film water. This distribution also underlies the P–θ curve (soil moisture characteristic) and the two curves can generally be related. The light textured soil has a higher proportion of macro pores through which water is able to flow rather easily, resulting in a higher K-saturated. At low moisture content the finer textured soil has a higher K-value because it has a larger proportion of interconnected water filled micro pores and filmwater than the lighter soil.

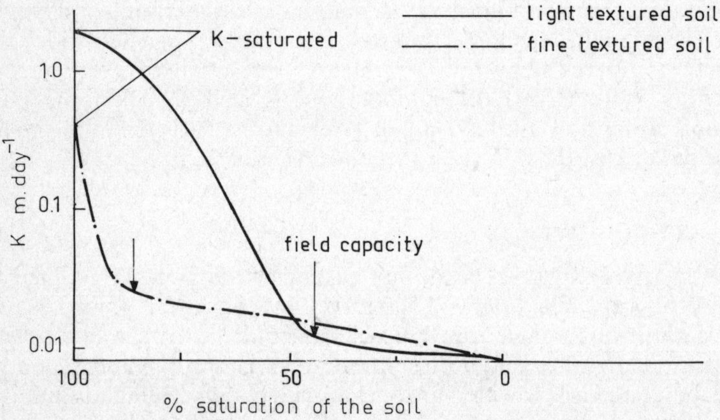

FIGURE 1.11 *Hydraulic conductivity of soils as a function of the soil moisture content*

The hydraulic conductivity generally drops to a very low value when most of the macro-pores have been drained and the soil moisture content has decreased to the extent that the water in the soil no longer forms an effective hydraulic continuity. This situation occurs roughly at field capacity (see section 1.9). Although under these conditions water movement over long periods (some months) may still be considerable, over short periods (few days) it is insignificant.

1.8 UNSATURATED FLOW

Darcy's Law applies also to the movement of soil moisture (unsaturated flow). This is illustrated for the case of steady state percolation of water through a soil profile to an underlying watertable as detailed in figure 1.12. The percolation rate is 20 mm.day^{-1} and the pressure in the soil moisture is measured at 20 cm intervals with tensiometers. K-values may be calculated for each interval by applying Darcy's Law:

$$v = -K\frac{\Delta\varphi}{L} \quad \text{or} \quad K = -\frac{L}{\Delta\varphi}v$$

Where: $v = 20\,\text{mm.day}^{-1} = 2\,\text{cm.day}^{-1}$

$L = 20\,\text{cm}$

$\Delta\varphi = (\varphi_g+\varphi_p)_{n+1}-(\varphi_g+\varphi_p)_n = (z_{n+1}+P_{n+1})-(z_n+P_n)$.

For the interval between z = 80 cm and z = 60 cm this yields:

$$K = -\frac{20}{(60-30)-(80-32{\cdot}5)}.2 = \frac{20}{17{\cdot}5}.2 = 2{\cdot}3\,\text{cm.day}^{-1}.$$

Values for all other intervals can be calculated similarly:

Interval z = 120→100.... K = 2·0 cm.day^{-1}
z = 100→ 80.... K = 2·1 cm.day^{-1}
z = 80→ 60.... K = 2·3 cm.day^{-1}
z = 60→ 40.... K = 2·7 cm.day^{-1}
z = 40→ 20.... K = 4·0 cm.day^{-1}
z = 20→ 0.... K = 8·0 cm.day^{-1}

Actually, as K varies continuously along the flow path, Darcy's Law should strictly be applied in its differential form:

$$v = -K\frac{d\varphi}{dz}$$

However, as an approximation steady state unsaturated flow may be treated as a series-flow with K = constant per interval (implies uniform moisture content per interval).

The potential profile in figure 1.12 shows that in comparison with the equilibrium situation depicted in figure 1.10 higher (ie less negative) pressure potentials prevail above the watertable, creating the hydraulic gradient for downward flow. The hydraulic gradient increases with height above the watertable, compensating for the decreasing K-value. Well above the watertable the soil moisture content becomes uniform and the gradient converges to one (no pressure gradient; water moves by gravity only). The moisture content in this zone settles at such a value that its corresponding K-value equals the infiltration rate (v = Ki; for i = 1, v = K). The steady state percolation reaches its maximum attainable value when the soil profile has become saturated and the percolation rate equals K_{sat}. When the rainfall exceeds this rate surface ponding will occur.

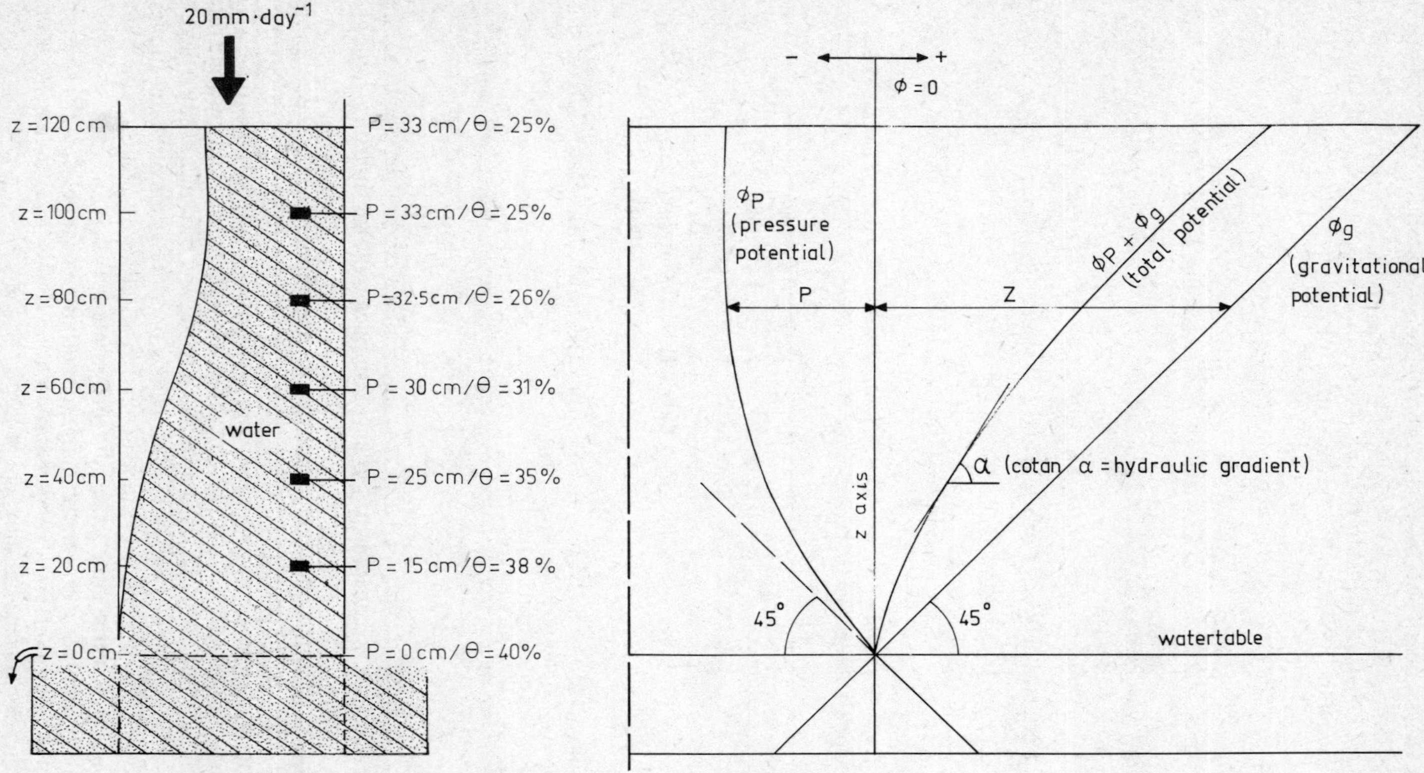

FIGURE 1.12 *Potential profiles for steady state percolation to the watertable*

1.9 Soil moisture profiles and constants

Soil moisture conditions vary greatly in time and space. The soil moisture profile shows the variation in soil moisture content with depth in the soil. Indirectly it expresses the prevailing pressures in the soil water at different depths since pressure and moisture content are related through the soil moisture characteristic. When the pressure changes, soil moisture content will adjust, and vice versa.

Three typical pressure profiles and corresponding moisture profiles are depicted in figure 1.13. In the static equilibrium situation as discussed in section 1.5 ($\varphi = 0$ at all depths), the pressure head at each point is equivalent to the height above the watertable. The moisture profile in this case has the same shape as the soil moisture characteristic plotted with a pressure scale in units of head.

The remaining profiles hold respectively for steady state downward flow to the watertable (percolation) and steady state upward flow from the watertable (capillary rise). The pressure profile for downward flow has been discussed in section 1.8. Suctions at all heights are lower (less negative) than in the equilibrium situation and the moisture contents are correspondingly higher. In the case of upward flow, suctions are higher and the profile is dryer than in the equilibrium situation.

Under the prevailing field soil moisture regimes, steady state flow conditions are the exception and non-steady state conditions are the rule. The corresponding moisture profiles may vary almost endlessly.

A number of special soil moisture situations are recognised, the corresponding indicative soil moisture parameters being known as the *soil moisture constants.* These situations are:

Field capacity (FC)

The soil is said to be at *field capacity* when after a good watering of the soil the deep percolation flow has virtually ceased while no significant evapo(transpi)rative soil moisture depletion has yet occurred.

When the watertable is at a shallow depth ($<$ 1–2 m below the soil surface), the field capacity situation approaches the equilibrium soil moisture profile. When the watertable is deep, an equilibrium profile never becomes established in the field since considerable evapo(transpi)rative depletion has already occurred by the time the deep percolation process has ended. In such cases field capacity refers to the situation where the water in the rapidly draining macro-pores in the upper soil layers (*rootzone*) has percolated through. In a readily draining soil profile this situation is reached within one day after rainfall or irrigation, leaving the upper soil layers at a moisture content roughly corresponding with $P = -200/300$ cm (pF $= 2{\cdot}2$–$2{\cdot}5$). In a slowly draining profile, field capacity is reached somewhat later and/or the corresponding suctions are somewhat lower ($P = -60/100$ cm; pF $= 1{\cdot}8$–$2{\cdot}0$).

Actually, the field capacity situation is only a momentary situation in the profile drainage process; this process continues after field capacity has been reached, although at a much slower rate (very low hydraulic conductivity, see figure 1.11). Its real significance is practical rather than physical: in most soils it constitutes a readily indentifiable starting point for soil moisture depletion calculations (see 'available moisture' below).

Wilting point (WP)

A soil is said to have reached *wilting point* when plants are no longer able to extract

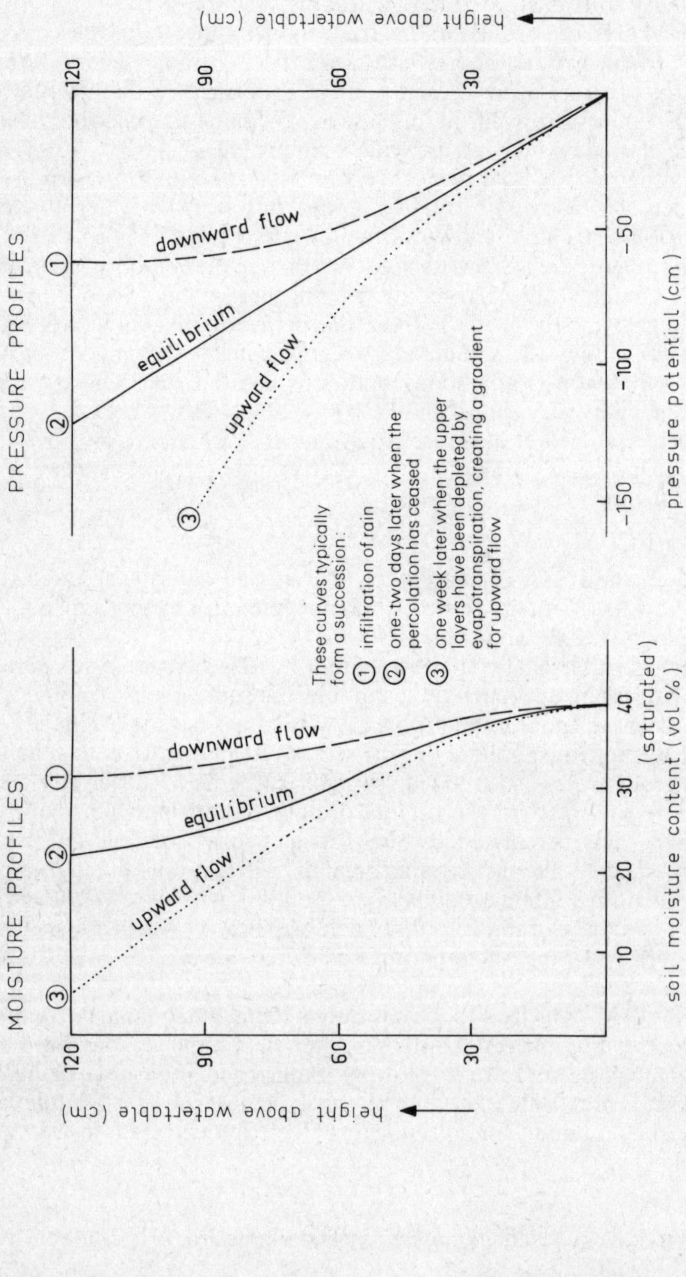

FIGURE 1.13 *Moisture and pressure profiles for three situations*

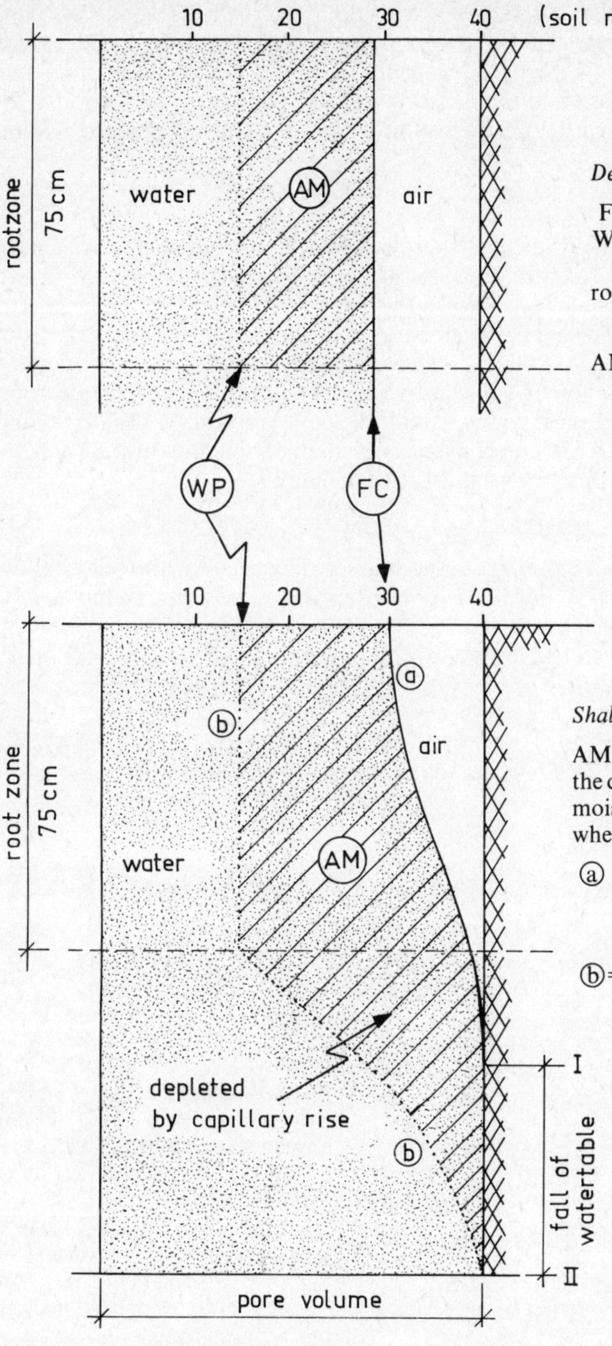

Deep watertable

FC = 29 vol. %
WP = 15 vol. %

rootzone depth = 75 cm

$$AM = \frac{(29-15)}{100} . 750 = 105 \text{ mm}$$

Shallow watertable

AM derived graphically as the difference between the two moisture profiles a and b, where:

ⓐ = equilibrium profile for a shallow watertable (at depth I)

ⓑ = profile pertaining to a rate of capillary rise to rootzone approaching to nil; in the rootzone θ = WP

FIGURE 1.14 *Calculation of the 'available moisture' in a soil profile*

enough water from that soil to survive. At wilting point the soil has lost all of its pore water and part of its film water, the remainder being held close to the colloid surface under pressures lower than the plant roots are able to overcome. For most crops this occurs when the suction has risen to -15 atm $= -15\,000$ cm (pF $= 4{\cdot}2$). Actually, crop growth is retarded well before the soil moisture content has reached wilting point.

Available moisture (AM)

The quantity of moisture held in the rootzone between field capacity and wilting point is referred to as the *available moisture*. Water in the rootzone above FC and below WP is unavailable to plants as it will either drain to a lower depth or be held at too strong suction forces. The available moisture is held at a range of suctions (from $P = -100/300$ cm to $P = -15\,000$ cm) and only the water held at the more moderate suctions can be taken up sufficiently rapidly by the plants to sustain good growth. This part of the AM is called the *readily available moisture* (RAM). It is often assumed that RAM $= 2/3$ AM. The concept of available moisture and the type of calculations pertaining to it are further illustrated in figure 1.14.

Drainable pore space (μ)

When in a situation with an equilibrium moisture profile existing above a shallow watertable, the watertable is lowered, some pore space will drain and a new equilibrium profile will develop (see figure 1.15). A similar shifting of the equilibrium profile occurs when the watertable rises and pore space is filled. The pore space which drains/fills with a fall/rise of the watertable is called *the drainable pore space* (μ).

Typical μ-values for field soils vary from 2 to 10%. When the value is eg 4% the watertable falls or rises by 10 cm for each 4 mm of water extracted from or

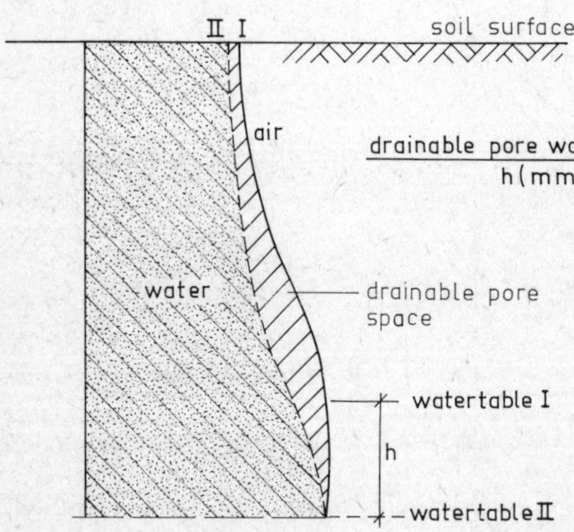

$$\frac{\text{drainable pore water(mm)} \times 100}{\text{h(mm)}} = \mu\ (\%)$$

*Indicative values**
- dense clay $\mu =$ 1– 2%
- well structured loams, clay loams and clays $\mu =$ 4– 8%
- fine sand $\mu =$ 15–20%
- coarse sand $\mu =$ 25–35%

FIGURE 1.15 *Illustration of the drainable pore space*

* The drainable pore space and the hydraulic conductivity both depend primarily on the pore geometry of the soil and their values are generally correlated. For rough estimates use may be made of the formula: $\mu = \sqrt{K}$, with μ in % and K in cm.day^{-1}.

respectively added to the groundwater. Watertable responses to additions/extractions are small when μ-values are large, responses being great when μ-values are small. The drainable pore space can be measured in the field by observing the drain discharge and the watertable recession over a period when the evaporation is insignificant. Alternatively, the μ-value may be estimated as the percentage air in the soil at field capacity as read from the moisture characteristic.

Lower plastic limit (LPL)

The deformation characteristics of the soil depend mainly on soil texture and moisture content. Fine textured soils are easily moulded without breaking up when their moisture content is above LPL, and the soil is in the so-called *plastic state of consistency*. In this state the soil easily smears, puddles or compacts when driven or walked on by cattle or when the soil is tilled, resulting in a poor soil structure for plant growth. After rain the upper soil layers will drain to field capacity. For most fine textured soils, however, FC > LPL, which means that after drainage, further evaporative drying is required to reach the workable stage (see figure 1.16). This takes longer the further LPL is on the dry side of FC. The FC–LPL relationship of a soil may thus be used to assess the scope for drainage to improve soil workability.*

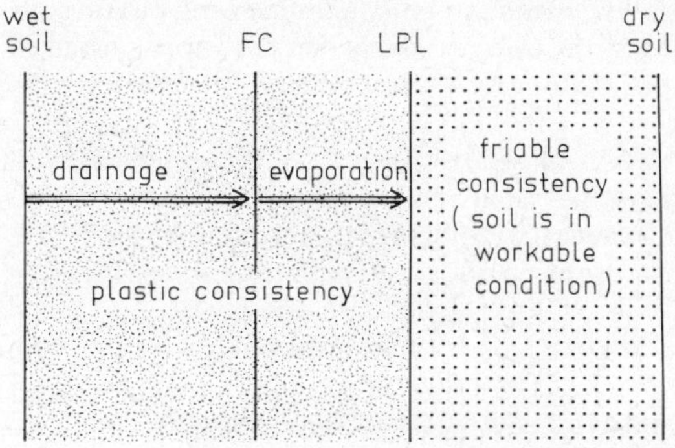

FIGURE 1.16 *The relationship between LPL and FC and their effect on soil workability*

Soil management, however, also has an influence. Smearing and compaction of the soil resulting eg from working the soil under too wet conditions, destroys macropores and as such tends to increase FC. Addition of organic matter to the soil tends to increase the LPL value of the soil. The first activity thus has a negative effect on soil workability; the second a positive effect.

1.10 Infiltration and percolation

The maximum rate at which water can infiltrate into a dry soil decreases normally from an initially high value to a much lower value as the infiltration process continues, becoming nearly constant some 1 to 3 hours after the start. The value of the nearly constant final infiltration rate reflects the pore-geometrical situation in

* *Workability = tillability + trafficability*; although criteria for soil tillage and vehicular movement over the soil differ, present knowledge does not justify such a refinement in criteria.

the topsoil which varies with the soil texture but is also very much influenced by the current state of the soil structure. Under normal conditions the following orders of magnitude apply:

	Total infiltrated water after 3 hours	*Infiltration rate after 3 hours*
Coarse textured soils	150–300 mm	15–20 $mm.hr^{-1}$
Medium textured soils	30–100 mm	5–10 $mm.hr^{-1}$
Fine textured soils (not cracked)	30– 70 mm	1– 5 $mm.hr^{-1}$

NB: initial soil moisture content of topsoil approaching WP (time for next irrigation).

A cracked clay soil may absorb almost instantaneously 100 to 200 mm of water but once the cracks have been filled and closed as a result of the swelling of the soil, infiltration virtually stops. In compacted or otherwise densely structured soils, very little water is able to infiltrate, and rainfall easily results in ponding. Soils may also disperse under the impact of rainfall and clog the surface pores (*surface sealing* or *capping* of the soil); this happens particularly in many silty and fine sandy loam soils in the semi-arid zones at the beginning of the rainy season when they are exposed to intense storms with little vegetative protection. As a result much of the rainfall runs off the land.

Darcy's Law also applies to infiltration of water into soil. Water enters the soil from the surface under the combined gravitational and pressure gradients in the soil water. The latter is very strong during the initial infiltration of water into a dry soil when at the surface the pressure is atmospheric (P = 0) while in the soil just below the surface it is strongly negative (P = −5 000/−10 000 cm). The resistance for the entry of water in the soil, however, is also high as the hydraulic conductivity of the dry soil is very low. Most of the available head is dissipated in the advancing wetting front where the resistance is highest and as a result this front becomes very sharp (see figure 1.17). In a moist soil, the wetting front is much more diffuse.

As the infiltration process continues, the flow paths become longer and consequently the pressure gradient becomes smaller. This partly explains the decline of the infiltration rate in time. By the time the infiltration rate has become constant, a pronounced transition zone has established with nearly uniform moisture content close to saturation. Pressure differences in the zone are small and the water movement is dominated by the gravity force. The final infiltration rate thus becomes approximately equal to the K_{sat} of the soil (compare to section 1.8).

Percolation refers to the downward movement of excess water in the soil profile to underlying layers. In the process the upper layers drain to field capacity. The post-infiltration percolation in a dry soil (*redistribution*, see figure 1.17) is a special case. Moisture profiles during percolation to a shallow watertable have been discussed in sections 1.8 and 1.9.

In field soils the downward flow may be impeded due to the occurrence of poorly permeable dense layers while stratification in general may impede downward movement of excess water. This may lead to perched groundwater and similar cases of near-saturated moisture conditions above the watertable. Where this occurs, horizontal gradients for water movement may exceed the vertical gradients,

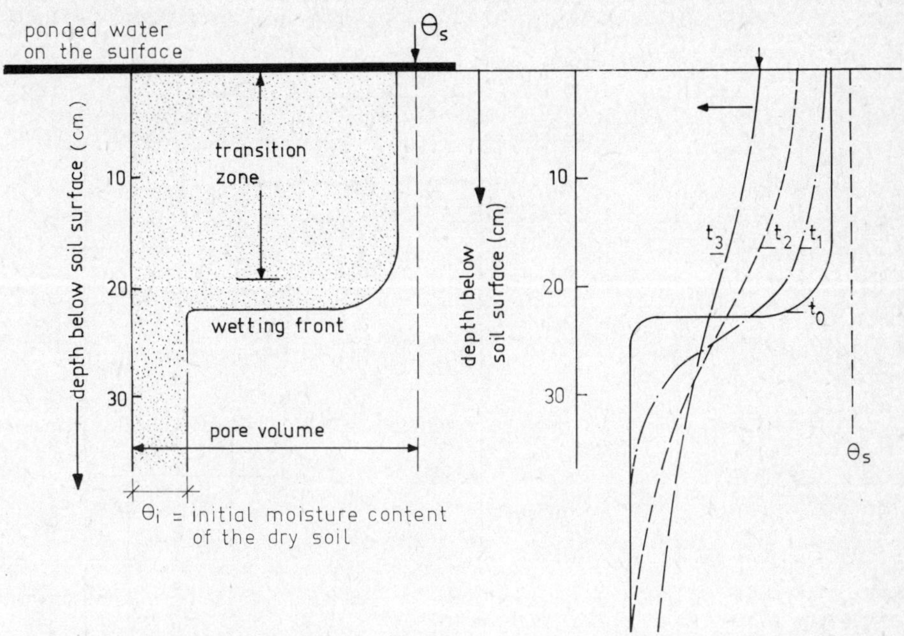

FIGURE 1.17 *Soil moisture during the infiltration of water into a dry soil and the post-infiltration redistribution of the infiltrated water*

resulting in lateral water movement above the true watertable, referred to as *interflow*. Conditions under which interflow occurs are often ill-defined, this type of flow rarely lending itself to the sort of physical-mathematical analysis used for other types of soil water movement. Interflow, however, plays an important role in the functioning of shallow drainage systems discussed in chapter 4.

1.11 GROUNDWATER FLOW

Groundwater flow refers to saturated flow below the watertable. The rate and direction of flow are controlled by differences in soil water potential (combined gravitational and pressure potential). Consequently, groundwater flow may be described by the classic equations of potential flow such as the *Laplace Equation.* This equation combines the two basic laws of groundwater flow, *viz* Darcy's Law and the Law of Conservation of Mass (also known as the *Continuity Equation*). For two dimensional flow these equations are:

Darcy's Law : $v_x = -K\dfrac{\partial \varphi}{\partial x}; \quad v_y = -K\dfrac{\partial \varphi}{\partial y}$

Continuity Equation: $\dfrac{\partial v_x}{\partial x} + \dfrac{\partial v_y}{\partial y} = 0$

Laplace Equation : $\dfrac{\partial^2 \varphi}{\partial x^2} + \dfrac{\partial^2 \varphi}{\partial y^2} = 0$ ......(eq. 1.5)

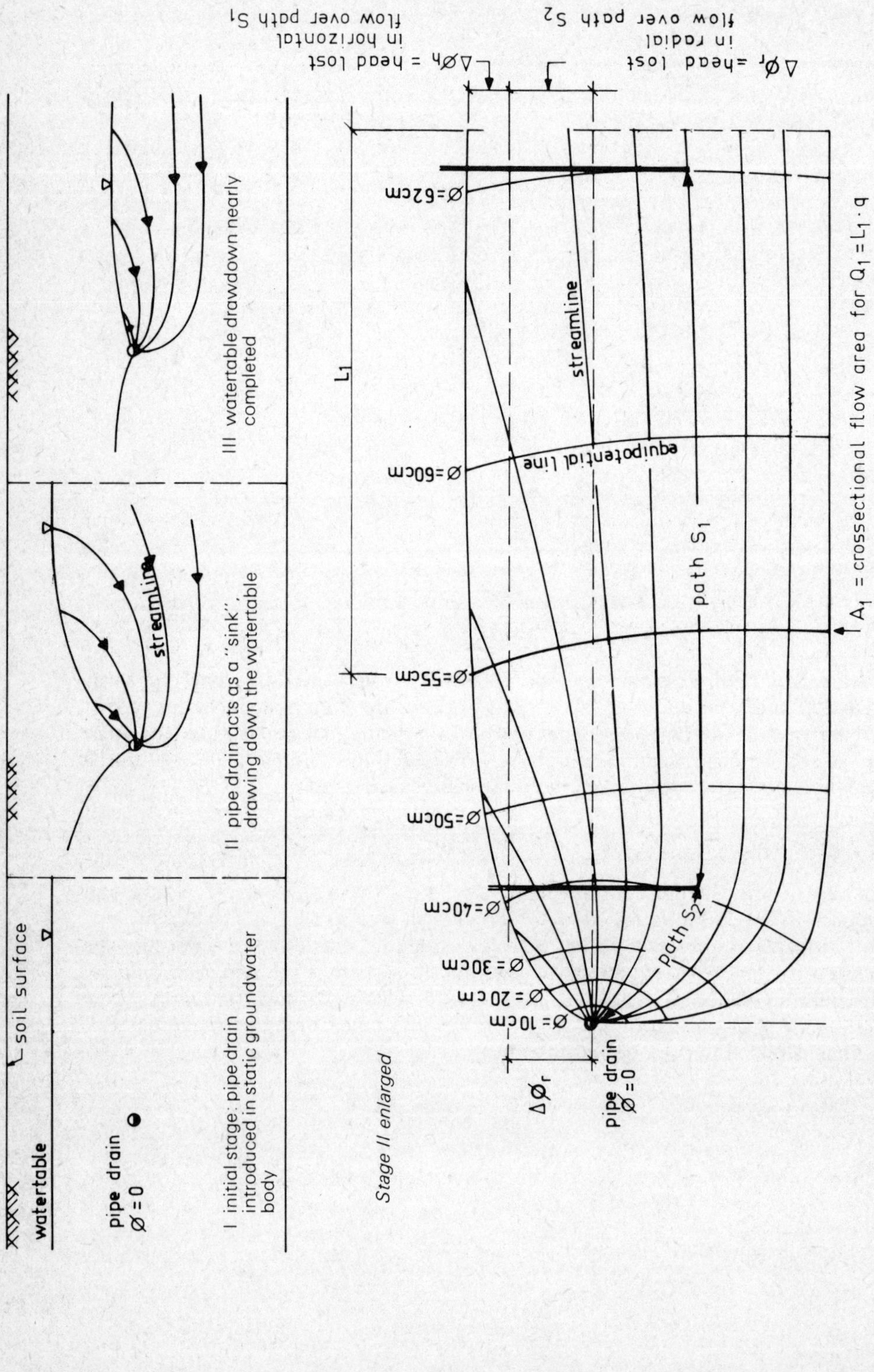

FIGURE 1.18 *The functioning of a groundwater drainage system*

Groundwater flow is a major feature of the functioning of the groundwater drainage systems described in chapter 3. In these systems, drains (ditches or underground pipe lines) are installed to some depth below the soil surface, reaching into the groundwater. The water potential in the drains is zero ($\varphi = 0$), so all the groundwater above this drainage base is under gradient to flow towards and into the drains (these drains constitute so-called *sinks*).

Figure 1.18 depicts the two-dimensional flow in a cross-section across a drain in a groundwater drainage system. Stage I represents the initial high watertable situation while stage III represents the final situation when most of the drainable water above the drainage base has been drawn in by the drain, lowering the watertable in the process. The watertable will settle at drainage base level (or even fall below due to evaporation or natural drainage of groundwater) and remain there until the groundwater is recharged again by deep percolating rain or irrigation water. When the watertable rises above the drainage base, the drain resumes its 'sink' function. The flow to the drain will increase as the watertable rises higher above the drainage base (more head), until the steady state situation is reached where the recharge to the groundwater equals the discharge by the drain. Steady state situations are rare and of short duration. With the tailing off of the recharge the watertable will fall again.

The *streamlines* in figure 1.18 indicate the paths along which water moves towards the drain. Along the *equipotential lines* the total potential (= gravitational + pressure potential) is constant. The flow of water occurs in response to potential differences and is always in the direction of the steepest gradients. As a result streamlines are perpendicular to the equipotential lines.

Far away from the drain the flow is mainly horizontal, the streamlines being essentially horizontal and the equipotential lines vertical. The flow pattern becomes radial as the flow approaches and converges onto the drain.

Piezometers show the pressure potential (as head equivalent) at different points in the flow system. The total potential decreases in the direction of flow towards the drain, becoming zero in the drain. The gradients in the potential field are indicative of the resistance met by the flow. These show that very little potential is lost in the horizontal flow zone where the water is able to move through extensive strata and the discharge is still well below its maximum value. The discharge increases towards the drain while the cross-sectional flow area beneath the declining watertable decreases. The gradients become steeper as is evident from the strong curvature of the watertable near the drain.

The Laplace Equation can be readily solved analytically for simple drainage situations with steady state flow and uniform soil conditions. For more complicated situations (ie stratified soil, three-dimensional flow, etc) this may require considerable schematisation. Many of these cases, however, can be solved numerically with the aid of a computer or by means of electrical analogue models. Most standard groundwater drainage situations have been thoroughly analysed and relationships between the different relevant parameters have been reduced to simple design formulae (see chapter 7).

References and further reading

Bouma, J, 1977: 'Soil survey and the study of water in the unsaturated zone', Soil Survey Paper no 13, Netherlands Soil Survey Institute, Wageningen

BOUWER, H, 1978: *Groundwater Hydrology*, McGraw-Hill, New York

CHORLEY, R J (ed) 1969: *Introduction to Physical Hydrology*, Methuen, London

DUNNE, T and L B LEOPOLD, 1978: *Water in Environmental Planning*, W H Freeman, San Francisco

HARTGE, K, 1978: *Einführung in die Bodenphysik*, Ferdinand Enke, Stuttgart (in German)

HILLEL, D, 1971: *Soil and Water: physical principles and processes*, Academic Press, New York

MARINO, M A and J N LUTHIN, 1982: *Seepage and Groundwater*, Elsevier, Amsterdam

MARSHALL, T J and J W HOLMES, 1979: *Soil Physics*, Cambridge University Press, London

WINTER, E J, 1974: *Water, Soil and the Plant*, MacMillan, London

YONG, R N and B P WARKENTIN, 1975: *Soil Properties and Behaviour*, Elsevier, Amsterdam

Chapter 2 Land drainage for agriculture – an introduction

The main objective of agricultural land drainage is to remove excess water in order to improve the profitability of farming the land.

There are periods on most land when excess water occurs. However, these need not be too harmful provided the quantities are small, the periods of occurrence are of short duration, or the excess occurs during a non-critical season (eg outside the main farming season). Most land also has some natural drainage which assists in the removal of a certain amount of the excess water. It is only when large quantities occur for prolonged durations at critical periods, that its removal by artificial means may be feasible.

2.1 Drainage problems

Excess water may occur on the land surface (surface ponding, often combined with waterlogging of the topsoil) or deeper down in the soil profile (waterlogging of the rootzone due to impeded percolation or due to high watertables). The adverse effects on farming may be broadly classified as:

(a) *Impaired crop growth*: most crops respire by gaseous exchange in the rootzone – the process whereby roots absorb oxygen (O_2) from the soil atmosphere and release carbon dioxide (CO_2) back into it. Roots are also able to absorb O_2 dissolved in the soil water but this capacity is very limited. Rice is unusual in being able to transfer air taken in through the stomata to the roots; as a result it is able to grow well even though its roots are submerged.

In waterlogged soils, the air content of the soil is low because most pores are filled with water. Moreover, the exchange between the remaining air in the soil and the air in the atmosphere above (O_2 moving into the soil, CO_2 moving out) is very restricted by these conditions. In consequence, respiration is restricted by the oxygen deficiency while at the same time the carbon dioxide accumulates to toxic levels, directly impairing the root growth and the roots ability to absorb nutrients. Anaerobic conditions in the soil may also result in toxic concentrations of reduced iron and manganese compounds, sulphides and organic gases.

Rootzone aeration generally becomes inadequate when the effective air-filled pore volume in the main rootzone falls below 5–10%. However, the duration of waterlogging and its timing in relation to the activity and stage of development of the crop also have a considerable influence. Waterlogging of the entire rootzone for a period of two to three days can be fatal when it occurs during the seedling stage, whereas a well-developed crop is likely to suffer relatively little damage from a similar incident. Furthermore, a vigorously growing healthy crop is able to withstand waterlogging better than a poor one. Most crops have a large proportion of their roots in the topsoil and so early restoration of an aerated topsoil is of particular importance.

Crops suffer much more from waterlogging under warm than under cold weather conditions. The reasons for this are that growth increases at higher temperatures, increasing the oxygen consumption (of the crops but also of the soil flora and fauna) and leading to earlier deficits. The lower solubility of oxygen in water at higher temperatures also plays a role.

In temperate climates, waterlogged soils often remain cold for too long in spring for good crop growth.

Waterlogging also *indirectly* affects plant growth by its adverse effects on soil biological life and on the structure of the soil. Thus, waterlogging during the winter period in Northern Europe is known to impair mineralisation and nitrification by microbes, resulting in nitrogen deficiency to crops in the spring. This, however, may be partially offset by increasing the nitrogen fertilizer application. It may also cause the soil structure to disintegrate or prevent it being restored by the action of frost. Drainage in temperate climates is therefore largely done to overcome these adverse indirect effects. Similar effects may also be expected in the humid tropics from excessive and prolonged waterlogging during the rainy season.

(b) *Impaired farm operations*: excess water on or in the soil adversely affects the soil workability. There are fewer workable days on poorly drained land and essential farm operations, ie seedbed preparation, planting, weeding, spraying and harvesting may be critically delayed. If through necessity these operations are not delayed but go ahead under unsuitably wet soil conditions, compaction, puddling and smearing of the soil is likely to occur, and the soil structure may seriously deteriorate. Besides affecting future yields, a poor soil structure also hampers the infiltration and percolation of rain water into and through the soil, leading to further reductions in the number of workable days. As explained in section 1.9, the soil will drain to FC (field capacity) but can only be worked safely when drier than LPL (lower plastic limit). Soils in which FC < LPL are thus workable sooner after rain and will generally permit more working days than soils for which FC > LPL.

The economic significance of the effects of excess water on farm operations depends on the type of farming; modern mechanised farming, for example, being much more affected than traditional peasant farming.

Watertable control

Impairment of crop growth and farm operations is directly related to the soil moisture conditions in the upper soil layers but since these are closely dependant upon the watertable depth below the soil surface, this readily determined depth is often used as the relevant diagnostic parameter. The relationship between watertable depth and crop yield is particularly well documented. The examples shown in figure 2.1 were derived in experimental fields in which the watertable during the growing season was narrowly controlled within a certain depth range below the soil surface. Under the variable watertable regime that is much more typical of the conditions in a field, crop responses vary with the pattern of the regime (the duration and timing of the high watertables are especially important) and the usefulness of curves such as in figure 2.1 is limited.

Relationships may also be expected to exist between watertable depth and farm operation parameters such as tillage costs, and the number of workable days. Available data generally indicate a continuous improvement with increasing watertable depths up to 50–100 cm for light soils and up to 150–200 cm for heavy soils, then remaining constant.

Other types of drainage problems: scope of the book

These *wet land type* of drainage problems are of course most prevalent in humid

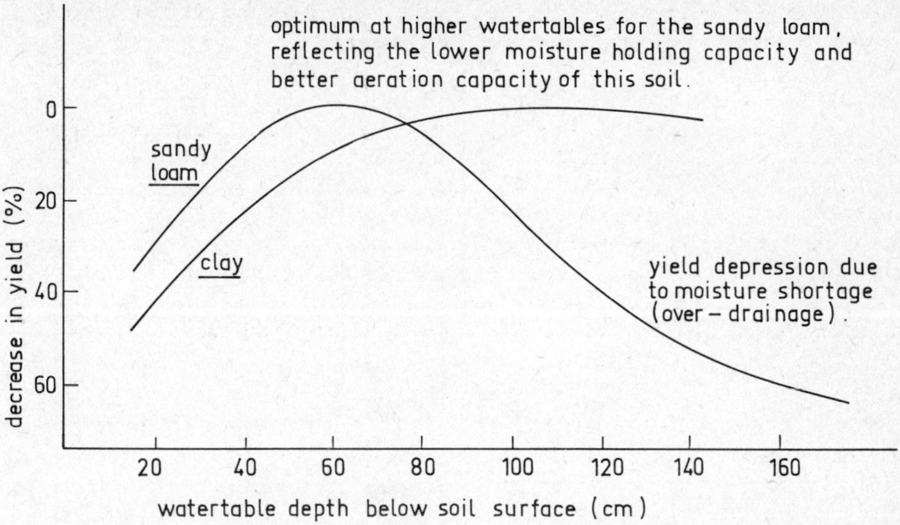

FIGURE 2.1 *Crop yield – watertable depth curves for two different soils* (Netherlands)

climates, whether temperate or tropical, and occur more on flat than on sloping land. Three other types of drainage problems occur:

1 *Flooding*: river and coastal plains may be flooded during high river or sea levels, impairing the agricultural use of the land. The solution to these problems requires in the first place flood control measures such as endiking, river regulation, or flood retention. These are problems which are outside the scope of the present book.
2 *Erosion by runoff*: this problem is to be expected on sloping land under intensive rainfall when the land has insufficient retention/detention capacity to prevent the rainfall that has failed to infiltrate from running off at high rates. As well as eroding the land on the slope, such runoff may cause considerable flooding of lower areas. Although there are many other influencing factors (soils, climate, land use, etc), erosion problems generally become serious when land slopes exceed 1–2%. The control of erosion and runoff can only be adequately discussed within the context of the entire field of soil and water conservation, which again is outside the scope of this book (this is, for example, treated by Hudson 1982).
3 *Soil salinisation*: inadequate drainage of the land is a main cause of accumulation of salts in the rootzone. The problem is most prevalent in (semi) arid climates, especially when the land is irrigated. Although the damage done to farming and the drainage requirements of the land in this case are quite different, the methods and techniques of drainage for soil salinity control are often similar to those applied for solving the wet land type of drainage problems. The specific aspects of drainage for soil salinity control are discussed separately in Part IV of this book.

2.2 SOURCES OF EXCESS WATER

Direct rainfall constitutes by far the major and most common source of excess water. However, another major source of excess water in many cold and moderate climates

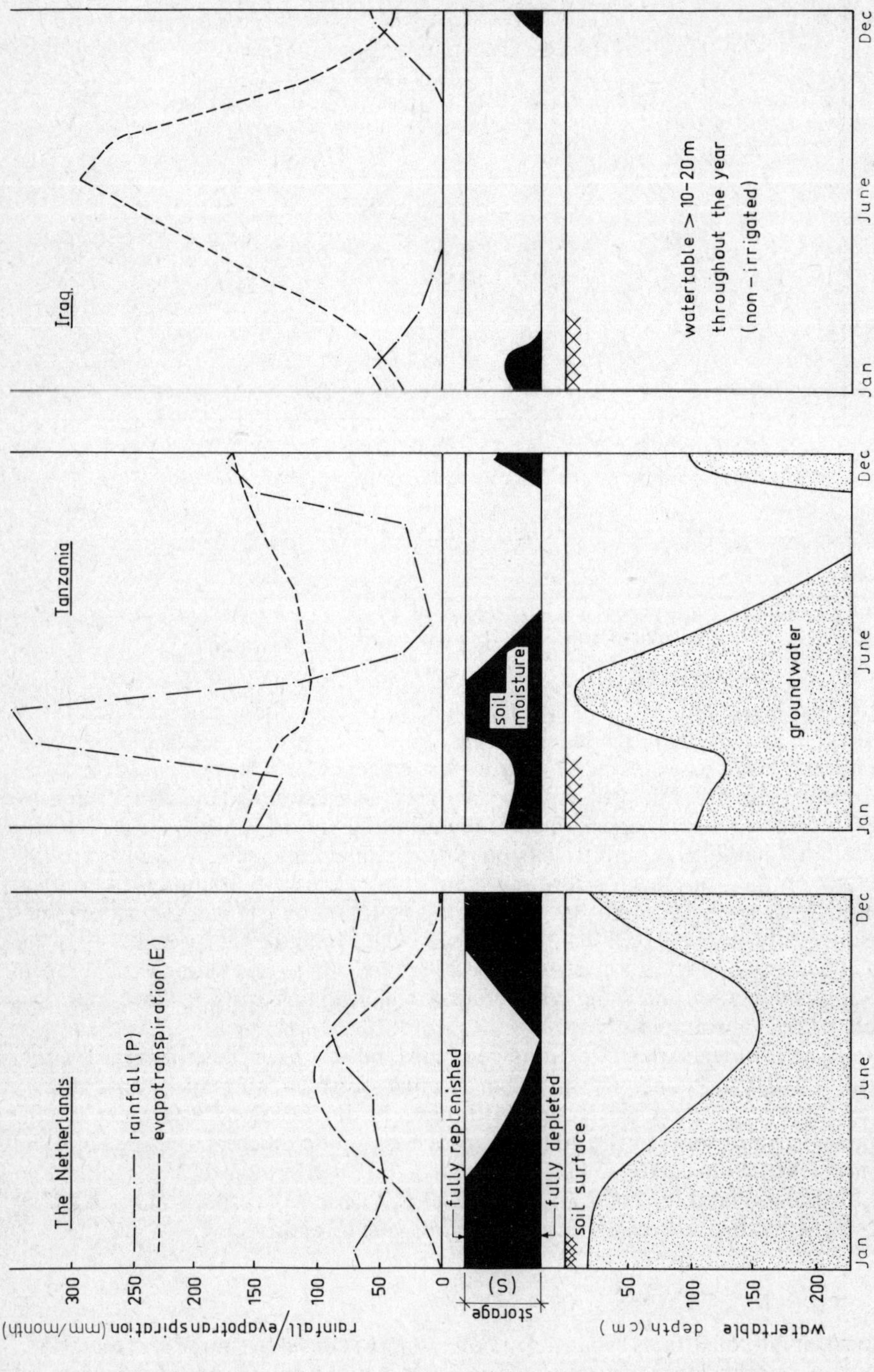

FIGURE 2.2 *Water balances and agrohydrological conditions*

is snowmelt water in the spring. Other sources such as irrigation, seepage, runoff and flood water are mostly of local or minor importance.

The occurrence of excess rainfall applies especially to humid climates. However, it may also occur in (semi) arid climates following the common type of intense, heavy storm or in general during the rainy season. The drainage load from rainfall not only depends on the amount of rainfall (P) but also on the storage capacity of the soil (S) and on the rate of evapotranspiration (E). Part of the rainfall may be stored beneficially in the soil profile or be readily evaporated so that only the remaining excess needs to be removed from the land.

Drainage problems for three quite different climatic conditions are illustrated in figure 2.2.*

1 *The Netherlands*: here P is in excess of E throughout the winter (November to March) while the summer is characterised by a small deficit of rainfall. Watertables are high during the winter and the soil moisture content even in the topsoil is close to saturation during much of the time. In the early spring, as E rises above P, the soil begins to dry and watertables fall. Crops compensate for the lack of adequate summer rainfall by depleting the soil moisture storage. For most soils the latter is adequate to meet the water requirements of the crops. When in late autumn P once again exceeds E, the excess rain first replenishes the depleted soil moisture storage and then any further excess percolates downwards to recharge the groundwater, causing watertables to rise.

2 *Tanzania*: during the early part of the rainy season P and E are more or less in balance. The soil moisture storage which was thoroughly depleted during the preceding dry season, is able to accommodate most of the short-term excess rain. Major drainage problems do not tend to occur during this period. However, when the rainfall increases in March, the soil moisture storage is rapidly replenished and watertables rise high, remaining so until June when the rainy season nears its end and P is again overtaken by E.

3 *Iraq*: under non-irrigated conditions high watertable problems are not to be expected as $P \ll E$ almost all year round. Recharge to groundwater from rain is practically nil. It is only during short periods in the rainy season that excess rainfall may occur and soil moisture storage may then become partly replenished. Some surface drainage problems may occur as rain often falls in intense storms.

Very little cropping can be done without irrigation. The introduction of irrigation, however, changes the hydrological balance. More water will percolate downwards to the groundwater as frequent irrigation keeps replenishing the soil moisture storage, and watertables may rise to critical levels causing salinisation of the rootzone.

2.3 Drainage systems

A typical agricultural drainage system has the following main components (see figure 2.3).

(a) *Field system*: this gathers the excess water from the land by means of a network of field drains, and where necessary these are supplemented by measures which promote the flow of excess water to these drains. The systems used depend mainly upon the drainage characteristics of the soil. Two principal types may be distinguished:

* The use of soil moisture calculations in drainage studies is discussed in further detail in section 15.1.

1 *groundwater drainage systems*: these are used in soils in which the excess water is able to infiltrate and percolate through the main rootzone to the subsoil/substratum and then move as groundwater flow to the drains;

2 *shallow drainage systems*: these are used where the infiltration or percolation of excess water is impeded at the surface or at shallow depth in the rootzone due to the presence of poorly permeable layers. The excess water either ponds on the surface or flows laterally above the impeding layer to the drains, either as *overland flow* or as *interflow* (perched groundwater flow). A number of measures may be introduced to promote this shallow lateral flow (land forming, subsoiling, moling, etc). The term *surface drainage* applies to situations in which the discharge to the drains occurs predominantly in the form of overland flow. On most flat to gently sloping land both overland flow and interflow occur and are difficult to separate.

(b) *Main system*: this receives water from the field systems and conveys it to the outlet. A main system is usually composed of ditches and canals of differing order (tertiary, secondary and primary). The smaller tertiary canals (collector ditches) are usually aligned along the field boundaries towards which the field drains flow. The tertiaries discharge into the secondary canals which discharge in turn into the primary canals.

(c) *Outlet*: this is the terminal point of the whole system at which it discharges into a major element of the natural open water system of the region (river, lake or sea). The waterlevel at the outlet constitutes the *drainage base* for the area concerned. This level, relative to the land level, governs the amount of hydraulic head available for the drainage flow. It determines how far the waterlevels/watertables may be lowered below the land surface. It also determines whether the area can be drained by gravity or requires pumping.

2.4 DRAINAGE AND FARMING; DRAINAGE REQUIREMENTS

The measures taken to improve the drainage conditions of an area ultimately depend upon the benefits these are expected to yield when compared with the costs. For agricultural drainage projects the benefits and the costs are mostly straightforward (farm) economic quantities. The drainage planner/engineer should of course carefully evaluate all the consequences of the proposed drainage, particularly the possible effects on the environment, and point out where these may be harmful so that alternative solutions may be considered.

Benefits from drainage of agricultural land may accrue from the improved crop growth conditions created by drainage (earlier, higher, more dependable, better quality yields; wider range of more demanding/attractive crops; etc) or from the improved soil workability conditions (earlier planting, more workable days, lower draught requirements, less damage to soil structure by farm machinery, increased scope for spreading and planning of the work, etc). All in all, land drainage may be expected to result in better yields at lower costs, in other words higher net returns to the farmer. These returns should always be considered in relation to the costs of the drainage improvements (see chapter 17).

Drainage is of course just one of the factors that determines the returns; others are cropping pattern, fertilizer application, irrigation, machinery use, management skill, etc. The *Law of the Minimum* applies and a significant increase in the returns due to improved drainage can only be expected when there are no major constraints imposed by these other factors. High investments in drainage, as required for

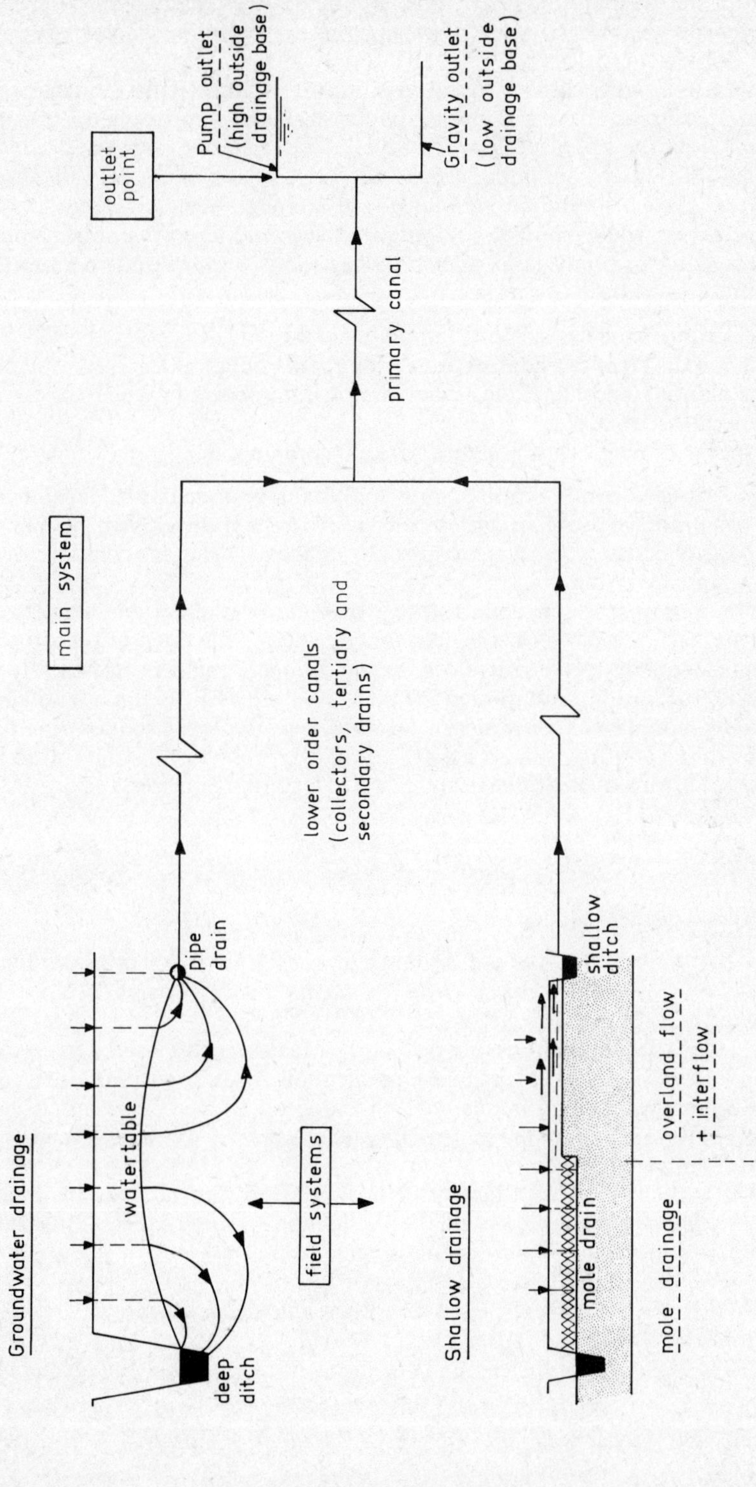

FIGURE 2.3 *Main components of a drainage system*

example for the installation of a pipe drainage system, may therefore not be warranted in an area with 'low level' farming (this applies to many drainage projects in the developing countries).

In general, soil and climatological conditions together with the farm system employed, determine the economic scope for drainage improvements. On many soils, notably on the so-called *heavy lands*, a high degree of excess water control is technically difficult or even impossible to achieve – at any event only at very high cost. On such soils a combination of improved drainage, achieving a moderate but significant excess water control, and adapted farming, imposing some moderate constraints on the use of the land, often provides a satisfactory solution. Examples of adaptations in farming in this respect are:

- crop selection on the basis of the drainability of the land.
- planning of the farm calendar so as to avoid performing critical farm operations, such as planting and harvesting coinciding with a period in which excess water may be expected to occur.
- use of special practices (low pressure tyres, air spraying, etc).

The most suitable combination of improved drainage and adapted farming in the solution of a drainage problem, varies from case to case. It should depend mainly on how the costs of better excess water control by improved drainage compare with the 'costs' of adapted farming.

The foregoing discussion demonstrates that many interactions exist between drainage and farming as well as the need for considering the integration of drainage with farming. It is important therefore for the drainage engineer to be thoroughly familiar with the farming situation in the area. This will assist him in identifying the constraints imposed by excess water on farming, to evaluate the economic impact of these constraints and to visualise how the drainage problems may be solved by an integrated set of improved drainage and adapted farming measures.

References and further reading

CANNELL, R Q 1979: 'Effects of soil drainage on root growth and crop production'. In: *Soil Physical Properties and Crop Production in the Tropics*, R Lal and D J Greenland (eds), Wiley, Chichester

ERIKSON, J 1979: 'Soil functions and drainage'. In: *Proceedings of the International Drainage Workshop*, J Wesseling (ed), publication 25, International Institute for Land Reclamation and Improvement, Wageningen

FAO/UNESCO 1973: *Irrigation, Drainage and Salinity; an international source book*, Hutchinson, London

VAN DE GOOR, G A W 1972: 'Plant growth in relation to drainage'. In: *Drainage Principles and Applications*, Volume I, publication 16, International Institute for Land Reclamation and Improvement, Wageningen

HUDSON, N W, 1982: *Soil Conservation*, Batsford, London

INRA 1981: *Drainage Agricole*, Chambre Régionale d'Agriculture de Bourgogne, Dijon (in French)

MAFF 1977: 'Drainage in the economy of the farm', Technical Management Note no 23, Ministry of Agriculture, Fisheries and Food, London

REEVE, RC and NR FAUSEY 1974: 'Drainage and timeliness of farming operations'. In: *Drainage for Agriculture*, J van Schilfgaarde (ed), Agronomy Monograph no 17, American Society of Agronomy, Madison, Wisconsin

RUTHENBERG, H 1980: *Farming Systems in the Tropics*, Oxford University Press

SMITH, LP and BD TRAFFORD 1975: 'Climate and drainage', Technical Bulletin 34, Ministry of Agriculture, Fisheries and Food, London

WESSELING, J 1974: 'Crop growth and wet soils'. In: *Drainage for Agriculture*, J van Schilfgaarde (ed), Agronomy Monograph no 17, American Society of Agronomy, Madison, Wisconsin

PART II
DRAINAGE SYSTEMS

Chapter 3 Groundwater drainage systems*

These systems normally consist of a network of deeply installed field drains, establishing a deep drainage base in the soil, well below the main rootzone. In the drains the soil water potential is effectively zero so that all groundwater above the drainage base has a higher potential and is under gradient to flow towards and into these drains. Low watertable depths may be maintained by selecting a suitable drain spacing. Provided the downward movement of excess water is not impeded, the overlying soil profile will drain to field capacity.

The field drains may either be pipes or deep ditches. Pipe drains are also frequently referred to as subsurface, buried or covered drains, in contrast to open ditch drains. Low watertables may also be maintained by a network of pumped wells. This method of groundwater drainage is referred to as *vertical drainage* in contrast to the *horizontal drainage* provided by pipe drains or ditches. Vertical drainage has a limited applicability (see section 12.2). The horizontal method is by far the most common.

Groundwater drainage is applicable in soils where:

(a) the rootzone is underlain by strata of reasonable hydraulic conductivity K and/or thickness D (the KD-value of these strata should be reasonably high);
(b) the excess water on or in the soil is able to infiltrate and percolate through the rootzone to the underlying watertable at reasonable rates.

Figure 3.1 illustrates the flow pattern of the excess water to the field drains for the case of a parallel pipe drainage system (the most common type of groundwater drainage system – see section 3.1). When the pipe drains are underlain by deep permeable soil, as is assumed in respect of figure 3.1, the flow of groundwater will extend well below the drains. The height 'h' of the watertable above the drainage base midway between two drains constitutes the head driving the groundwater towards the drains. When the KD-value of the soil at and below drainage base depth is high, the excess water may be discharged with a low head and so low watertables can be maintained even with rather wide drain spacings (> 50–100 m). When the KD-value is low, either more head must be provided or a smaller drain spacing must be used to generate the same discharge. In soils with poor infiltration or percolation characteristics and/or with poorly permeable substrata, excess water control in the rootzone is both technically and economically impossible, and other drainage measures have to be used (see chapter 4).

3.1 Pipe drain systems

Most groundwater drainage for modern farming in temperate climates is done by means of pipe drain systems. These systems are also extensively used for salinity control in irrigated areas. Very little pipe drainage has as yet been installed in the

* Also referred to as *underdrainage* and *subsurface drainage*. The term *groundwater drainage* is, however, preferred because it emphasizes the primary function of these systems, ie to control the watertable by discharging groundwater.

(semi) humid tropics. The reasons for this are partly economic (the present low output level does not warrant a high investment in drainage as required for pipe drainage) and partly technical (open ditches are often preferred as these drains can also serve to evacuate excess surface water).

A great deal of research has been done on the functioning of the pipe drain, the materials used and the installation methods, all of which has added to the sophistication and dependability of the pipe drainage system.

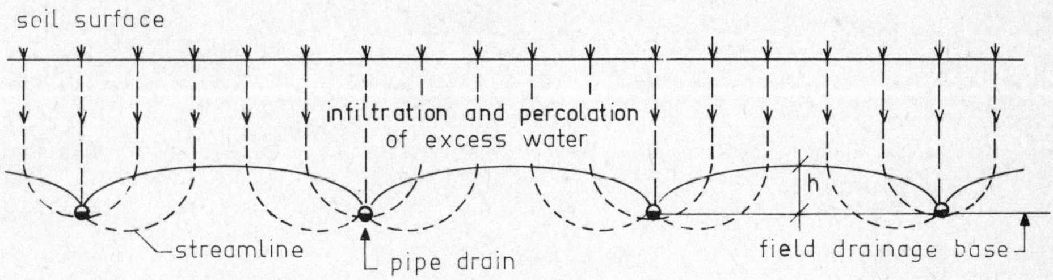

FIGURE 3.1 *Typical flow pattern to parallel pipe drains*

3.1.1 Layout patterns

The alignment of field drains and the collector drains into which they discharge are mutually dependant, their alignment, however, in the first instance being determined by the topography of the land to be drained. Drains are most effective when they are sited so as to pass through the lowest areas in the land since these are the positions to which water gravitates naturally.

For example, a field in which waterlogging is confined mainly to a number of depressions, is drained most efficiently by aligning the drains so that they pass through and connect these depressions. The irregular pattern of field drains which results is termed a *natural system* (see figure 3.2a). As a drain running through a depression (*depression drain*) is liable to collect some surface runoff and interflow as well as groundwater flow, an open ditch is often used in preference to a buried pipe.

As the size of the depressions decreases and their number increases, the natural layout increasingly loses its advantage compared to a regular network of field drains uniformly covering the field. Common examples are the *parallel grid system* (see figure 3.2b) and the *herringbone system* (figure 3.2a).

In the herringbone system the collector drains are aligned down the main slope and the field drains (*laterals*) are aligned across the slope but at a slight angle to the contours, so that the pipes slope downwards towards the collector drain but remain at a constant depth below the surface. In a parallel grid system, the laterals are given slope by increasing the installation depth towards the collector. As typical slopes are 5–10 cm per 100 m length and the maximum drain length is seldom in excess of 250 m, the difference in drain depth below the soil surface between the ends is only 10–20 cm. The depth uniformity, therefore, is hardly a consideration in favour of the herringbone layout. This also applies to the alleged advantage of an oblique entry by laterals into the collector drain in the herringbone system. The choice of one system in favour of another depends mainly on the field situation. In flat plains where large rectangular fields are the norm, the field drainage systems are commonly of the grid type. Herringbone systems have an application in more undulating terrain where

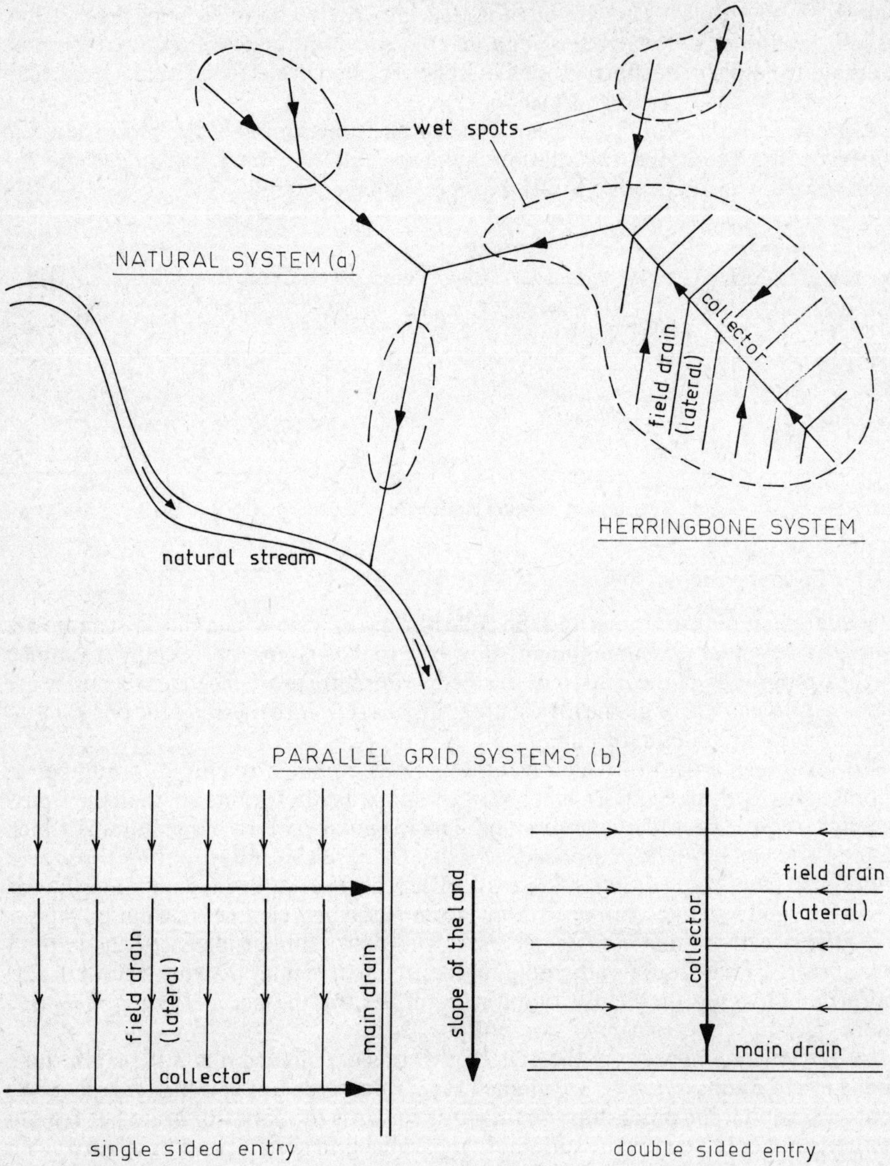

FIGURE 3.2 *Field drainage layout patterns*

only part of a field requires drainage. The situation depicted in figure 3.2a is typical, with the collector passing through a depression whilst laterals cover the lower slopes.

In situations where there is a distinct groundwater flow in a certain direction, field drains are best aligned as *interceptors* across the direction of flow. In sloping fields this means that the laterals should be aligned across the slope more or less parallel with the contours with the collector running down the slope (see section 13.1.3).

Alignment with the collector across and laterals parallel to the contours is also the most suitable choice on most flat land since this enables collectors to be entered by laterals from both sides. When collectors run parallel to the contours, laterals are normally only able to enter from one side (see figure 3.2b), unless land slopes are minimal. Single sided entry of laterals requires twice as many collectors as double sided entry.

3.1.2 Singular v composite systems

Pipe laterals may either discharge into a ditch collector or into a pipe collector. In the first case the system is referred to as a *singular system*; in the latter as a *composite system* (see figure 3.3).

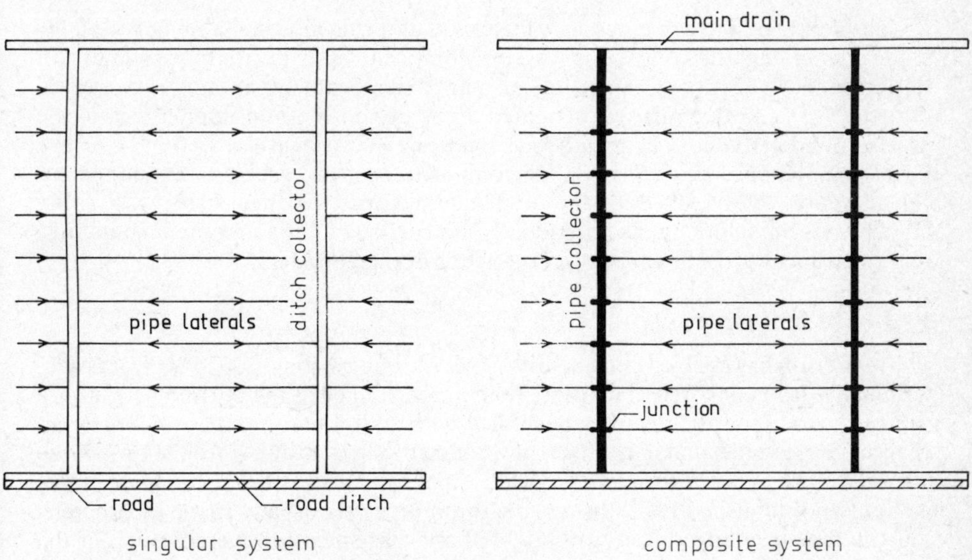

FIGURE 3.3 *Singular and composite pipe drain systems*

With respect to the choice between the two systems, the following considerations apply:

(i) *surface water*: a ditch collector (singular system) provides an outlet for excess surface water;

(ii) *field size and land loss*: pipe laterals as a rule should not be made longer than 300 m, limiting field width in the singular system to 300 m (single sided entry), or 600 m (double sided entry). Land loss by ditches in a singular system may add up to some 2–3%;

(iii) *blockage*: the outflow of a pipe drain into a ditch collector (singular system) is easy to inspect and malfunctioning easy to localise. Moreover, blockage affects a small area only. In a composite system, blockage may affect a large area (the more so the closer to the discharge point the blockage occurs) while it is liable to continue over a longer period as malfunctioning is not so obvious;

(iv) *maintenance*: ditch collectors require much more maintenance than pipe collectors (once or twice a year compared to once per five to ten years);

(v) *outlets*: the many pipe outlets in the singular system represent weak spots as they are easily damaged, eg during ditch maintenance. Moreover, they hinder (mechanical) ditch maintenance;

(vi) *hydraulic gradient*: a pipe collector requires about five to ten times as much gradient as a ditch collector;

(vii) *costs*: installation costs for composite systems are considerably higher than for singular systems. Costly provisions are required at the various junctions to allow for inspection and maintenance. However, considering the lower maintenance requirements of the composite system, total annual costs differ little, although normally remaining lowest for singular systems.

Considering all factors, singular systems are as a rule most suitable in flat plains under humid climatic conditions whereas composite systems are more suitable in sloping land. The latter systems often also have significant advantages in respect of drainage for salinity control in irrigated areas due to the non-interference of the underground drainage system with the irrigation canal system (see section 12.1). The lower maintenance requirements of composite systems are also an important consideration. Not only does the ditch maintenance become costly with rising labour costs but in practice it is frequently not carried out up to required standards, to the detriment of the functioning of the whole drainage system.

3.1.3 The drain pipe

The pipe drain has evolved considerably since its first use some 200 to 300 years ago. Originally these consisted of no more than a trench filled at the bottom with stones or brushwood (and then further backfilled with soil). Later, regular underground channels were made, first in the form of wooden box spouts but, since the turn of the century, mostly in the form of clayware pipes (also called *drain tiles*). In places without an established tile industry, the drain pipe was usually made of concrete. Plastics pipe was introduced around 1960 and has since conquered much of the drain pipe market in many countries. Figure 3.4 illustrates some of the more commonly used pipes whilst relevant and up to date information on them has been compiled below (for standards see ISO 1979).

(a) *Clay tile pipe*: standard sizes vary between countries although typical pipe sections are 30 cm long and have internal diameters (ID) equal to 5, 6·5, 8, 10 up to 20 cm. Special pipes with collars are available for use in soils in which consolidation is likely to occur, eg peats or non-ripened marine soils, although these are 25–30% more expensive than conventional straight pipes.

The clay tile is highly resistant to deterioration in aggressive soil conditions.

Pipe sections are abutted against each other and water enters through the joints that exist as a direct result of the imperfect fit between the ends of the pipe sections, see figure 3.5a.

(b) *Concrete pipe*: mostly medium to large size with diameters of 15–20 cm or more and section lengths of 30 cm for small diameter pipe and up to 50 cm for large

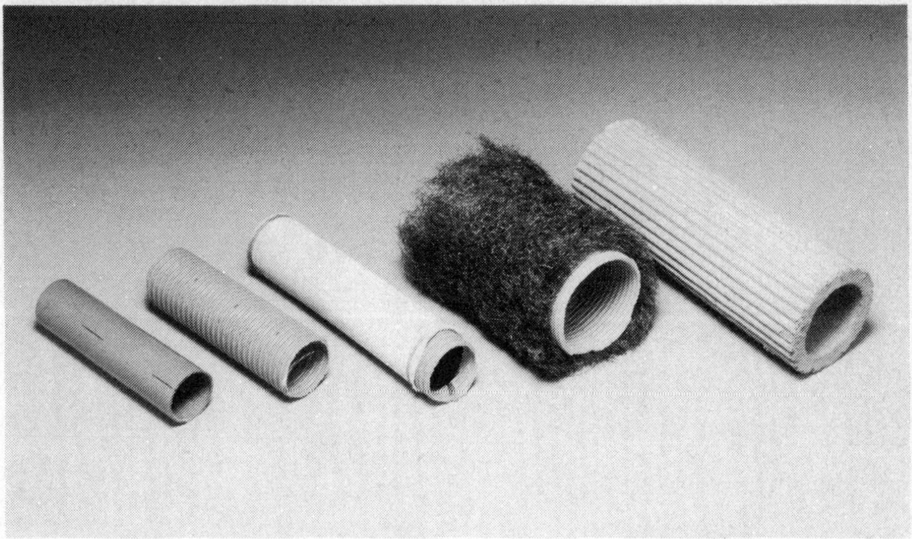

FIGURE 3.4 *Widely used types of drain pipe*

diameter pipe; water entry occurs through the joints. Pipes made with ordinary (Portland) cement are liable to deteriorate in acidic or salty soils (especially attacked by sulphates) and in these circumstances special resistant cements should be used.

(c) *Plastics pipe*: may be made from PE (polyethylene) or PVC (polyvinyl chloride) of which the latter is the most common (higher bearing capacity, lower costs). It is very durable but subject to deterioration by long exposure to the ultra-violet radiation of strong sunlight. PVC becomes brittle with freezing temperatures and can then easily fracture.

Most plastics drain pipe is nowadays supplied as corrugated pipe. Corrugated pipes have a higher hydraulic roughness and require a 20% increase in diameter to carry the same flow as smooth pipes. The cost per metre length of smooth and corrugated pipes having similar water transport capacities, however, is about equal, since more material goes into a smooth pipe compared to a corrugated pipe of the same diameter.

Plastics pipe should be properly embedded to avoid excessive deflection which may cause failure.

Standard corrugated drain pipes usually have outside diameters (OD) of 40, 50, 65, 80, 100 and 125 mm, whereas $ID \sim 0{\cdot}9 \times OD$. At small diameters the pipe is flexible and is delivered in coils of 30–50 kg, containing up to 200 m of pipe. The price of plastics pipe increases sharply with increasing diameter (a 10 cm diameter pipe is for example roughly four times as expensive as a 4 cm pipe). Large diameter pipe ($>$20 cm) as required for collector drains, is available in plastics but is seldom competitive with concrete pipe.

The pipes are perforated for water entry as is shown in figure 3.5a (0·6–2·0 mm wide slots, usually on the grooves of the corrugations). The perforations may be arranged in any pattern which provides an even distribution around the whole circumference. The open area should be minimally 800 mm^2 per m pipe length.

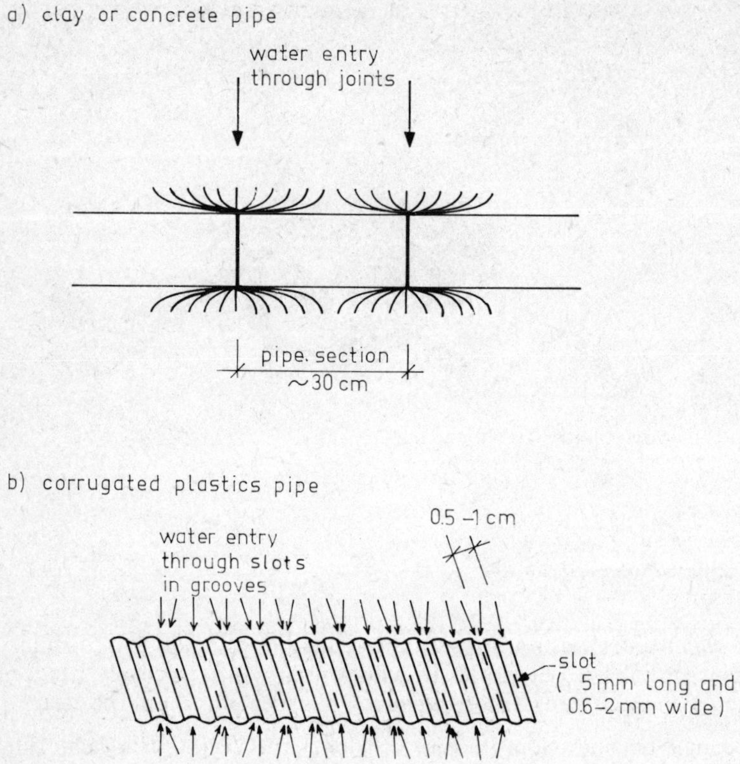

FIGURE 3.5 *Entry flow pattern for clay/concrete pipe and for corrugated plastics pipe*

Entry losses

Head may be lost as the water flows towards and through the rather limited open areas into the pipe. This head loss constitutes the entry loss and it follows that, for the same inflow, pipes with a very small entry area will have a much higher head loss than pipes with a large entry area. Entry characteristics of different pipes may be compared using a standardised *resistance factor* α which is related to the head loss at entry as (ILRI 1979):

$$h_e = \alpha \frac{Q}{K} \quad \text{(compare to Darcy's Law, section 1.6)}$$

Where: h_e = head loss at entry, m
α = resistance factor, dimensionless
Q = the inflow rate to the pipe per m length of pipe $m^3.m^{-1}.day^{-1}$
K = the hydraulic conductivity of the material enveloping the pipe, $m.day^{-1}$.

The ranges of variation of α are:

clay and concrete $\alpha = 0{\cdot}4 - 2{\cdot}0$

smooth plastics $\alpha = 0{\cdot}4 - 0{\cdot}6$
corrugated plastics $\alpha = 0{\cdot}05 - 0{\cdot}1$

Clay and concrete pipes generally have a higher entry resistance than other pipes due to the wide spacing between points at which water is able to enter the pipes. Corrugated plastics pipe offers the least resistance which is at least partly due to the fact that it tends to have a considerably greater slot area than the International Standard minimum normally adopted for smooth plastics pipe. These values of α are of importance in selecting a suitable type of envelope for a drain pipe (see section 3.1.4).

Selection criteria

As far as their practical drainage performance is concerned, there is very little to choose between the different types of drain pipe, provided they have been installed correctly. Selection, therefore, is mostly based on cost comparison and on local availability. In addition the following considerations may be relevant:

(i) where pipe drains are not locally available, local manufacture of concrete pipes is the most straightforward and easiest to organise. It requires little skill and can be economically done even on a very small scale. Local manufacture of clay tile drains requires a considerable investment and skill and is economically feasible only for large quantities.
Plastics occupy an intermediate position; local manufacture from imported (plastics) material is feasible for reasonable quantities;
(ii) where all types are available the use of corrugated plastics pipe often has distinct advantages. These are due to its light weight and delivery in long lengths, making transport costs (to the site as well as on site) relatively low and on-site handling relatively easy (can be carried by hand, as may be required under wet conditions);
(iii) plastics pipe has the general advantage that performance is less affected by poor installation.
Plastics pipe is particularly suited for machine installation;
(iv) prices of small diameter pipe ($\emptyset \leqslant 10$ cm) are usually of the same order of magnitude for tile, concrete and plastics.
For large diameter pipe, concrete is normally the cheapest and plastics the most expensive.

3.1.4 The pipe envelope

A pipe envelope is the material placed around the pipe to serve one or both of the following functions:

(a) *hydraulic function*: to reduce the head loss at entry by facilitating the flow of the water to the entry points;
(b) *filter function*: to prevent the entry of fines into the pipe, mostly by reducing the velocities of the flow converging onto the entry points, to the extent that very little sediment is actually transported by the water.

Envelopes used in practice have their strong and weak points and selection should be based on the function that is most critical and which is best served by a particular envelope. The filtering function is especially important during the first period after installation, until the backfill around the pipe has become stabilised, although in unstable soils the filter function is more or less permanent.

Most traditional envelopes provide for a thick layer of highly permeable material around the drain pipe. This has two beneficial effects. Firstly, the head loss at entry is considerably reduced, the resistance factor α becoming generally very much less than 0·1. Secondly, the rate of flow of water leaving the soil and entering the surround is several hundred times less than the rate at which it would flow directly from the soil into the slots of a pipe, without a permeable surround. This reduction in velocity lessens the risk of siltation occurring as a direct consequence of erosive flow velocities. These two objectives are generally sufficiently achieved when the envelope has a minimum thickness of 5–10 mm and a K-value at least ten times greater than that of the soil (ILRI 1979). The materials that may be considered as fulfilling this condition are coarse sands and gravels, organic materials such as peat litter, coconut fibre, flax, straw or thick synthetic materials such as propylene mats or polystyrene granules. The organic materials are only semi-durable and straw or chaff in particular decay rapidly. Under anaerobic conditions which can occur, decomposition will produce humic substances which may adversely affect the hydraulic conductivity of the envelope.

The filter function is important in situations where the soil is liable to be transported into the pipe, causing blockage. In order to fulfill this function the pore size of the filter has to be carefully designed to match the particle size of the protected soil. A perfect filter prevents all particles from passing but experience has demonstrated that a successful field drain filter must allow some fine material to pass right through to the drain. These fines generally remain in suspension and are discharged at the drain outlet.

Soils for which a filter is required are those which are most readily eroded and transported by water. Experience indicates that the most critical soils in this respect are uniformally graded coarse silts and fine sands, having median particle sizes in the range 20–100 microns. The entry flow rates are seldom high enough to transport larger sized particles whilst very fine particles <2 microns have an inherent stability due to their cohesion. Instability can also be caused by the chemical status of a soil, eg due to salinity, or it may result from very high erosive forces occurring during irrigation or intense rainfall. The erodibility of a soil may be roughly evaluated by observing the disintegration of the soil when left under water for some hours/days.

Gravel envelope

Design criteria for granular drain envelopes have been developed by both the US Soil Conservation Service and the US Bureau of Reclamation through careful experimentation. The emphasis is on the filtering function (the term *filter* is used rather than envelope). The USBR criteria allow use of somewhat coarser material but both have important restrictions on the quantity of fine materials permitted in the filter.

The US Soil Conservation criteria are*:

$$\frac{D_{50F}}{D_{50S}} = 12-58; \quad \frac{D_{15F}}{D_{15S}} = 12-40$$

* D_{50} = sieve mesh size through 50% of the material passes whereby S refers to soil material, F to filter material.

For stability of the filter:

$$\frac{D_{15F}}{D_{85S}} = \leqslant 5; \quad D_{100F} \leqslant 38\ \text{mm}; \quad D_{90F} \leqslant 20\ \text{mm} \quad \text{and}$$

$$D_{85F} \geqslant \text{half the width of the pipe perforation.}$$

Because the control of fines is of paramount importance, it has recently been recommended that:

$$D_{5F} \geqslant 0{\cdot}42\ \text{mm} \quad \text{(FAO, 1980)}$$

The US Bureau of Reclamation criteria are set out in table 3.1 whilst figure 3.6 illustrates the procedure used to design a filter for a siltation prone soil.

TABLE 3.1 *Gradation relationship between base material and filter* (US Bureau of Reclamation 1978)

Base material 60% passing (diameter of particles, mm)	*Gradation limitations for filter (diameter of particles, mm)*											
	Lower limits						*Upper limits*					
	Percentage passing						*Percentage passing*					
	100	60	30	10	5	0	100	60	30	10	5	0
0·02–0·05	9·52	2·0	0·81	0·33	0·3	0·074	38·1	10·0	8·7	2·5	—	0·59
0·05–0·10	9·52	3·0	1·07	0·38	0·3	0·074	38·1	12·0	10·4	3·0	—	0·59
0·10–0·25	9·52	4·0	1·30	0·40	0·3	0·074	38·1	15·0	13·1	3·8	—	0·59
0·25–1·00	9·52	5·0	1·45	0·42	0·3	0·074	38·1	20·0	17·3	5·0	—	0·59

required gradings for $D_{60S} = 0{\cdot}075$ mm

$D_{60S} = 0{\cdot}075$ mm

lower limit $D_{60F} = 3$ mm

upper limit $D_{60F} = 12$ mm

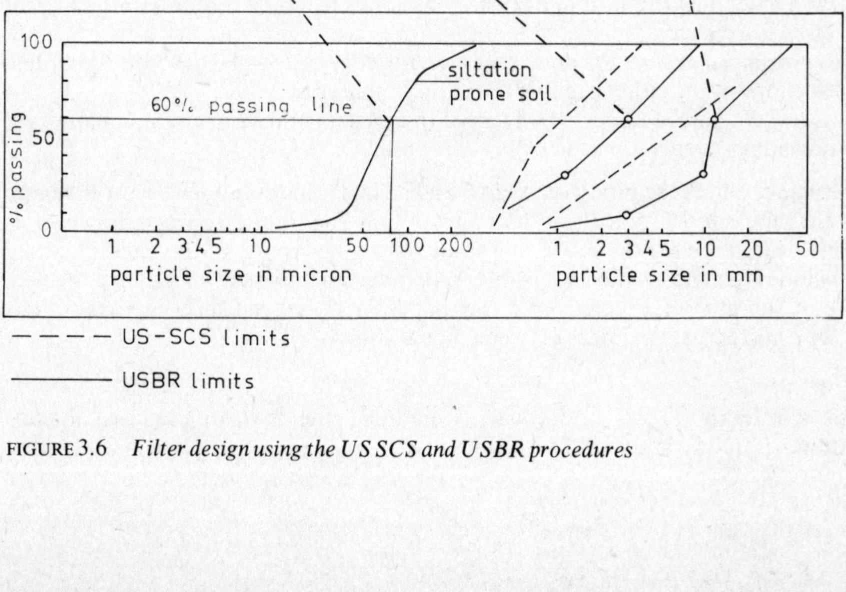

FIGURE 3.6 *Filter design using the US SCS and USBR procedures*

Synthetic and organic envelopes

Synthetic substitutes for graded gravel filters have been developed in recent years. These range from thin woven or non-woven fabrics to thicker felt materials. Design procedures have also been developed which seek to design the pore size of the fabric so that they correspond to the critical pore sizes that exist in a comparable graded gravel filter.

The thin fabrics generally have a high hydraulic conductivity across the fabric, $K > 20\,m.day^{-1}$ but the transmissivity in the plane of the fabric is very low. Thin woven fabrics have a uniform pore opening size whereas non-woven fabrics have a distribution of pore sizes around a mean. Both types of material are satisfactory as a filter but their application as a surround to smooth plastics, concrete or clay pipe cannot be recommended. The reason for this is that the only functional part of the fabric lies directly above the slots which in effect reduces the slots open area and leads to increased entrance resistance. Ideally the thin fabrics should be used to protect a coarse gravel envelope from ingress of loose trench backfill. They are also useful as sleeves for corrugated pipes which are slotted on the grooves. The tight fitting sleeve bridges the corrugations, leading to a considerable reduction in the already low entrance resistance.

The thicker felts, coconut fibre or propylene mats have the advantage of possessing transmissivity in the plane of the fabric so that the whole of the fabric is involved in the transfer of water from the soil to the entry slots in the pipes. The pore size distribution of felts will, however, change due to compression by the soil backfill, a fact which should be allowed for in evaluating this type of material.

Synthetic filters can be susceptible to clogging by iron, manganese or humus, and some fibres can swell in the presence of water so it is advisable that performance tests should be carried out in the field before a fabric is specified.

Methods of application of envelope material

The envelope material may be applied by different methods depending mostly on the envelope material and pipe installation method used:

(a) *in bulk*: the material is spread loosely in the trench, usually on top of the pipe (giving an approximate 75% envelope). A complete envelope is possible also with machine installation. Bulk application is common for materials like peat litter, gravel, etc;
(b) *pre-wrapped around the pipe*: common for corrugated plastics pipe; complete envelope, using either organic or synthetic materials;
(c) *as a strip*: placed under or on top of the pipe (synthetic or organic material); nowadays mostly replaced by pre-wrapping.

Most water enters the pipe from below and so from this point of view an envelope below the pipe is more important than one on top. The siltation hazard, however, is often greatest in the region above due to the initial instability of the backfill, and a full envelope is generally most advisable. Where severe ingress of backfill is expected, the top of the pipe may be covered (tar paper or polythene sheet for example is sometimes laid across the joints between pipe sections).

Selection criteria

Besides considering costs and local availability, the following technical considerations should be taken into account:

(i) in stable, well structured soil (at drain depth) an envelope may be omitted when it is assured that the pipes are installed under dry conditions and backfilling is done carefully, using stable, well structured topsoil around the drain. When there is no such assurance, an envelope should be provided (if only as a precaution against poor installation);
(ii) a relatively thick (1–2 cm when compressed) permeable envelope, with good filtering properties (eg gravel or a good synthetic substitute) should be used where pipes are installed in weak structured, easily dispersed soil, or where this type of soil is used for backfill around the pipe. This applies to most silty and fine sandy loams, silty sand and unstable sodic soils;
(iii) a (cheap) thin synthetic fabric, woven or non-woven, may be used where the envelope functions mainly as a screen but these should only be used on corrugated pipes (eg installation in, or backfill with, very permeable sandy soil). With smooth plastics, tile and concrete pipes the use of such an envelope may lead to high entry resistances;
(iv) in situations in which the flow to the drain is likely to be very high (wide spacings and/or high drainage rate), good envelopes are both technically and economically justified.

3.1.5 Structures in pipe drain systems

Most structures are of a simple type and occur mainly in the larger composite systems. Typical examples are shown in figure 3.7.

Outlet of a pipe drain into a ditch or canal

This structure should meet the following requirements:

- be reliable: malfunctioning affects the performance of the entire drain;
- prevent erosion of the side slope;
- preferably not interfere with ditch maintenance:
- prevent the entry of animals (applies to pipes with ID > 5 cm).

For singular field drains draining areas up to 1·0 ha, 150–200 cm long plastics pipes (half enveloping the drain pipes, half protruding into the ditch) are suitable. Where protrusion of the pipes seriously hinders ditch maintenance, removable or telescopically insertable pipe (commercially available), or a non-protruding pipe with side wall protection (chute) may be used. Collector pipe drains warrant more expensive outlets (see figure 3.7).

Junctions and inspections chambers

These are used in composite pipe drain systems to connect lateral pipes with the collector drain pipes. Most connections are made using 'blind junctions', which cannot be inspected after installation (figure 3.7). In many locations, however, a combined junction/silt trap should be provided, with access for inspection and for cleaning the drain from time to time. This type of inspection chamber is fairly expensive and so it should only be used in vital locations, eg where the slope of a collector drain flattens out (liability to silting) or on long lengths of drain (every 250 m on long runs). Experience has shown that above ground inspection chambers are liable to be vandalised or damaged by farm implements and so there is an increasing tendency to install buried chambers with some marker (eg metal covers used in combination with a metal detector) to enable their location to be pinpointed at a later date.

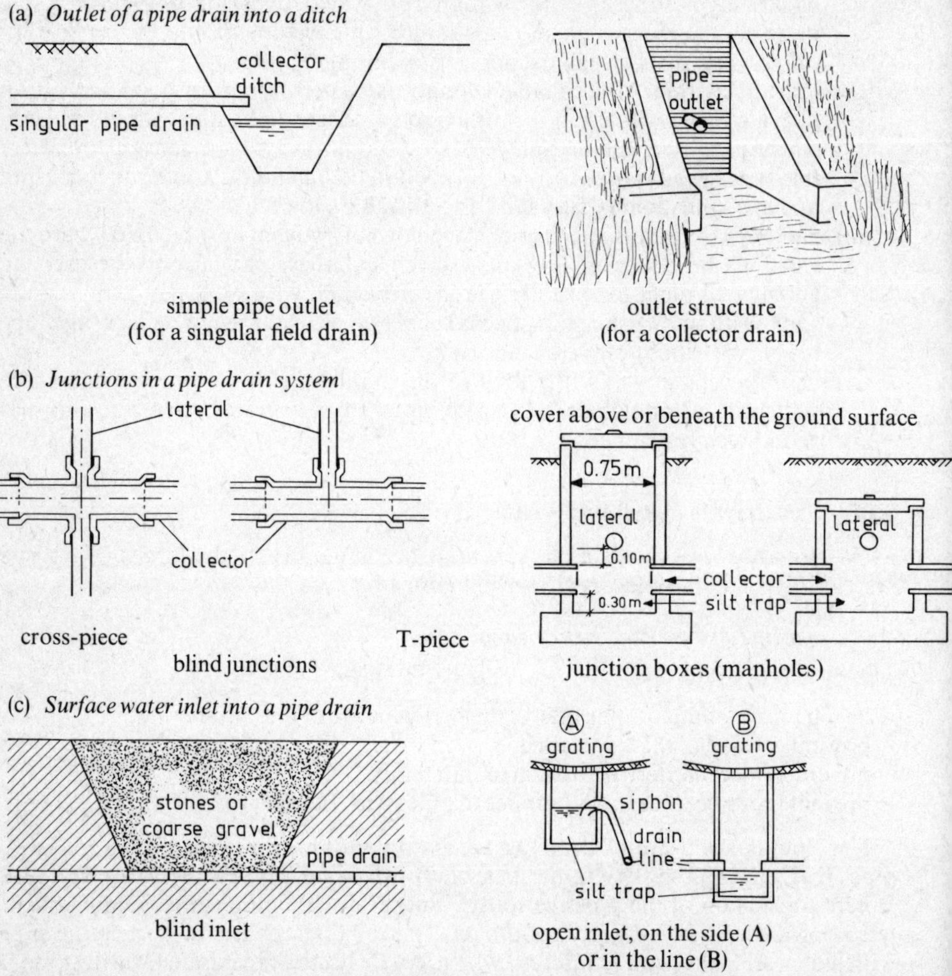

FIGURE 3.7 *Structures in a pipe drain system* (Cavelaars ILRI 1974)

Surface water inlets

Used for incidental discharge of surface water, eg where localised areas of excess surface water occur. Provision of proper silt traps is essential for open inlets; an inlet located to the side of the drain line, providing a safeguard against poor maintenance, is to be preferred in this respect.

Bedding

The soil in which the pipes are laid, generally has enough strength to support the pipe. Pipes laid in a trench are usually embedded in a V-notch or other form-fitting groove excavated in the bottom of the trench; (most machines form such a groove which also facilitates alignment of the pipes). A gravel envelope provides very good bedding conditions for the pipe, although this is seldom the sole reason for it being used. It may, however, be an additional reason for using a gravel envelope where easily deflecting corrugated pipe is to be used under uncertain bedding conditions.

So-called 'drain bridges' are used where the pipe drain has to pass through a soft spot for example a recently in-filled ditch. If the pipe is not supported it is likely to subside, resulting in mis-alignment or breakage. Wooden planks laid beneath the pipe or a continuous length of rigid pipe may be used for this purpose.

The special problems encountered with pipe installation in soft, unripened soils, are given in chapter 14.

3.2 DITCH SYSTEMS

These consist of ditches laid out in various patterns. In order to effectively act as groundwater drains, ditches must be rather deep (water level ~ 1·0—1·5m below soil surface). The ditches normally have a trapezoidal section with side slopes varying from 1:1 to 1:2, depending on the soil conditions.

A number of factors enter into the choice between pipe drains and ditches as the most suitable field drain for an area. The influence of each of these factors depends very much on the prevailing local conditions, although the following main considerations apply:

(i) *surface water*: ditches can also collect excess surface water, a function which pipe drains are only incidentally able to perform;
(ii) *land loss*: can add up to 10% for ditches (when narrowly spaced and constructed with gentle side slope) as compared to no loss in the case of pipe drainage;
(iii) *hindrance to field operations*: ditches restrict machine manoeuvrability, resulting in higher farm costs compared to pipe drainage;
(iv) *maintenance*: ditches require frequent maintenance (a few times a year as compared to once every few years for pipe drains). In practice maintenance is either not done or it frequently fails to meet required standards, resulting in poor performance of many ditch drain systems;
(v) *installation*: pipe drains can be installed by machine under good grade, guaranteeing good discharge performance. Similar machines are not available for ditch construction, and ditches often have adverse bed grades resulting in poor discharge and functioning;
(vi) *soil conditions*: may favour or prevent the use of ditches or pipe drains (caving in of ditches; clogging of pipes by iron compounds; etc).

Generally pipe drains result in better drainage than ditches due to better installation and mostly trouble free functioning, while the fact that pipe drains do not hinder (mechanical) farm operations is often of overriding importance. However, the use of ditches may well be preferable when:

- the land in question requires both surface and groundwater drainage (applies to large areas of land in the humid tropics);
- ditches do not severely hinder farm operations as for example would be the case where ditches are widely spaced, where the land is used for perrenial crops like sugarcane, bananas, etc, or where farm operations are mostly done manually;
- standard of drainage required is not high (applies to many grassland areas);
- cost advantages overshadow all disadvantages.

Many peat soils are best drained by ditches, as pipes are liable to become clogged by iron compounds, become mis-aligned due to uneven subsidence and chemically deteriorate. Newly reclaimed, non-ripened marsh and sea or lake bottom soils are also best drained by ditches; after some years when the soil has ripened sufficiently, the ditches may be replaced by pipe drains (see also chapter 14).

Most factors set out in section 3.1.1 with respect to alignment and layout of field drains are equally applicable to ditches. As noted, ditches are often well suited to a natural system of field drainage but are also commonly used in parallel systems. The field ditch system presented in section 4.2.1 as a shallow drainage system for heavy land, may also function as a groundwater drainage system by making the ditches deeper than usual (75–100 cm). This system would for example be applicable where intense rainfall occurs on poorly infiltrating soils underlain by a fairly permeable subsoil/substratum.

Widely spaced (100–200 m), deep (2·0–3·0 m) ditches are also used in irrigated areas to provide groundwater drainage for salinity control. In recent times, however, ditches are being replaced by pipe drains or (less frequently) by pumped wells (see chapter 12).

3.3 Construction of pipe drain systems

The construction of a pipe drain line normally involves:

(a) setting out of alignments and levels
(b) installation of drain pipe in the soil.

The main technical aspects of these works are discussed in the following sections while some relevant information and guidelines on their costing is presented in Annex 3.1.

3.3.1 Setting out

The principles are illustrated in figure 3.8. The beginning and end point of each drain line is marked by stakes (for long lines intermediate stakes are installed). The stake levels are adjusted to establish a sightline parallel with the grade line (= trench bed slope) at a convenient height above the soil surface. Boning rods (also called *travellers*) are used as necessary to establish the correct level of the bed of the trench.

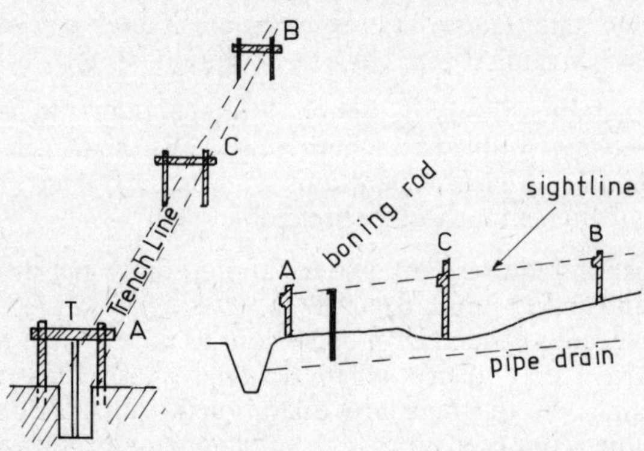

FIGURE 3.8 *The staking of a pipe drain line*

3.3.2 Installation methods

Pipes may be installed in a trench, excavated by hand or by machine, or they may be installed directly in the soil without first excavating a trench. The principal types of machines used for installation are illustrated in figure 3.9.

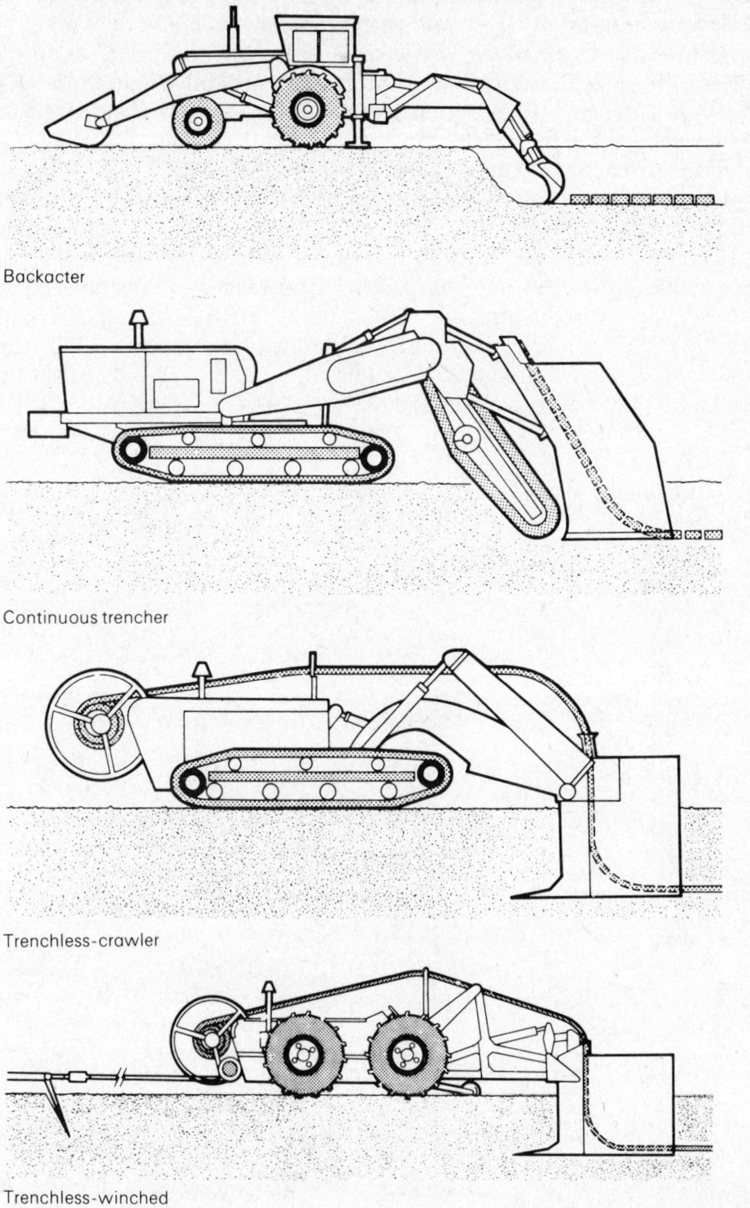

FIGURE 3.9 *Principal types of drainage machines* (BAILEY and TRAFFORD 1978)

Trench installation involves the following activities:

- the excavation of a trench;
- the laying of the pipe, with or without an envelope;
- the backfilling of the trench.

The excavation of the trench, the installation of the pipe and the envelope can be done in a series of continuous operations by machines. However, manual installation may still be justified where either only a very small area needs to be drained and there is no machine locally available, or in general where there is a preference for labour intensive work. Considerable skill, however, is required to match the accuracy and overall performance of the available machines.

The combination of machine trenching using a backacter and hand laying is frequently employed on small areas particularly within stony soils, and for the installation of large collector drains.

The trenchless method of installation was introduced around 1960 and has since developed into a viable, although somewhat restricted, alternative to installation in a trench. Figure 3.10 illustrates the principle of the method. To avoid compression of the soil around the pipe, leading to high entry resistances, it is essential that the opening in the soil where the drain pipe is to be placed, be created mainly by lifting the overlying soil rather than by forcing the soil to flow around the implement.

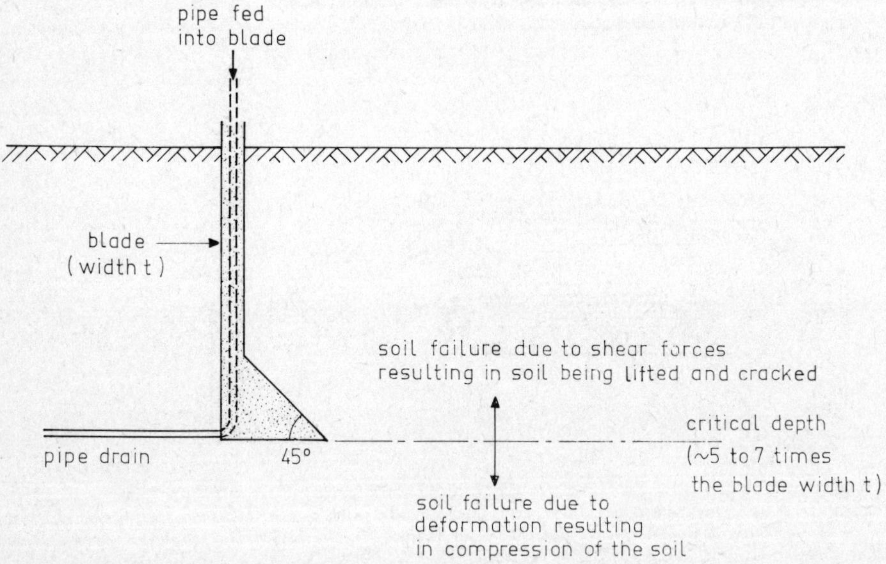

FIGURE 3.10 *The concept of critical depth and its application to trenchless drain pipe installation*

The principles of soil failure by the passage of a vertical blade are set out in some detail in section 4.3. From this it follows that the soil above a certain 'critical depth' is lifted and suffers shear failure which cracks the soil and improves its hydraulic conductivity. The soil below the critical depth fails by plastic deformation which compresses the soil and reduces its hydraulic conductivity. It therefore follows that the pipe should be installed at, or above this critical depth. The critical depth depends mainly upon the aspect ratio (working depth of implement/width of base of implement) as well as upon soil conditions, especially the soil moisture content. In wet soils critical depth is always much shallower than in dry soils. The geometry of the foot of the blade used for opening up the soil (or the foot of the plough if the blade is of the plough type) also has an influence. The foot should be inclined and be rather wide, although the latter of course increases the draught requirement. High installation speeds are also unfavourable. All in all these considerations limit safe installation depth to 1·5–2·0 m in fairly dry soils and much less in wet soils. It has, however, been observed that initially high entry resistances, due to installation below critical depth, often reduce to normal within a few years of installation, probably due to the cracking and other structure formation processes around the pipe, induced by the improved drainage.

3.3.3 Drainage machinery

The technical capabilities of the principal types of machines are detailed in table 3.2. This indicates that the backacter is the slowest machine. The trenchless machines have the maximum working speed capability although speed of operation is usually dependant upon and limited by the efficiency of the support services available; the continuous availability of pipe, envelope, spare parts for machine and facilities for on-site maintenance.

The large trenchless machines also have the largest power requirements of any machine.

TABLE 3.2 *Principal types of drainage machinery and their characteristics*

Machine type	*Power requirements kW(HP)*	*Maximum digging depth m*	*Trench widths m*	*Working speed* $km.hr^{-1}$
Backacter	30–50 (40–65)	up to 4·0	0·20–0·60	0·2
Continuous crawler trenchers				
light (8–10 tonnes)	60–90 (80–120)	1·5–2·0	0·20–0·30	0·5–1·0
medium (13–14 tonnes)	100 (150)	2·5	0·20–0·45	1·0–2·0
heavy (16–23 tonnes)	150 (200)	3·0–3·5	0·20–0·45	up to 3·0
Trenchless, winched	50–80 (65–110)	1·2–1·7		1·7–2·0
Trenchless, crawler	100–150 (135–200)	1·6–2·0		up to 5·0

The selection of a machine depends upon the technical suitability for the conditions likely to be encountered and upon the total costs involved.

(a) *The backacter* is a very popular choice of machine because it is relatively cheap and technically versatile. Its main disadvantages are that it is difficult to form a uniformally graded bed on which the pipe can be laid, and it is slow. It is, however, particularly suitable for working on small drainage schemes in hilly areas where the limitations of the grade are overcome by adequate natural slope. The machines work well in stony soils and its slow speed can be turned to advantage in situations where careful exploration of the soil is required. In this way services (water mains, electricity cables) can be exposed before they are damaged; old drainage systems may be intercepted and linked with the new system and the location of springs may be explored and thus intercepted.

(b) *The continuous trenchers* have a wide range of trenching capacities. The majority of these have a series of chain driven cutters but these wear very rapidly in stony soils and for these conditions a bucket wheel trencher is more suitable. Clay tile or plastics pipe may be laid using these machines and gravel, fed from a hopper, may be placed around the drain pipe as part of the continuous sequence of operations.

Light bulk envelope materials are often applied by hand after the passage of the machine and strip type filters may be fed from a roll and be either wrapped around the pipe as it passes through the machine or placed on top of the pipe or gravel in the trench.

The speed of operation of light drainage machines can be up to 1 000 $m.hr^{-1}$ although 600 $m.hr^{-1}$ is more typical. The net output depends on installation depth, soil type, field dimensions, project size, weather conditions and work organisation. For medium textured soil, 1·0–1·2 m installation depth and 200 m long drain lines, a normal output would be 300 m per gross workable hour, reducing to 200 $m.hr^{-1}$ in heavy clay soils. For large machines working at 2·5–3·0 m depths, the output is much less.

Trenchers normally require a crew of five men, comprising the operator, the assistant-operator (who also does the staking) and three labourers who supply the pipe, put in the envelope material, install the outlet pipe and backfill the trench (using a tractor with blade and trailer). With pre-wrapped plastics pipe, the number of labourers may be reduced to two. For normal output, this involves some 10–20 manhours/1 000 m drain length (compared to 250–300 manhours/1 000 m for manual installation).

(c) *Trenchless machines* have the potential to work faster and more economically than trenching machines (up to 2–3 × as fast ≃ 5 km/working day) but, here again, continuity and reliability of support services are essential requirements.

Trenchless machines are ideally suited to the installation of corrugated plastics pipe (60–180 mm diameter). They can be used in stony soils; a large machine having the capability to push stones up to 1 tonne out of the way. Envelopes can also be installed although they cannot be inspected easily.

The winch type of trenchless machine operates by winching itself along cables towards a ground anchor, normally fixed to a support tractor some 50 m ahead of the plough. In practise the main problem encountered with this type of machine is one of ensuring adequate anchorage.

Trenchless machines require more power than comparable trenchers especially when working at great depths ($>2{\cdot}0$ m) in dense dry soils (eg in heavy clay soils). The power requirements (P) roughly increase with depth (D) as D^3 and a major problem experienced with the powerful crawler mounted machines is one of securing adequate traction under wet surface conditions.

Depth and and grade control

Installation depth and grade may be controlled manually by the machine operator keeping a reference mark on the digging part of the machine fixed on a sightline such as that shown in figure 3.8. With hand installation, boning rods are used for this purpose.

Modern methods employ laser equipment. Essentially, this consists of an emitter, located at the edge of the field, which establishes a sloping reference plane over the field by means of a rotating laser ray. A receiver on the machine visualizes this plane of reference and transmits the signal via the regulating system to the machine's control hydraulics. The reference plane can be set to a grade with an accuracy of 0·5 cm per 100 m (0·005%). About five depth checks per second are normally made (rotation speed of the laser beam).

Standards of depth control vary somewhat between different countries though there is general agreement that the position of the drain pipe should not be allowed to deviate by more than half the pipe diameter from its designed depth. For grades the maximum permitted deviation is normally half the pipe diameter per 100 m. These standards can be achieved by skilled operators using manual control while machines fitted for laser control can easily lay claims to even more exacting standards. With trenchless machines working at high speed and/or in non-uniform soils, there is a danger that installation depths may become irregular and so extra care needs to be taken to maintain standards.

Timing of installation

Ideally drain installation should be carried out during the dry season when the ground is able to support the heavy machines and the continual passage of heavy wheeled tractor and gravel trailers. Drains should be installed even through a growing crop (causes ~5% crop loss) in preference to installation under wet conditions. The serious compaction of the soil by the passage of heavy machinery on wet land should not be underestimated. The greatest damage results from the passage of wheeled vehicles. The provision of crawlers leads to a lower and more even distribution of pressure (0·2–0·3 $kg.cm^{-2}$) although the larger trenchless crawler machines often have such high ground pressures that they are still unable to operate on soft soils.

3.3.4 Backfill

The initial backfilling should be done with topsoil (blinding) since this is generally more stable than the subsoil. This may be done by hand or by means of a blade mounted on the machine, although the main backfilling utilises graders, scraper blades or even screw augers mounted on the trenching machine. In humid temperate climates it is common practice to run the tracks of the trencher back along the trench line to consolidate the apparent excess fill. In irrigated areas serious problems can arise when water moves rapidly through the unconsolidated trench fill causing severe erosion (piping) and in extreme cases washing much of the trench backfill into the collector ditches. This problem can be overcome by a combination of trench backfill consolidation, preventing standing water coming into direct contact with the trench and by commissioning the work after the first few irrigation applications so that any voids appearing may immediately be backfilled with soil. The level of consolidation may be varied between the critical 10 m of trench adjacent to its outlet and the main length of the drain. This first 10 m of trench should contain an unperforated drain pipe and the trench should be backfilled in 0·3 m deep layers.

Each layer should be sprinkled with water so as to raise its moisture content to the optimum for maximum compaction by hand tamping. The remaining length of trench should be consolidated using a carefully controlled stream of water fed along a V ditch formed in the excess trench fill, major voids being filled as they become evident.

Any remaining fill should finally be formed into a low bund to prevent water ponding above the recently filled trench. Piping normally ceases to be a problem after one season, and irrigation may then proceed without regard to the position of the drain.

Similar problems may occur with trenchless installation. The crack formation which is considered a desirable feature in humid zones, can lead to piping under irrigated conditions. The passage of the plough will also cause the surface to heave and whilst this settles over a period of time, it presents problems for the mowers used on intensive grassland.

3.3.5 Installation below the watertable

Modern pipe laying machines are capable of installing drain pipe beneath the watertable but the attendant risks are that the soil around the pipe will become somewhat puddled during excavation or backfilling, resulting in high entry resistance. Use of good, thick envelopes reduces this risk and for these conditions gravel is often most suitable since lighter material may float.

A special problem occurs with the installation of drain pipe under 'quick conditions' (occurrence of unstable fine sandy/silty or sodic material at drain depth, below the watertable), a situation which is occasionally encountered in the deep drainage for salinity control of alluvial soils. The provision of suitable guard plates on the trencher will ensure that the trench will remain sufficiently open to allow machine installation to be used. The problems are most acute when installing collector pipes which are either too big or too deep to be installed by machine. This applies to most collector pipelines in irrigated areas, which are, therefore, normally installed by hand. Under 'quick conditions' extensive dewatering has to be used to keep the trench dry and prevent it collapsing. This may be done by conventional well-pointing along the trench (on one or both sides, depending upon the severity of the caving in problem) although horizontal dewatering provides a suitable alternative system. This latter method involves pumping water from a pipe (usually an ordinary corrugated plastics drain pipe) installed lengthwise below the bottom of the trench. Special machines have been developed which are able to install these horizontal dewatering pipes at great depths (5 m and more; some types of conventional drainage machines can also be adapted for the installation of horizontal dewatering pipes up to a depth of 3·0 m).

3.4 Maintenance of pipe drain systems

Pipe drains need to be cleaned regularly when there is a danger of deposition of fines or of iron compounds (ochre) or when the flow becomes restricted for any other reason.

Roots of normal field crops rarely enter drain pipes installed at normal depths. Problems may, however, occur where drain lines pass beneath hedges or pass near trees. At these locations unperforated pipes may have to be installed.

Clogging by ochre

This problem is to be expected when soils which contain iron sulphates are drained for the first time. Typical soils so affected are lowland peats, recently reclaimed marine soils, and soils derived from iron pyrite bearing rocks. Virtually all soils contain appreciable quantities of iron. Generally in a well drained, aerated soil it occurs as water-insoluble ferric oxide (Fe_2O_3) which remains dispersed and immobile in the soil profile, causing no problems. However, in some badly drained soils the iron occurs as iron sulphide (FeS_2) which is oxidised, when the soil is drained to form the water soluble ferrous sulphate ($Fe\,SO_4$) and sulphuric acid (H_2SO_4):

$$2Fe\,S_2 + 7O_2 + 2H_2O \rightarrow 2Fe\,SO_4 + 2H_2SO_4.$$

The ferrous sulphate is unstable at near neutral or alkaline conditions (pH 6·5 or above) such as exist in calcareous soils and it soon oxidises further to insoluble ferric oxide. The orange staining that this results in is normally observed as localised mottling well distributed throughout the soil. If a lot of iron sulphide is present the sulphuric acid produced, lowers the pH and stabilizes the soluble ferrous sulphate which then seeps in solution to the drains. At low pH ($<4{\cdot}5$) certain acidophilic soil bacteria catalyse the oxidation of iron sulphide which then accelerates the production of soluble ferrous sulphate in a self promoting cycle. The soluble ferrous sulphate is liable to oxidise to insoluble ferric hydroxide, $Fe(OH)_3$ either by direct contact with atmospheric oxygen or by bacterial action, the latter resulting in the formation of slimy filamentous deposits in and around the drain. The insoluble ferric hydroxide eventually turns into the more stable ferric oxide either by drying or ageing.

The severity of the problem varies with the pH of the groundwater and its divalent iron content and, as table 3.3 shows, is most acute in low pH, Fe^{++} rich conditions.

TABLE 3.3 *The variation of the risk of clogging by iron ochre with divalent iron content,* Fe^{++} *and* pH (Eggelsmann 1978)

Fe^{++} *concentration mg/litre*		*Iron ochre clogging risk*
pH<7 *Acidic conditions*	pH>7 *Alkaline conditions*	
<0·5	<1·0	hardly any
0·5–1·0	1·0–2·5	slight
1·0–2·5	2·5–5·0	medium
2·5–5·0	5·0–7·5	pronounced
>5·0	>7·5	very pronounced

In extreme cases the pipe and envelope can become totally clogged very rapidly and there is at the present time no forseeable economic cure to the problem.

The evidence available indicates that the leaching of iron rises to a peak rate some two to three years after the installation of drains, subsequently falling off until eventually it ceases to present problems. In these circumstances drainage should be undertaken in a series of carefully planned stages using ditches initially, since these may be cleaned easily, and pipes for the later stages when the risks of clogging have been minimised. In areas where there is only a slight risk of ochre clogging, pipes may be installed immediately provided they are frequently cleaned by flushing.

Pipe cleaning

Deposits of fines and ochre may be removed from the pipe by flushing. Flushing used to be done in combination with rodding (pushing a jointed bamboo rod through the pipe to loosen the deposits), but nowadays pipe cleaning is mainly done using pressurised water jetted from a self-propelling nozzle (figure 3.11). Water pressures at the nozzle of 10–15 atm suffice to remove loose deposits but fail to remove deposits that have become consolidated (eg due to lack of regular maintenance). Such deposits may be removed by flushing at higher pressures but this can disturb the stabilised soil around the pipe and lead to renewed entry of fines. High sediment loads often occur during the initial period after installation before the soil above and around the pipe has consolidated. Usually the pipe needs to be cleaned soon after installation (one to two years) although later a frequency of five to ten years is normally sufficient.

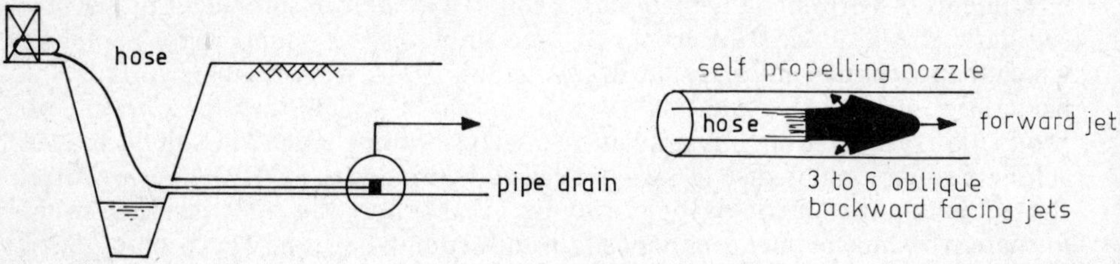

FIGURE 3.11 *Drain pipe cleaning by means of a jetting nozzle*

Most commercially available drain pipe flushing machines have a maximum 250–300 m reach. Their working capacity ranges from 400 m.hr^{-1} under normal conditions to 200 m.hr^{-1} in highly silted pipes. The cleaning may be done from the pipe outlet into a ditch (as in figure 3.11) or from a junction box. In composite systems with long drain lines, special access points may have to be provided for cleaning purposes.

References and further reading

BAILEY, A D and B D TRAFFORD 1978: 'Drainage construction techniques in England and Wales', Tenth Congress on Irrigation and Drainage, Question 34.2, International Commission on Irrigation and Drainage, New Dehli

BUREAU OF RECLAMATION 1978: *Drainage Manual*, US Department of Interior, Washington DC

CAVELAARS, J C 1974: 'Subsurface field drainage systems'. In: *Drainage Principles and Aplications*, volume IV, publication 16, International Institute for Land Reclamation and Improvement, Wageningen

EGGELSMANN, R 1981: *Dränanleitung*, Paul Parey, Hamburg (in German; also in English translation: *Subsurface Drainage Instructions*, 1978)

FAO 1973: 'Drainage machinery', Irrigation and Drainage Paper no 15, Food and Agriculture Organisation of the United Nations, Rome

FAO 1980: 'Drainage design factors', Irrigation and Drainage Paper no 38, Food and Agriculture Organisation of the United Nations, Rome

ILRI 1979: *Proceedings of the International Drainage Workshop*, J. Wesseling (ed.), publication no 25, International Institute for Land Reclamation and Improvement, Wageningen

ISO 1979: 'UPVC pipes and fittings for use in sub-soil drainage-specifications', Publication ISO/TC 138 WG 1 N431, International Standard Organisation. DIN Deutsches Institut für Normung c.v. Berlin 30, (In draft)

LUTHIN, J N 1966: *Drainage Engineering*, Wiley, New York

VAN SCHILFGAARDE, J 1974 (ed.): *Drainage for Agriculture*, Agronomy Monograph no 17, American Society of Agronomy, Madison, Wisconsin

ANNEX 3.1 Cost calculations

Costs of pipe drainage systems vary from project to project and are valid for a limited period only. Therefore emphasis is placed upon methods of cost calculation and on the structure of the total and component costs. For some items the range of comparative costs is indicated.

1 *Purchase costs of drainage machinery*

The purchase costs of important drainage machinery in western Europe are presented relative to the purchase costs of the backacter:

– backacter	100
– continuous trencher	150 to 400
– trenchless machine, winch type	150 to 300
– trenchless machine, crawler	450 to 600
– tractor and pipe loader	75 to 100
– tractor and gravel trailer	75 to 100
– digger/loader (tractor mounted)	75 to 100
– mole plough or subsoiler with matching tractor	100 to 150
– laser equipment	75 to 100
– pipe cleaning equipment	20 to 30

2 *Cost structure for pipe drainage construction*

The breakdown of the total costs of pipe drain construction given below shows that the machinery cost component is rather small in spite of their high purchase costs. This is due to the high efficiency and high capacity of these machines. The materials represent well over half of the costs when a good envelope is needed (for example a gravel envelope):

	% of total costs
– machinery costs	10 to 15
– costs of material (pipe and envelope)	40 to 65
– labour costs (W. Europe)	10 to 15
– overheads (including design and supervision)	15 to 25

This costs breakdown applies to singular drainage. The example of cost calculation presented further on gives an idea of the cost structure for a composite pipe drainage system.

3 *Guidelines for cost calculations*

General guidelines are presented together with some standards which enable rough cost estimates to be made. Final estimates should always be based on locally applicable standards.

Drainage machines

- *annual running time*: 1 000–1 500 hours in humid climates, 2 000 hours in arid climates; *annual output* for a light machine 250–400 km (humid climate) to 500–600 km (arid climate), for trenchless machines 500–700 km (humid) to 1 000 km (arid); output of heavy machines more variable, very much depending on drain depth;
- *fuel consumption*: engines assumed to be running on average at half their rated power, consuming then about 0·3 litre diesel oil per HP-hour;
- *oil and lubrication*: about 20% of the fuel costs;
- *depreciation*: five and seven years periods respectively for trench and trenchless type (10000–15000 hrs);
- *maintenance and repairs*: about 15% of the purchase price.

At 10% interest, the total of depreciation, interest, maintenance and repair costs add up to an annual cost factor of 35% of the purchase cost for a trench type machine and 25% for a trenchless machine.

Labour costs (four to five men crew)

Taking total employer's costs as 150% of the salary costs and assuming the salaries of respectively the operator, assistant-operator and labourers to relate as 300:200:100, the total annual labour costs comes to roughly 10 to 12 × S where S = annual labourer's salary.

Materials

The comparitive costs of different types of drain pipe detailed below are indicative only, actual costs depending very much on the locally available materials and production facilities:

Diameter[1]	*Clay pipe*	*Corrugated plastics pipe*	
		Plain	*With coconut fibre envelope*[2]
50–60 mm	30–35	30–35	60–65
80 mm	40–50	50–60	110–120
100 mm	70–80	100	150–165
150 mm	100–125	250	
200 mm		400–450	

[1] ID for clay pipe, OD for plastics pipe
[2] thick synthetic envelope 50–100% more costly than the organic equivalent

For a gravel envelope some 0·02 to 0·04 m^3 per meter drain length is needed (including wastage). Costs depend very much on the availability and quality of the local gravel material.

For the on-site transport of materials and for the backfilling of the trench a tractor with a trailer and a blade are required, costing per annum about 50% of the purchase price of this equipment.

4 *Example cost calculation*

- drainage of irrigated land, Iraq 1980;
- composite pipe drainage system with laterals 220 cm deep, 40 m spacing and 250 m length, consisting of 8 cm corrugated plastic pipes with gravel envelope; collectors 250 cm deep and 1 000 m length consisting of 20 and 30 cm concrete pipe;
- laterals laid by machine (trench type 175 HP machine costing ID 40 000 cif Basra; annual output 2 000hours × 150 m = 300 km; fuel consumption 30 l/hr at ID 0·02 per litre;*
- labourer's salary ID 600.— per annum.

Laterals: pipe		ID 0·250 /m
gravel, supply on site	ID 0·100	ID 0·150 /m
installation	ID 0·050	
drainage machine: depreciation, interest, maintenance and repairs	ID 0·047	
fuel, oil and lubrication	ID 0·005	ID 0·076 /m
labour	ID 0·024	
other costs		ID 0·024 /m
		ID 0·500 /m

Collectors: pipe	ID 0·500 /m	ID 1·500 /m
installation	ID 1·000 /m	

Junction: ID 50/piece
Outlet : ID 35/piece

Construction costs:

	Units	*Requirements per ha*	*Unit cost ID*	*Total costs/ha*
Laterals	m	250	0·500	125·––
Collectors	m	20	1·500	30·––
Junctions	piece	0·5	50·––	25·––
Outlets	piece	0·02	35·––	0·7
				180·7
Overhead (15%) and engineering (20%)				63·3
				ID 244.––

* Local price, below world market price; IDs = Iraqi Dinar (1 ID = 3.50 US $, 1980)

Annual costs:	capital annuity	ID26·88
	(10%interest, 25 years life time)	
	maintenance, 2%of construction costs	ID 4·88
		ID 31.76
	(10%–15% of the construction costs)	

Chapter 4 Shallow drainage systems

These systems apply to land in which groundwater drainage systems cannot be used because of the inadequate infiltration or percolation characteristics of the main rootzone and/or inadequate hydraulic conductivity in the substrata. This situation is typical in so-called *heavy land*. It occurs in many parts of the world, representative examples being the basin clay soils of the fluvial deposits, the pseudo-gley soils of mid- and eastern Europe, the planosols of the semi-humid/semi-arid tropics, the vertisols of the semi-arid tropics ('black cotton' soils) and the glacial till soils ('boulder clays') covering much of northern America, United Kingdom and northern Europe. The prevailing profile characteristics of these soils are shown schematically in figure 4.1.

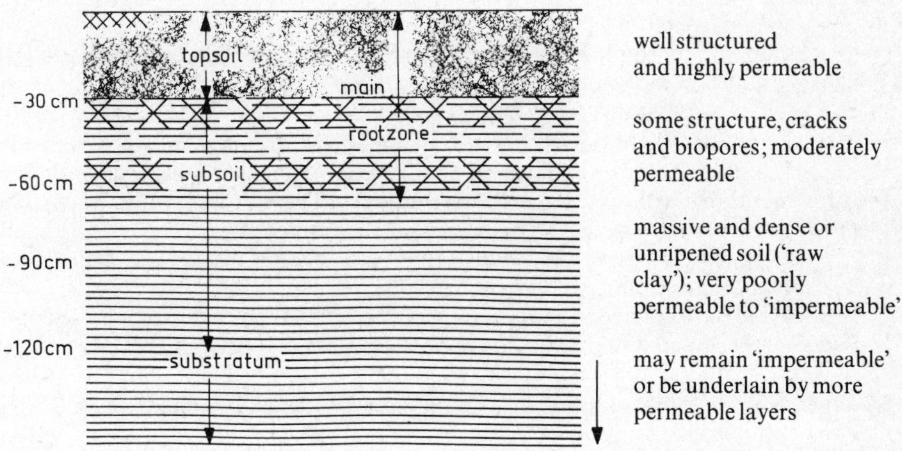

FIGURE 4.1 *'Heavy land' soil profile*

Whereas in these soils the excess water is unable to drain downwards through the profile, it will drain away by lateral flow. Thus water ponded on the surface will be drained by lateral overland flow whilst water impeded at some depth in the rootzone will be drained by lateral interflow over the impeding subsoil layer. The depth of the drained soil will extend down to the impeding layer and because hardly any roots will penetrate into the layer, it will coincide with the main rootzone depth. As this type of subsoil normally has a low drainage pore space ($\mu < 2$–3%), there is little to be gained by deep drainage unless it is combined with subsoil improvement.

Surface ponding is promoted also by flatness and unevenness of the land surface, for under these conditions the water will only run off slowly. The excess water will collect in the depressions leading to an unequally distributed infiltration load. Moreover, conditions for infiltration in depressions are poor as the soil structure deteriorates under the frequent ponding and the pores become clogged by the sediments carried into them by the water. The ponding thus becomes self-promoting. Better grading and smoothing of the land may solve part of the drainage problem.

The rainfall intensity and magnitude also have an influence. Under conditions of high, intense rainfall groundwater drainage may well fail to provide adequate drainage in moderately permeable soils which could otherwise be drained under less severe climatic conditions (relevant to much of the humid tropics).

Finally, surface ponding and impeded percolation are also frequently due to poor soil management. Modern mechanised arable cropping systems severely tax the soil structure, on the one hand by adding very little organic matter to the soil and, on the other hand by working the soil with heavy machines, frequently under adversly wet conditions. As discussed in section 1.9, soils with a low organic matter content are rather sensitive to the compacting, puddling and smearing effects of cultivation, resulting in poor soil structure and in 'impermeable' layers (*pans*) at some depth below the seedbed. Of course, where pans occur in a soil having otherwise adequate hydraulic conductivity they should be broken up by subsoiling, and their re-establishment prevented by improved soil management so that normal groundwater drainage may be used.

As a rough guide, shallow drainage systems are applicable when the impeding subsoil lies within some 60 cm below the soil surface, and the hydraulic conductivity of the subsoil is less than 0·1 m.day^{-1}. When the 'impermeable' subsoil occurs deeper, say at 80–100 cm depth, normal groundwater drainage may be used. The drains in this case should be laid just on top of the impermeable subsoil since all flow towards the drains will take place above this layer. These drains would have to be installed at rather narrow spacings in view of the limited flow zone and the small head (mid-spacing watertable height) that can be permitted. Nevertheless a reasonable rootzone drainage can be achieved.

Of course it is no coincidence that much of the heavy land in the temperate zones is used as grassland and in the warmer zones as rice land. Rather, it is a fine example of adapted farming in the sense discussed in section 2.4. The heavy lands as a class, however, are far from uniform and many can be drained sufficiently well with shallow drainage systems to allow a number of other field crops to be grown successfully.

4.1 Bedding systems

Bedding is the classical drainage method for flat, heavy land in humid climates. The excess surface water is drained by lateral overland flow towards the field drains. However, where a permeable topsoil exists, much of this water will also drain by lateral flow through the perched groundwater zone forming itself in the topsoil on top of the 'impermeable' subsoil (interflow, see figure 4.2). This '*topsoil drainage*' helps to remove ponded water and also promotes early restoration of the aeration of the upper rootzone (usually the main rootzone in this type of soil). Topsoil drainage is enhanced by maintaining a deep plough layer on the beds (30–35 cm).

For the typical conditions under which bedding is used, flat beds should generally be no wider than 10 m to obtain good drainage. However, narrow beds are not conducive to mechanised farming. Crowning of the beds promotes overland flow and interflow enabling wider beds to be used (up to 20–30 m), commonly being referred to as *cambered beds*. The crowning may be achieved in the course of time by deliberate 'uphill' ploughing towards the centre of the beds, or by construction using grader and/or scraper.

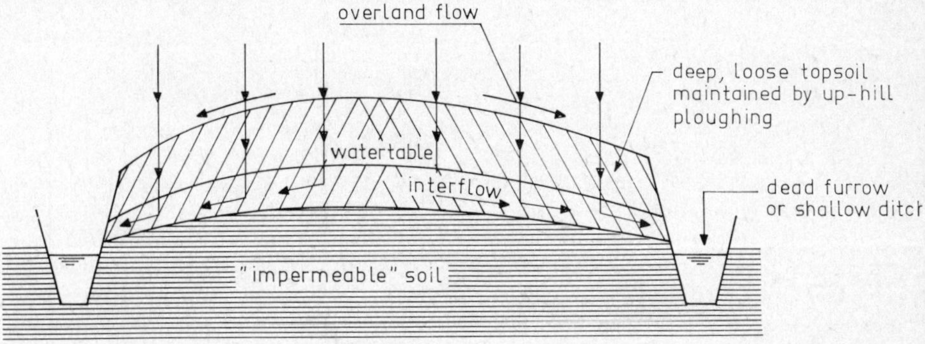

FIGURE 4.2 *Drainage by overland flow and perched groundwater flow in a cambered bedding system*

Bedding is most appropriate for crops grown 'on the flat' like grassland, other foddercrops and various grain crops. Ridge cropping, although in itself a desirable practice on heavy soils, also has its drawbacks. Ridges aligned along the length of the beds, parallel with the field drains, prevent lateral surface runoff and should only be used when there is a good deep topsoil capable of allowing considerable interflow to pass beneath the ridges. Small furrow drains may be made across the ridges to permit ponded water, standing in the furrows between the ridges, to flow to the field drains (these small drains, usually made by hand, are termed *quarter drains* or *ridge cuts*).

The field drains separating the beds, may vary in form from simple dead furrows left by the uphill ploughing of the beds, to small ditches constructed by hand or with a ditcher. For unobstructed topsoil drainage the field drains should be incised well into the subsoil so that waterlevels at normal flow remain below the topsoil (a total depth of some 40–50 cm will generally suffice). They should have a continuous gradient towards the outlet. In practice, this latter condition is not always fulfilled. Many field drains are constructed without staking and suffer from impeded discharge due to adverse gradients. This also applies to machine constructed field drains, as the ditchers used for this type of work are generally not provided with sighting equipment. Ideally the land should have some slope so that field drains can run down the slope, guaranteeing good discharge (this applies especially to the bedding system shown in figure 4.3B).

Correct maintenance of the field drains is a time-consuming task due to their great number, the required frequency, and the fact that an established crop may only permit handwork. In practice, the poor construction and/or maintenance of the field drains proves to be the main constraint to the proper functioning of the bedding drainage system.

4.2 SHALLOW DITCH SYSTEMS

These systems comprise shallow field ditches laid out in certain patterns. The ditches are usually too shallow and the subsoil too impermeable to achieve much groundwater drainage. Their main function is to collect surface runoff and provide shallow profile drainage, mostly through interflow.

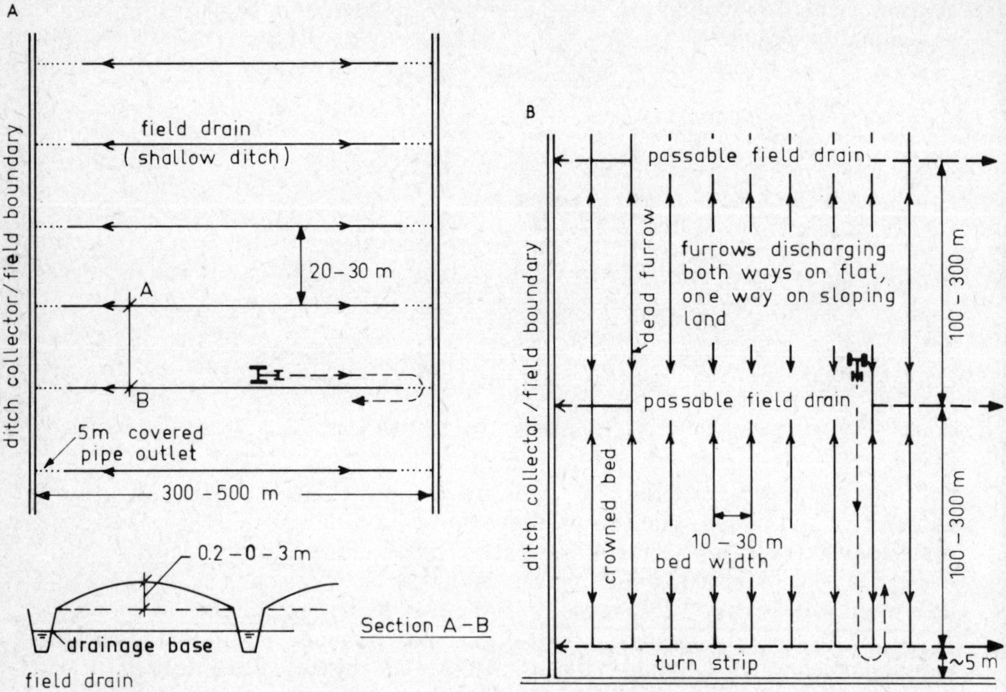

A Bedding system designed to achieve good topsoil drainage. The beds are formed by ploughing and grading, and are then maintained by 'uphill' ploughing. The dead furrows between the beds are excavated well into the poorly permeable subsoil (depth about 50 cm; bottom width about 30 cm).

The non-passable field drains discharge both ways and can only be crossed at the ends. The farm operations are carried out parallel with the field drains. The system is suitable for flat or gently sloping land; in the latter case field drains are best laid out parallel to the contours (in which case the field drain can discharge both ways).

B Bedding system with a shallower drainage base than the first system, achieving less topsoil drainage. The crowned beds are established and maintained by ploughing. Crown height is about similar to the first system although the dead furrows are shallower. The furrows discharge into passable field drains, the latter having side slopes of not more than 1:6.

FIGURE 4.3 *Examples of modern bedding systems, suitable for mechanised farming*

4.2.1 Types of shallow ditch systems

Four types of field ditch system may be distinguished on the basis of their layout pattern, type and main function.

1 *The depression ditch system* (figure 4.4)

This system is applicable to fields in which a limited number of pronounced, elongated depressions exist lending themselves to drainage by means of ditches. The depressions may be drained individually or be linked. The ditches discharge towards the field boundary into the main system.

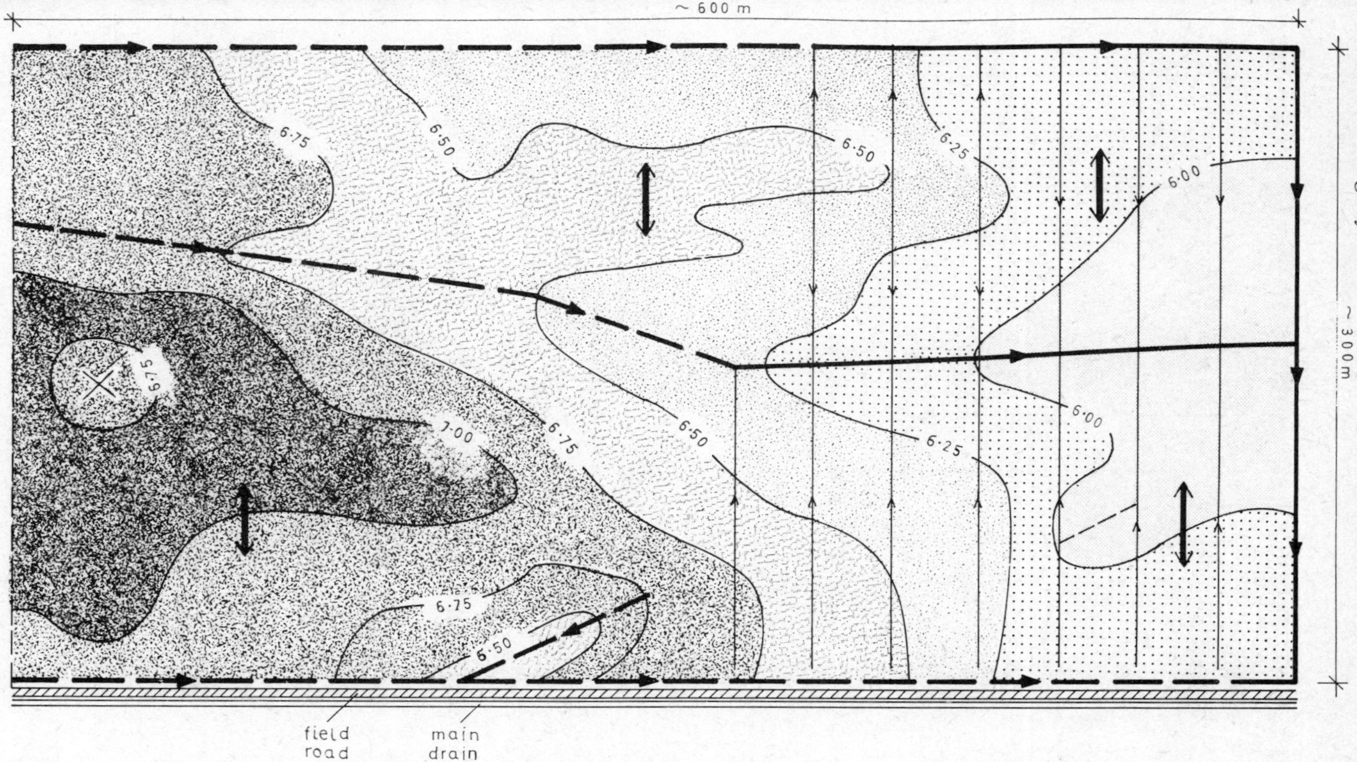

Legend (also for figure 4.5)

collector ditch (deep, non-passable)

collector } ditch (shallow, passable)
field

field ditch (60–75 cm deep, non-passable)

quarter drain/ridge cut (passable)

direction of farm operations

slope to be corrected (grading)

6·25 contour line

FIGURE 4.4 *Examples of shallow ditch systems (I)* (with explanatory text)

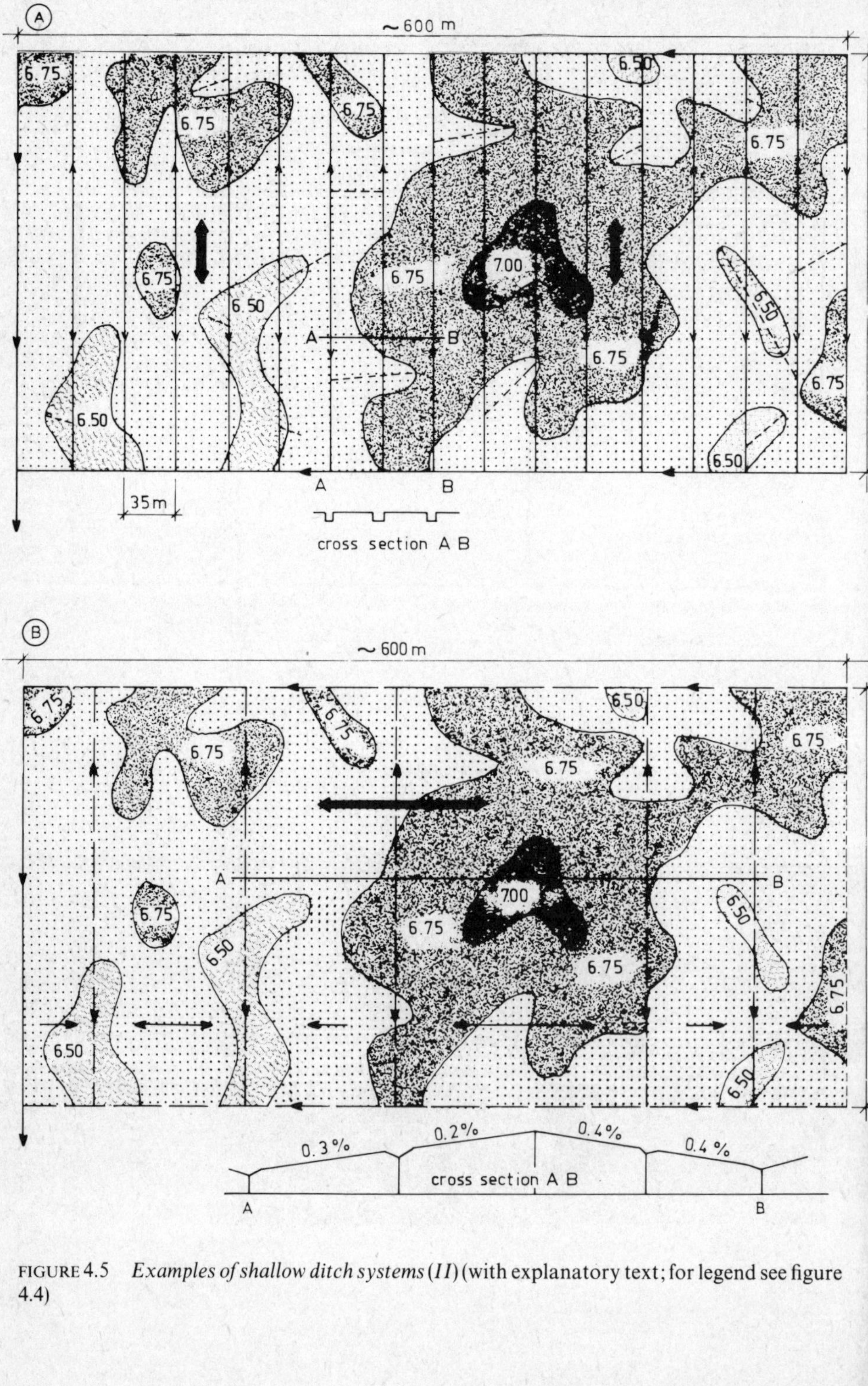

FIGURE 4.5 *Examples of shallow ditch systems (II)* (with explanatory text; for legend see figure 4.4)

Explanatory text for figures 4.4 *and* 4.5

Figure 4.4:

The soil and hydrological conditions in this field are such as to require surface drainage only in the upper half of the field while both shallow and groundwater drainage are required in the lower part. For the latter area a parallel non-passable ditch system has been planned. The ditches are made 0·75 m deep to make them reasonably effective as groundwater drains and their spacing is 30 m. Surface water will collect in the ditches partly by lateral overland flow and partly by lateral interflow drainage. Small cross drains (passable) may be installed in depressions (quarter drains/ridge cuts). The ditches discharge into (deep) collector drains which either follow pronounced depressions or field boundaries.

In the upper part of the field surface water collects in depressions or border ditches. These ditches may be rather shallow and are made passable where desirable. The land is carefully graded towards the ditches to provide good conditions for overland flow. No watertable control is required in this part of the field (watertables kept adequately low naturally).

Figure 4.5:

Two alternatives have been presented for a field with a rather irregular topography.

The parallel non-passable ditch system as planned in figure 4.5A would be applicable when:

- the field requires both shallow and groundwater drainage;
- very little earth movement in the field can be tolerated;
- the field will be cropped in such a manner that the ditches will not interfere unacceptably with the required field work.

The planned system requires only minimal earth movement. Only a few depressions have to be filled-in (using the spoil from the ditch excavations). To be effective groundwater drains, the ditches should be at least 75 cm deep. Culvert-outlets may be used to enable the ditches to be crossed by tractors. Portable bridges may also be used or a short length of the ditch may be temporarily filled in. Ponded water can enter the ditch by lateral overland flow and/or by lateral interflow. Under ridge cropping, a good number of ridge cuts should be installed to allow water collecting in the depressions within furrows to escape to the nearest ditch.

The parallel passable ditch system in figure 4.5B presents an alternative system which is more compatible with the field accessibility and manoeuvrability requirements of modern mechanised farming. The system, however, provides surface drainage only; no water is drained from the soil profile and fields may be expected to dry up slowly and crops may suffer from poor aeration unless there is adequate natural drainage. Ridge cropping is desirable in this situation in that it provides the means for good row drainage while at the very least the soil in the ridge bodies gets reasonably well drained.

High quality grading of the land towards the ditches is required which may involve considerable earth movement and locally may leave very little (valuable) topsoil.

Only the main depressions should be provided with ditches as too many infield ditches hinder farm operations. Small and/or inconveniently located depressions should be filled in with, for example, the spoil from the excavated ditches. An infield ditch system connecting a number of isolated, widely distributed depressions is commonly referred to as a *random ditch system* which is comparable to the natural system for groundwater drainage depicted in figure 3.2.

Because of their location, depression ditches are effective in drawing excess water from the surrounding, higher land, an advantage that applies to water on as well as to water in the soil. Where the overland flow rate towards these ditches is too slow, smoothing of the land, land forming, row drainage, etc, may be used to promote the flow (see section 4.2.2).

2 *The parallel non-passable ditch system* (figures 4.4 and 4.5A)

This system is applicable to fairly regular, even-surfaced, flat land and also to land with a highly uneven ('pocketed') surface. The internal field topography of this type of land does not indicate a preferential alignment of the ditches and so a parallel alignment which least hinders farm operations, is often the most appropriate. The ditch spacing may vary within a field although where conditions do not differ greatly, the spacing is best kept uniform.

The ditches are well incised and definitely aim to achieve considerable profile drainage as well as to collect excess surface water. The system differs from the flat bedding system in that the drains are deeper and wider spaced and, to be successful, the soils should allow somewhat better and deeper profile drainage than is possible on the poorly drainable soils suited to bedding. By promoting lateral interflow, eg through subsoiling, and by promoting surface flow through smoothing and careful land preparation, bedding may often be replaced by a parallel ditch system.

As with bedding, ridge cropping is on the one hand desirable whilst on the other hand it prevents lateral overland flow and so should only be used where there is good profile drainage. Ridge cuts should be made to evacuate any ponded water from within the furrows.

3 *The parallel passable ditch system* (figure 4.5B)

In this system the dilemma noted earlier with respect to ridge cropping is overcome by making the ditches passable by machine. Ridges are laid out, and indeed all farm operations are done, across the ditches. The land between them is carefully graded towards one or both of the ditches. With ridge cropping, good row drainage can thus be achieved, enabling the intercepting ditches to be spaced rather widely. Hardly any profile drainage is however achieved (not even topsoil drainage).

4 *The cross-slope ditch system* (figure 4.6)

This system is applicable on gently sloping land where on the one hand sufficient runoff may occur to require attention but on the other hand the water may also remain ponded on the land. Under this system cultivations are carried out with the slope (straight down or at an angle to the slope) with runoff being intercepted at regular, short intervals. Slopes are generally not less than 0·5% and not more than 2% for row crops and 4% for close growing uniformally broadcast crops. On steeper slopes more emphasis should be placed on detaining/retaining the excess water by practising contour farming and by constructing various types of terrace.

Ditch spacings should not exceed the maximum safe (from erosion point of view) length of run. In practice, spacings vary from as narrow as 20–30 m to as wide as 50–70 m. A number of factors influence the spacing, the effect of the most important ones being indicated qualitatively below:

	Narrow spacing	*Wide spacing*
– soil type: infiltrability	low	high
erodibility	high	low
– slope of the land	steep	gentle
– intensity of the rainfall	high	low
– crops to be grown	row type	close growing

As the influencing factors interact in various ways, required spacings are difficult to predict and are best determined by trial and error.

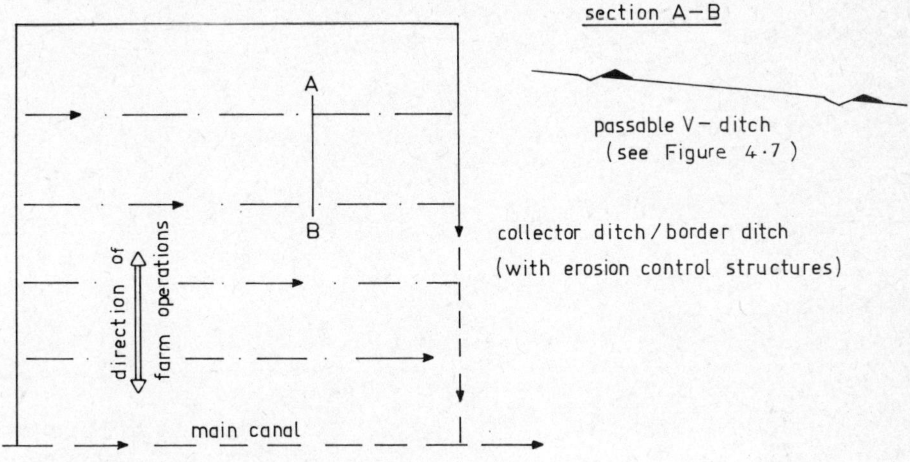

FIGURE 4.6 *The cross-slope ditch system*

4.2.2 Some technical aspects of shallow ditch systems

Field and collector ditches

The ditches may be made passable or non-passable examples of the first type being presented in figure 4.7. The prevailing conditions will generally dictate which type is most suitable. Depression ditches which for example divide up the field rather badly, should preferably be made passable. This also applies to parallel ditches aligned across the direction of farm operations (as in figure 4.5B). However, where the ditches run parallel to the direction in which the farm operations are carried out non-passable ditches, occupying less land, may be used instead (as in figure 4.5A). The function of the ditch also plays a role: ditches which only collect surface water can be made passable rather more easily than the deeper ditches which are designed to function as (shallow) groundwater drains as well. Not only are the latter deeper and thus more expensive to make passable but they also dry up later.

Passable ditches are normally constructed using such commonly available earth movement machinery as bulldozers, graders and scrapers. Self propelling graders and scrapers may be used for large collector ditches whereas small ditches may be constructed using farm-type equipment (blades, drags, small scrapers, etc, drawn by farm tractors). Various types of ditching equipment are also available for constructing non-passable ditches such as the rotary ditcher shown in figure 4.8.

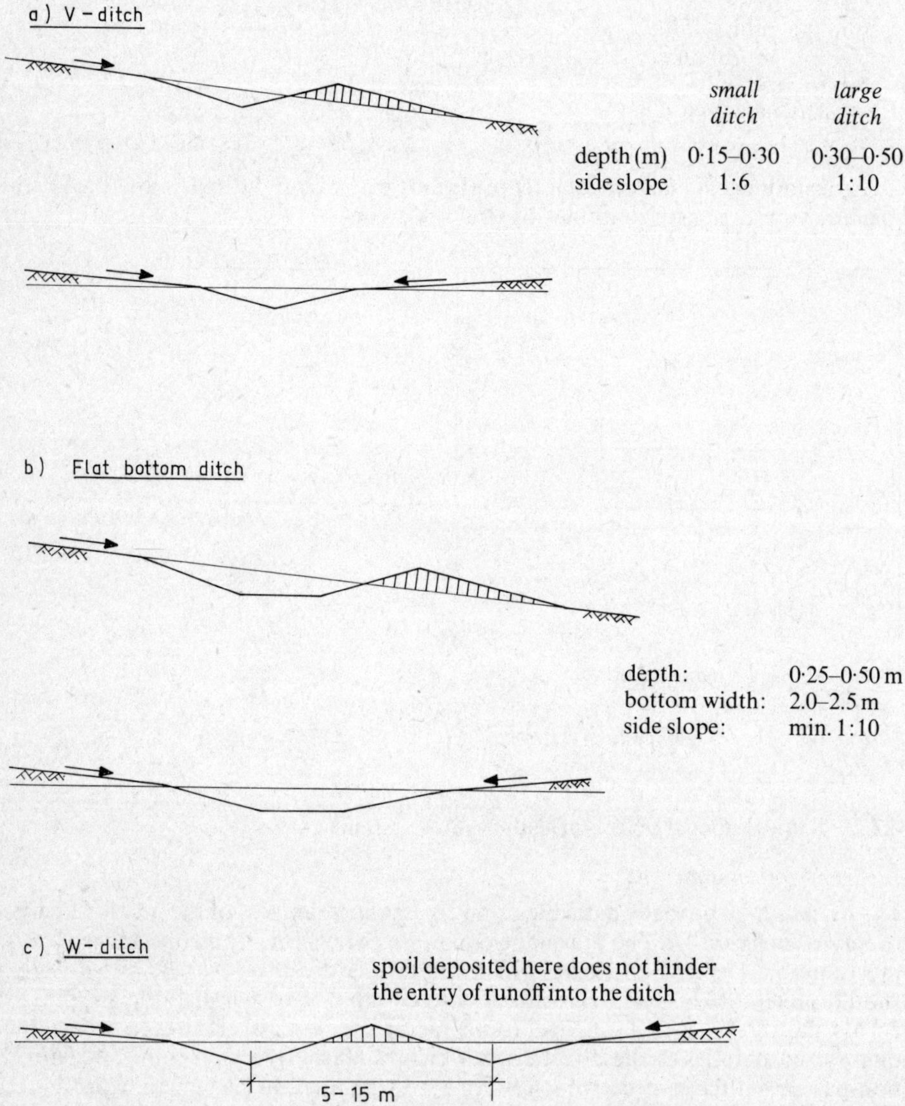

FIGURE 4.7 *Types of passable field and collector ditches*
(these ditches are generally farmed through and upper side slopes may well be planted. They should be cleaned before the drainage season, using eg a shovel or a V-drag. A small furrow drain is often installed in the centre to ensure that the ditch is dry in sufficient time for tractors to pass through

FIGURE 4.8 *Rotary ditcher suitable for use with* 40–50 *HP tractor*
(reproduced by courtesy of Pietro Dondi & Figli, Italy

Overland flow

Without good overland flow conditions, the excess surface water will not reach the ditches, rendering them ineffective.

Good overland flow requires primarily that the land draining towards a ditch has a continuous grade towards that ditch. Where the topographical conditions are unfavourable in this respect, earth movement should be undertaken to establish such grades artificially. This may involve correcting minor irregularities in the slope (*smoothing*) or extensively changing the natural lie of the land (*land forming*).

Under favourable conditions, field ditches may be widely spaced with travelling distances for the overland flow up to 100–200 m (see figures 4.4 and 4.5).

Row drainage

This refers to a special form of overland flow, *viz* the run-off through the furrows formed by two crop ridges. Given a well maintained furrow of a good grade, row drainage can be very effective; moreover, as mentioned earlier, ridge cropping is in itself a beneficial practice on poorly drained land.

Furrows should be aligned in relation to the slope of the land so as to achieve the most desirable furrow grade. For clean, well maintained furrows a minimum grade of 0·10–0·20% is required, but rough, trashy furrows require more slope to get a good discharge (minimum 0·20–0·25%). Long furrows are, of course, desirable from the farming point of view but are more liable to become blocked. A length of 150–200 m is normally a good compromise. The erodibility of the soil should also be considered. Few soils tolerate furrow grades in excess of 2% while on easily erodible soils the grades should generally not exceed 0·5% at normal furrow lengths. On steeper slopes, row length should be adapted to the degree of slope and the erodibility of the soil for which some general guidelines have been given in figure 4.9.

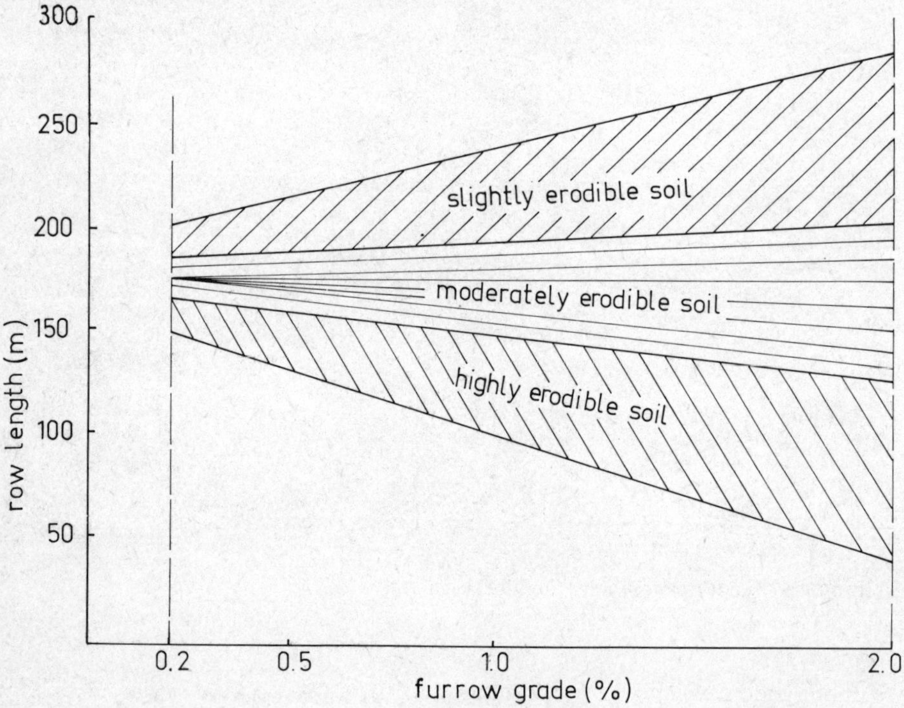

FIGURE 4.9 *Furrow grade/length relationships for row drainage*

The water draining from rows is best intercepted by field drains running across the slope or through depressions as for example in figures 4.4 and 4.5B. Where the ridges run parallel with the field drains as in figure 4.5A, the practice is usually chosen more for the benefits of cropping on ridges than of row drainage. Drainage in this situation is conceived as being mainly lateral (topsoil and shallow profile drainage) rather than longitudinal (row drainage). To provide an outlet for any flow that may occur along the rows, cross-drains (ridge cuts) should be made. These may be preplanned and installed at planting or they may be installed on an *ad-hoc* basis during the rainy season, when and wherever the need arises.

Spoil

A common cause of malfunctioning of ditch systems is that the spoil from excavation or from maintenance is not disposed off adequately. Often the overland flow into the ditches is blocked by spoilbanks.

The spoil from ditch construction should be used to fill in depressions on the slope or be feathered out by carefully grading it away from the ditch. Where satisfactory disposal of spoil from maintenance is difficult due for example to maintenance during the crop season, the installation of a W ditch, as depicted in figure 4.7, offers a solution.

4.3 Mole drainage systems

Here the lateral shallow flow through the soil is promoted by mole channels. These are installed with a mole plough (figure 4.10); a bullet (mole) attached to a vertical tine/blade is drawn through the soil, leaving a tunnel (mole channel). The mole is followed by an expander which expands and strengthens the channels somewhat. The mole channels thus formed normally have a diameter of some 5–10 cm and are spaced every 1·5–3·0 m. Installation depth is usually between 0·40–0·60 m below soil surface, well into the 'impermeable' subsoil.

Normal, though heavy and powerful farm tractors can be used for moling. The installation capacity under normal conditions is around 5 km.hr^{-1} (gross output) or some 5–10 ha per workable day.

Installation

The success of mole drainage depends upon the successful exploitation of the two basic modes of soil failure which occur when a tine is pulled through a cohesive soil. At shallow depths the soil is displaced forwards, sideways and upwards, failing in shear along well defined rupture planes which radiate from the tip of the tine towards the surface at an angle of approximately 45° to the horizontal (see figure 4.11). This results in the soil cracking and an increase in the soils porosity and hydraulic conductivity. When the installation depth reaches the so-called *critical depth*, the soil at the base of the tine begins to flow fowards and sideways, only creating compaction and a decrease in porosity and hydraulic conductivity (Spoor 1979).

The critical depth generally occurs at a depth corresponding to an aspect ratio (tine depth/tine thickness) of the order of 5:7, although it becomes shallower as the soils become more plastic (ie wetter) or the surface confining layers become harder and drier. Installation depth should be somewhat deeper than the 'critical depth'. Under good conditions, moling should form a channel just below the critical depth, linked to a zone of cracked soil which extends up to a band of disturbed soil at the surface. These cracks provide direct access for impeded water to the mole channels. The channels also establish a groundwater drainage base but when the surrounding soil is nearly impermeable, hardly any water will be drawn into the mole drains by normal groundwater flow.

Moling is best suited to clay soils with a minimum clay content of ~30%. When during installation the moisture content at mole depth is near the lower plastic limit,

Mole plough

Subsoiler

FIGURE 4.10 *Moling and subsoiling equipment* (reproduced by courtesy G Miles and Son and Ransomes Sims and Jefferies Ltd, England)

not too much smearing (hindering water entry) will occur. Channels installed at much drier conditions than the lower plastic limit are liable to cave in rapidly. Best installation conditions will often occur shortly after harvest, when the upper part of the soil profile has dried out.

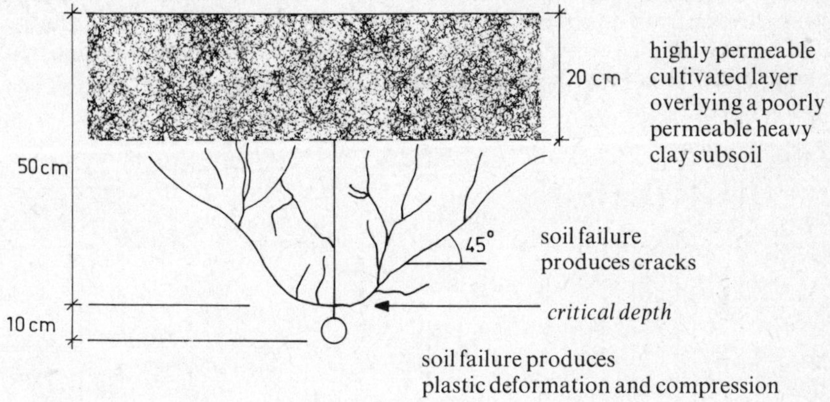

FIGURE 4.11 *Water movement to mole drains*

Length, slope and outlet

The length of mole channel between outlets and their effective life time are influenced by the inherent stability of the soil to repeated wetting, by the soil uniformity and by the conditions at the time of installation. Liability to blocking increases with increasing channel length. Safe lengths vary with soil type between 20 m and 80 m and should be determined on the basis of experience. Channels have been known to function for well over 20 years in the most stable, calcareous clays although it is sound practice to remole every three to seven years.

It is also important to prevent water stagnating in the mole channels since this will weaken the walls and lead to premature deterioration. Some slope and a good outlet are therefore important. Too much slope, however, may cause erosion in the channel, or lead to 'blow-outs' when combined with long lengths. Safe gradients are in the range of 0·2–3·0%.

Mole drains are usually aligned with the slope of the land where the latter is within the acceptable range. Few mole ploughs allow a very accurate grade to be maintained and when the land surface is rather uneven, some irregularities in the grade of the channels are to be expected. Mole channels may discharge directly into a ditch, either with an unprotected or with a protected outlet (the latter eg in the form of a 1–2 m long pipe inserted into the channel). A better system is often to intercept the flow in the mole channels at suitable intervals by means of drain trenches installed across the mole channels (see figure 4.12). This provides a dependable outlet and allows the use of short length mole channels in large fields. The drain trenches are installed first and then the mole channels are drawn across and through the trenches. The trenches may be parallel and regularly spaced or be laid out in close accordance with the topography, with the trenches mostly following the depressions in the field. The trenches are provided with a pipe drain and are backfilled to well above mole depth with gravel to allow the discharge from the mole channels to flow readily down to the pipe.

Mole drains generally respond rapidly to rainfall and their discharge rate can be substantial. Therefore it is essential that the hydraulic conductivity of the backfilled gravel should be high so as to minimise head losses at the discrete points at which water cascades from the mole channel into the gravel-filled trench. The gravel should be clean and have a minimum size of between 3 and 5 mm. The pipes should also be designed to cope with this rapid response to rainfall.

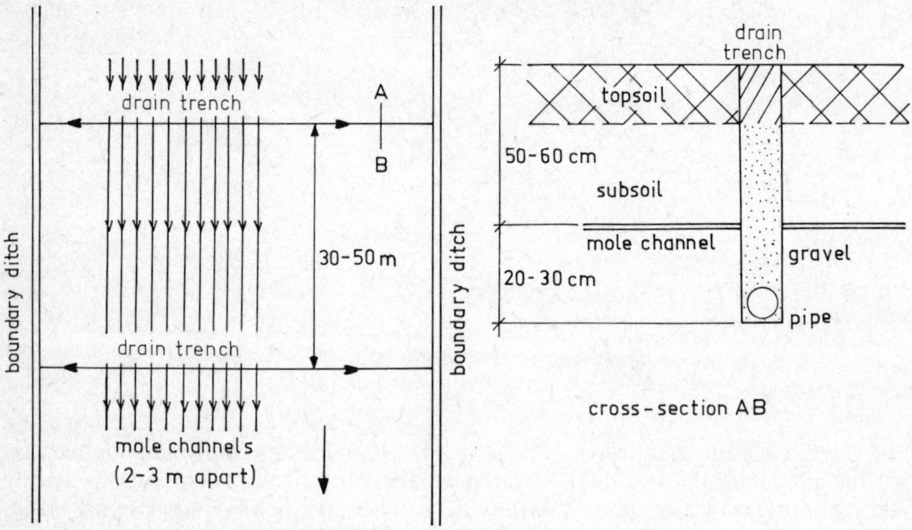

FIGURE 4.12 *Mole drains discharging into drain trenches*

Soil improvement

Mole drainage aims to establish a medium deep drainage base in the soil and although not all of the soil above this depth is well drained and rooted, opportunities for water movement and rooting are enhanced by the cracking of the soil and may further increase as the improved drainage induces more cracking and deeper, more thorough exploration of the soil by roots. This process of rootzone deepening may be assisted by the measures discussed in section 4.5. By gradually increasing the installation depth up to a maximum of 70 cm over say 20 years, a parallel deepening of the rootzone may also be achieved. Mole drainage should not be considered as a cheap, shallow, narrowly spaced, unlined version of pipe drainage: it is a *soil* and *drainage* improvement measure, whereby the latter supports the former and *vice versa*, based on cracking and subsequent improvement of the subsoil and on the provision of lateral outflow facilities for impeded water through the mole channel.

4.4 Pipe drainage systems

Pipe drainage systems are seldom feasible in heavy land as these derive their usefulness (and justify their high costs) from their groundwater drainage function. Owing to the immobility of the groundwater in the subsoil and the inhospitable nature of these strata for root development, groundwater drainage is in-effective in this case. Pipe drains thus serve mainly as (expensive) 'underground ditches' (figure 4.13), achieving essentially only the same kind of shallow drainage as the bedding and the shallow ditch systems described in sections 4.1 and 4.2, and this only for so long as the backfill material in the trench retains adequate hydraulic conductivity.

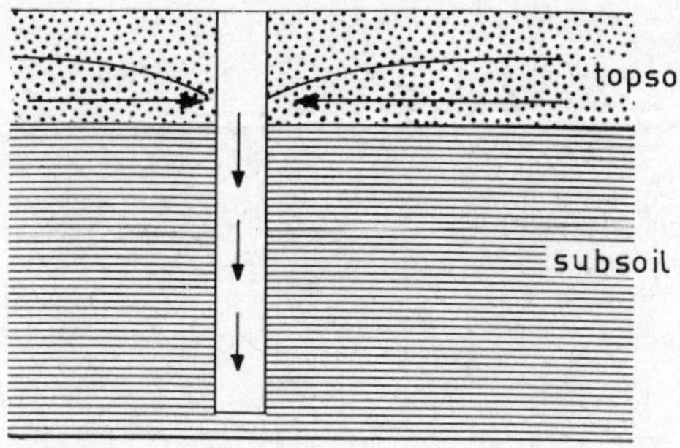

FIGURE 4.13 *The functioning of a pipe drain in heavy land*

Pipe drainage may however be used effectively in heavy land in the situation where the 'impermeable' subsoil is underlain within normal pipe drainage depth (1·0–1·5 m) by a reasonably permeable substratum, as sometimes occurs in alluvial basin clay soils (see figure 4.14). Pipe drain trenches reaching into this substratum perform the shallow drainage function described above and, in addition, establish a low hydraulic potential in the permeable substratum which induces vertical flow down through the overlying profile. This flow reaches its maximum value when the watertable is in the topsoil and the soil water pressure in the substratum at drain depth is about zero. This condition will be met when the drains are rather narrowly spaced, so that the amount of head lost in horizontal and radial flow in the permeable substratum will be insignificant. Under these conditions virtually all head is available for vertical flow through the impeding layer. This vertical flow will occur under a hydraulic gradient approaching its maximum attainable value of one with the result that the flow rate will be nearly equal to the hydraulic conductivity of the subsoil (eg in figure 4.14 about 3–4 $mm.day^{-1}$). This rate, although not very high, is similar in magnitude to the normal rates of evaporative drying of the soil and thus contributes as much to the early workability of the soil after rain as the latter. Also, this flow continues for a longer period than the shallow drainage and in time achieves a more thorough, deeper drainage of the soil than shallow drainage alone. Plant growth and workability conditions, generally improve sufficiently to warrant the extra costs of pipe drainage in situations such as that depicted in figure 4.14.

The lateral topsoil drainage may reach peak flows of 20–25 $mm.day^{-1}$ which imposes high demands upon the hydraulic conductivity of the trench backfill. For a normal 15–20 m spacing and 20 cm trench width, the minimum K-value in the trench should be of the order of 1–1·5 $m.day^{-1}$ for unrestricted conveyance of the shallow drainage water to the pipe. For recently excavated trenches this is generally no problem but the hydraulic conductivity of the trench generally decreases as the backfill consolidates. Being located above the pipe, drainage conditions in the trench are favourable and, as a result, a good soil structure will often develop in the backfill, and may even be promoted by adding lime to the backfilled soil. Experience shows that trenches are more permeable than the undisturbed soil. At wide spacings, however, the trench conductivity may well act as a bottleneck in the drainage of the topsoil. Trenchless installation of pipe drains obviously has no application with this sytem which relies on trench hydraulic conductivity.

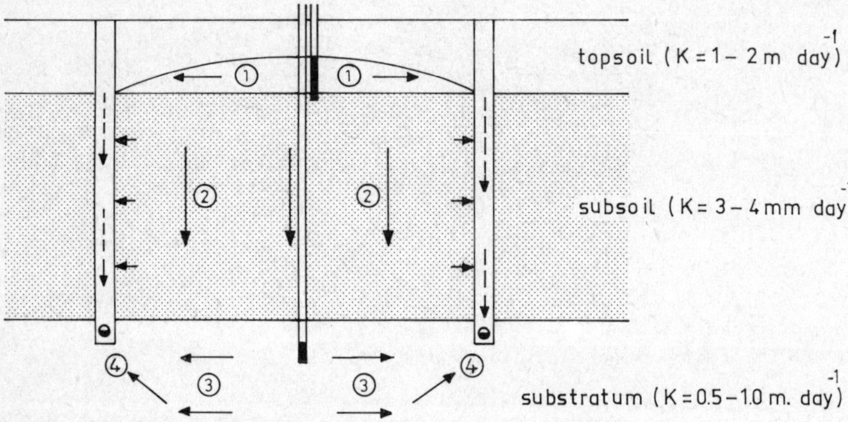

FIGURE 4.14 *Functioning of pipe drainage in heavy land with a well permeable substratum* (VAN HOORN 1974)

4.5 COMPLEMENTARY MEASURES

Measures may be taken to improve the flow conditions in the soil, thus adding to the effectiveness of the shallow drainage systems (and sometimes also of groundwater drainage systems). These measures are:

(a) *Subsoiling*

Moling and subsoiling both aim at loosening the soil although in the case of moling a stable channel should also be formed. This requires somewhat different equipment and working conditions to subsoiling (see figure 4.10).

Subsoiling is applicable to soils in which the hydraulic conductivity of the entire rootzone, or of layers within the profile, is very low. It involves pulling tined shanks through the soil, lifting, shattering and loosening the treated soil layers in the process (figure 4.15). These implements can be mounted on a frame or on a toolbar (see figure 4.10). For shallow subsoiling, for example, to break up a plough pan, 4–5 tines spaced 50 cm apart may be used; for a working depth of 60–80 cm (~ max. depth) 2–3 tines only, about 75 cm apart is normal. As shown in figure 4.15, the base (taken at soil surface) of the loosened triangular area has about the same width as the working depth. The tine spacing is usually about equal to the working depth which gives a considerable overlap of soil disturbance in the upper soil layer and adequate disturbance in the lower layer.

Subsoiling may be done instead of moling in heavy textured soils which have such a variable nature (eg containing pockets of sandy soil) that the integrity of the mole channels cannot be assured.

For subsoiling to be effective, the working depth should always be less than the 'critical depth' (see section 4.3) which depends upon the width of the tine and the installation conditions.

The tine should be rather thick and have an inclination of not less than 25–30° to produce a good lifting and loosening of the soil. Thin, shallowly inclined tines are less effective although they may be used for light work. As the tines become larger and more steeply inclined, power requirements increase. Power requirements also increase with the addition of side wings to the conventional chisel shaped foot of a subsoil tine (see figure 4.15) although this is generally more than compensated by the increased loosening which occurs.

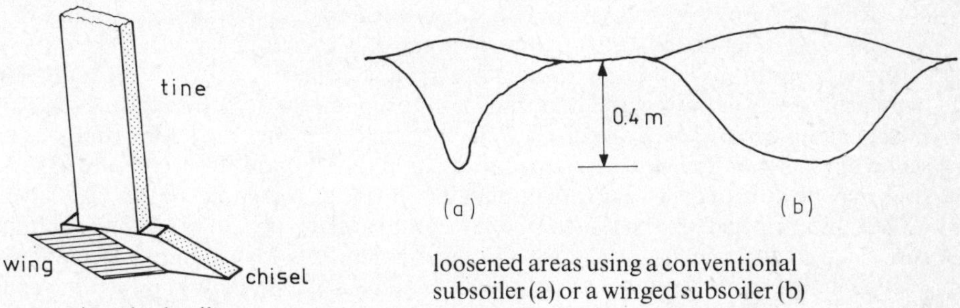

winged subsoiler

loosened areas using a conventional subsoiler (a) or a winged subsoiler (b)

FIGURE 4.15 *Patterns of soil failure created by conventional and winged subsoilers* (SPOOR 1979)

Subsoiling without field drainage may create a 'bath tub': the excess water readily sinks into the loosened soil but is unable to drain away. Both groundwater drainage and shallow drainage systems may be used in combination with subsoiling. The first system would for example be applicable where the soil below drain depth was reasonably permeable and the subsoiling would improve the vertical drainage in the soil all the way down to drain depth. Shallow drainage would be applicable where subsoiling has loosened the upper part of the soil while below the subsoiling depth the soil remains virtually impermeable. Subsoiling should, as a rule, be aligned across the field drains since, normally, flow conditions for water in the subsoiling direction may be expected to have been improved by the treatment. Subsoiling for example may be done across the pipe drains such as in figure 4.14 which would improve the lateral shallow flow (compare to moling across the pipe trenches illustrated in figure 4.12; subsoiling would be less effective as no defined channels are formed). The waterlevel in the field drain should of course be lower than the subsoiling depth.

The initial improvement in drainage and rooting conditions may be perpetuated and re-inforced by farming measures such as the growing of deep and vigorously rooting crops, careful soil management (avoidance of compaction), green manuring, liming, etc. The direct, immediate effects often last no more than one or two crop seasons.

On many heavy soils crops suffer from poor drainage on the one hand and drought on the other, both due to the limited rooting depth imposed by the poor drainage. Subsoiling helps to alleviate both of these problems.

(b) *Deep ploughing*

Deep ploughing, even more than subsoiling, is a total, integral soil profile improvement measure. It assists drainage by promoting the vertical flow of excess water through the soil but farming and crop growth normally also benefit from the improved moisture retention/supply, improved soil fertility, improved soil mechanical properties, etc. The combined benefits justify the high investment in deep ploughing. For maximum effect, deep ploughing should normally be combined with the farming measures discussed above under 'subsoiling'.

(c) *Use of chemical ammendments and organic matter*

These measures may be combined with subsoiling and deep ploughing but they may also be used independently. They improve the soil structure and as such improve drainage conditions although their effect is most pronounced in the topsoil. The beneficial effects of liming and of organic manuring on the structure of clay soils are widely recognised. Easily dispersed soils become less prone to slaking and surface sealing at a higher lime and organic matter content of the soil. The structure of sodic soils may be improved by gypsum application. Lime and organic matter also have a favourable influence on the FC/LPL relationship and as such on the soil workability (see section 1.9). It generally follows that well-limed soils with a good percentage of organic matter are less liable to compaction. In each of these cases infiltration and percolation of excess water into and through the soil improves.

(d) *Farming*

The relationships between drainage and farming have already been described in section 2.4. Drainage is assisted, and the harmful effects of imperfect drainage are partly alleviated by farming geared towards the maintenance of adequate organic matter in the soil. Soil management is able to enhance drainage or make it deteriorate. Different crops pose different drainage requirements. All this reflects on the farming system and cropping pattern to be adopted.

References and further reading

BOWLER, D G 1980: The Drainage of Wet Soils, Hodder and Stoughton, Auckland

CONCARET, J *et al* 1981: *Drainage Agricole*, Chambre Régionale d'Agriculture de Bourgogne, Dijon (in French)

DIVIS, J and J DVORAK (eds.) 1966: *Proceedings of the Symposium on Hydrological and Technical Problems of Land Drainage*, Czechoslovak Scientific and Technical Society, Agriculture and Forestry Section, Prague (two volumes)

EGGELSMANN, R 1981: *Dränanleitung*, Paul Parey, Hamburg (in German; English translation: *Subsurface Drainage Instructions*, 1978)

FAO 1971: 'Drainage of heavy clay soils', Irrigation and Drainage Paper no 6, Food and Agriculture Organisation of the United Nations, Rome

VAN HOORN, JW 1974: 'Drainage of heavy clay soils'. In: *Drainage Principles and Applications*, publication 16, volume IV, International Institute for Land Reclamation and Improvement, Wageningen

HUDSON, AW *et al* 1962: The draining of farm lands, Bulletin 18, Massey Agricultural College, Palmerston (New Zealand)

ILRI 1979: *Proceedings of the International Drainage Workshop*, J Wesseling (ed.), publication 25, International Institute for Land Reclamation and Improvement, Wageningen

LUTHIN, JN 1957 (ed): *Drainage of Agricultural Lands*, Agronomy Monograph no 7, American Society of Agronomy, Madison Wisconsin

RAADSMA, SR and FE SCHULZE 1974: 'Surface field drainage systems'. In: *Drainage Principles and Applications*, volume IV, publication 16, International Institute for Land Reclamation and Improvement, Wageningen

SCHWAB, GO, RK FREVERT, TW EDMINSTER and KK BARNES 1966: *Soil and Water Conservation*, Wiley, New York

SCHULTE KARING, H 1968: 'Die Unterbodenmelioration', Landes Lehr- und Versuchstation, Ahrweiler (in German)

SOIL CONSERVATION SERVICE 1971: *Drainage of Agricultural Land*, National Engineering Handbook, section 16, US Department of Agriculture, Washington DC

SPOOR, G 1979: 'Soil disturbance with deep working field implements in field drainage situations'. In: *Proceedings of the International Drainage Workshop*, J. Wesseling (ed.), publication 25, International Institute for Land Reclamation and Improvement, Wageningen

Chapter 5 Main drainage systems

A main drainage system receives water from the field drainage systems. In a small system, the field systems may be of a uniform type but large systems are likely to be fed by a variety of field systems and receive surface runoff, interflow as well as groundwater flow. The system is also likely to receive 'natural' drainage flow and it may also have to convey the discharge from urban drainage systems.

The main function of the main drains is to convey all this water to the outlet point. Main drains may, however, also collect a fair amount of excess water directly from the fields themselves, especially as these drains tend as a rule to follow the depressions or run along the lower sides of the fields.

5.1 THE GENERAL CASE

The situation shown in figure 5.1 is fairly typical of the layout and functioning of a main drainage system. The area encompassed (*drainage basin*) is about 8 000 ha of farmland which includes two small urban areas.

The layout of the system is based mostly on the geography and topography of the area, although historical and infrastructural developments have obviously also had a considerable influence. The latter is for example apparent in the interrelationships between road and canal systems and in the division between the two sub-basins formed by the main road crossing the area. The hierarchical composition of the canal network with small lower order canals combining into larger higher order canals in the process of centralisation of the flow and convergence towards the outlet, corresponds to the arrangement of most natural drainage systems. Designation of the different canal reaches and auxiliary structures may follow the format used in Annex 9.1. A systematic designation greatly facilitates the design of the canal system as well as the subsequent operation and maintenance of the system.

Agricultural fields (plots) may vary in size from < 1 ha up to some 25–30 ha. These are drained internally by the field drainage systems described in chapters 3 and 4. They will normally be bounded on one or more sides by a lower order main drainage canal (collector ditch or tertiary canal) into which the field system discharges. In figure 5.1 only the secondary drains are shown which typically serve areas from a few hundred up to several thousand hectares.

The hierarchical composition of the drainage network emerges naturally in a design based upon topography and is compatible with the aim of arriving at a well ordered, lucid system which can readily be managed. No hard and fast rules apply however, and where significant savings may be made, by for example allowing tertiary drains to discharge directly into a primary canal rather than via a secondary canal, this should be done.

In planning the layout of the drainage system, the basin should be perceived at different levels of generalisation. Initially, perception at a high level of generalisation is required to determine the division of the basin into sub-basins and the alignment of the primary drains serving these. Next, at a lower level of generalisation, the secondary networks and finally the tertiary networks should be planned. In this process various factors must be weighed and integrated: the topography of the area,

FIGURE 5.1 *Example of a main drainage system* (*scale* 1 : 50 000)

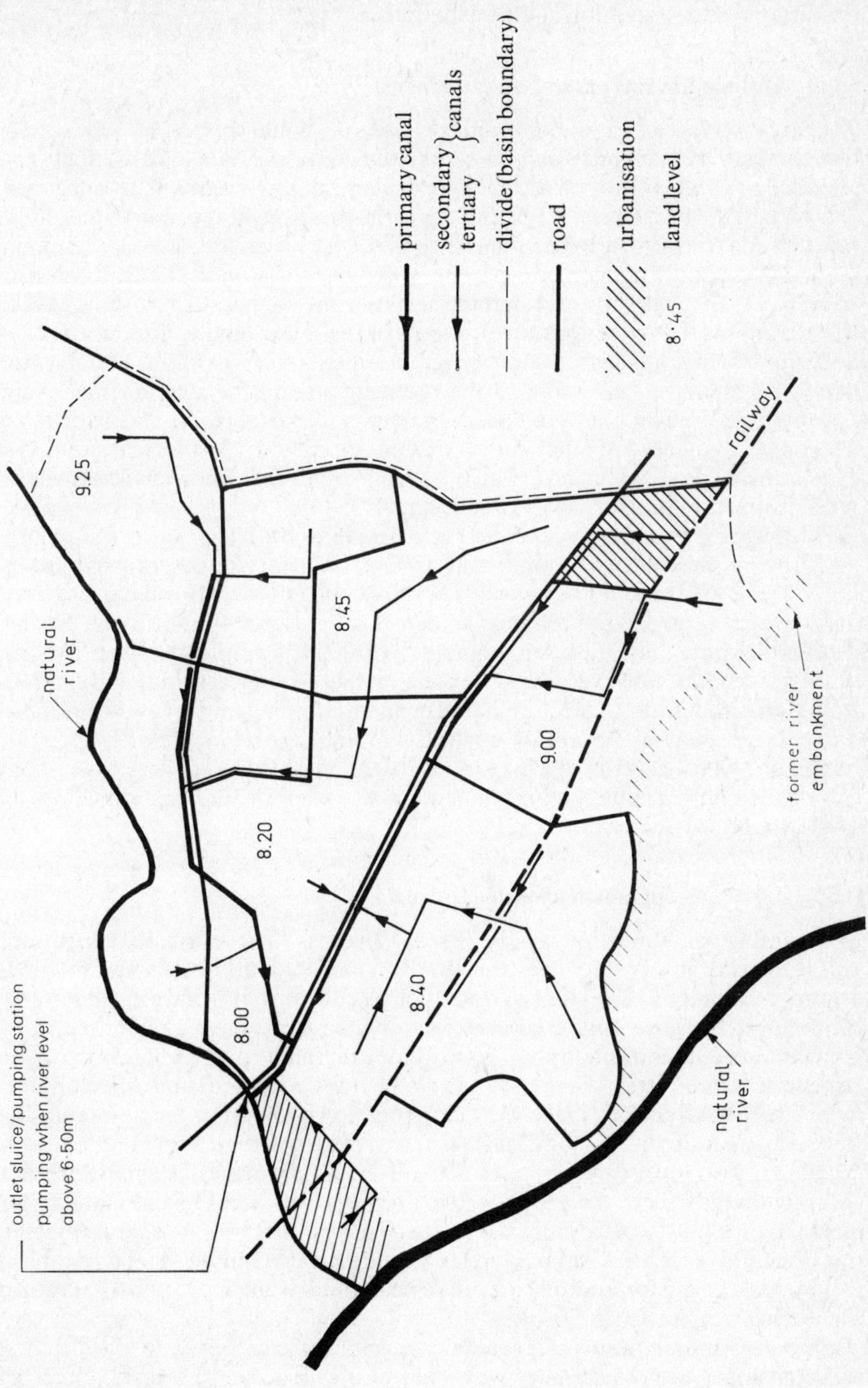

the present and planned infrastructure, property and administrative boundaries, soil conditions, costs, operation of the system, etc.

5.1.1 Drainage basin (watershed, catchment)

The area served by a main drainage system should preferably constitute an independant hydrological unit. If the system only serves a part of such a unit, remedial or compensatory arrangements will normally be required to safeguard the drainage of the part outside. Furthermore, the drainage of the area within may not be quite so straightforward as if the entire area had been included. For example, a system that is only concerned with the upper part of a natural basin, will usually have to pass through the lower part to the outlet and so special provisions have to be taken in order to ensure that the drainage of the lower area is not impeded. Main drainage systems, however, seldom cover an entire large river basin; rather it is more likely that the river will serve as the receiving stream for a number of drainage systems organised on a sub-unit basis. In river and coastal plains and, in delta areas, hydrological divides are often not very pronounced and hydrological sub-divisions can be changed or established relatively easily as desired. These divisions will often result in the formation of polders (see section 5.2).

Drainage basins may vary in extent from less than 100 ha (covering eg one large or a number of small farms) to more than 100 000 ha. They are generally largest in the flat lowland areas but it also depends on how much of the natural drainage system has been incorporated. Technically there are no strict limits to the basin size provided the area constitutes a suitable hydrological unit. Most large sized main drainage systems, however, are communal or public undertakings and limits arise from the consideration that the management of too small an area is inefficient while of too large an area it becomes unwieldy. Optimum sizes therefore range from a minimum 5 000/10 000 ha to a maximum 20 000/50 000 ha. Very large basins may be sub-divided into smaller management units under the overall direction of an authority/board (see section 5.3.2).

5.1.2 Types and alignment of drainage canals

Main drains will almost always be (open) canals as these normally constitute the most economical way to convey the relatively large quantities of water involved in drainage discharge. The storage potential of canals is also valued under certain conditions. Pipes are only used in exceptional circumstances as main drains, and then for lower order drains only. This may for example apply to a highly mechanised large scale farming area where collector ditches may be replaced by collector pipes to form larger fields and to make the field more accessible from the adjoining roads. Buried main drains may also be feasible in an irrigated area to avoid the complicated crossing of two open canal systems (irrigation and drainage systems). Such drains will usually only serve the groundwater drainage system and so a relatively small diameter pipe may be used since discharges will be rather low. A separate system for the discharge of excess surface water will then be required. Unfavourable soil conditions causing for example extensive instability problems in canals may also be a factor in favour of using pipelines.

Flow velocities in drains can generally be kept sufficiently low to allow the use of earthen canals, where necessary with bank protection. Smooth rectangular bends may well be used in the smaller tertiary canals (capacity up to 1–2 $m^3.sec^{-1}$). For larger canals, curves should be used. Recommended radii of curvature generally vary from 5 m for canals with a capacity less then 5 $m^3.sec^{-1}$ to about 5 to 10 times the bed width for large main canals with a capacity of over 10 $m^3.sec^{-1}$. Transitional sections as commonly used in road curves, are generally not needed for canals.

Two main considerations apply with respect to the canal alignment and layout, ie:

(a) main drains should, wherever possible, follow the depressions and downslope sides of the fields. Higher order canals should preferably run through the major depressions or otherwise follow the lowest sites in the area, the next lower order of canal following the next lower sites and so on. This alignment establishes the best gradients for the discharge of field and collector drains into the tertiary canals, tertiary into secondary, then into primary and onwards to the outlet. In addition, this approach will make the main drains themselves effective collectors of excess water. So, on the basis of this consideration, the general layout of the main system, as well as the alignment of the individual drains, will be largely dictated by the macro/meso relief of the land;

(b) field and farm boundaries will be largely formed by the lower order drainage canals towards which the field drainage systems discharge, and therefore these should preferably be laid out in a manner compatible with efficient farming. This calls for some regular pattern, rectangular or trapezoidal, of fair dimension, forming fields for example some 200–300 m wide and 300–500 m long.

These two considerations are obviously not always in harmony. However, an acceptable compromise is often possible where the relief of the land is fairly regular and even. A small number of depressions in a field can always be filled-in, using the spoil from the newly excavated ditches and canals.

However, where the macro/meso topography of the land is irregular, the choice must be made between either forming regular shaped, good sized fields (but containing a number of in-field depressions) or adapting the layout to conform to the relief of the land, resulting in an irregular layout, comprising a large number of odd shaped, small to medium sized fields (see figure 5.2). This dilemma applies mainly to the lower order canals; higher order canals should generally be aligned solely on the basis of topography.

Main drain alignments may also be dictated by the existing infrastructure (roads, railways, existing bridges and culverts, etc) and/or by administrative divisions (property boundaries, district boundaries, etc). There may be a preference to plan the new drains along existing roads as this facilitates inspection and maintenance and also avoids unnecessary further division of the land. In complete contrast, they may be sited away from the roads for reasons of safety, to avoid abuse, etc. The latter considerations also apply with respect to the alignment of main drains relative to settlements. Alignment of main drains along property and administrative boundaries divides the landloss evenly between the two parties and emphasises the communal character of these drains.

Wherever feasible, use should be made of the existing drains (excavated or natural). Natural drains as a rule follow the predominant gradient of the land, often through the depressions, and as such can normally be integrated rather well into the main drainage plan. Some minor canalisation and/or re-alignment can always be done to improve the fit of these natural drains into the desired layout. In general, the scope for incorporating the natural drainage system into the new plan will increase towards the outlet.

The location of the outlet is of course also a determining factor in the planning of the main drainage system. It determines the overall orientation of the system, in particular of the primary canals. Generally speaking, the shortest route to the outlet point should be sought.

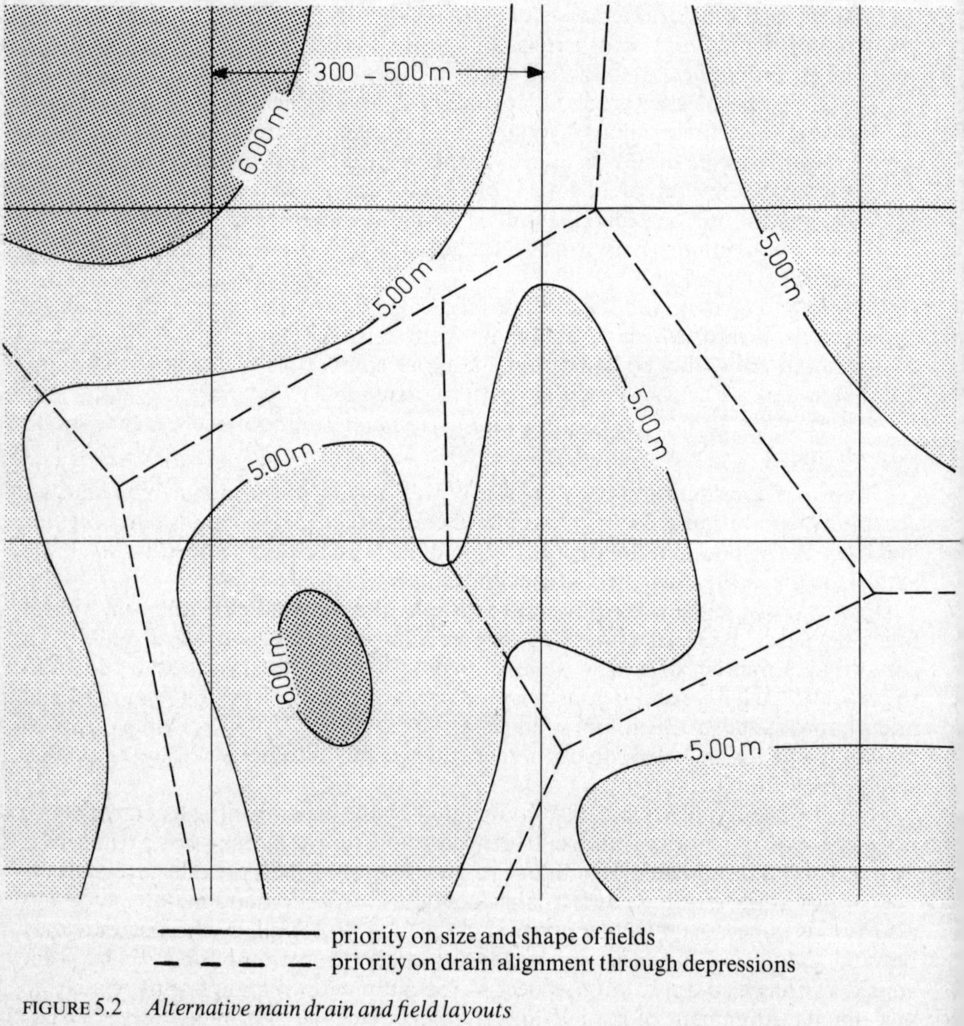

FIGURE 5.2 *Alternative main drain and field layouts*

5.1.3 Outlet and waterlevels

The outlet point of a drainage system will normally be a conveniently low point on a river, a lake, the sea or any other component of the hydrological system which may be suitable to act as a recipient of drainage water, eg a nearby (marshy) depression.

An outlet into a river will often be located sufficiently far downstream to create a good gradient in the main drainage system and to allow gravity discharge (see section 5.2.2). Prevailing wind directions, foundation conditions, liability to scouring/siltation, accessibility, proximity to powerlines, etc, should also be considered. In humid climates, the natural drainage system is usually well developed and so finding an outlet does not generally pose a problem. In arid/semi-arid areas where the runoff is much less and the natural drains are fewer and less well developed, outfall drains of irrigation projects may have to be extended far outside the area to end up at a suitable outlet point. In some cases the discharge may be led into an evaporation-pond or a vertical outlet may even be used, ie discharge into the underlying substrata. The latter solution is, for example, suitable where an enclosed area is underlain by a well-drained aquifer.

The level difference between the regional drainage base (= critical outer water level at the outlet point) and the field drainage base (= water level in the field drains) constitutes the total available head for drainage flow out of the area (P_1–P_3 in figure 5.3).

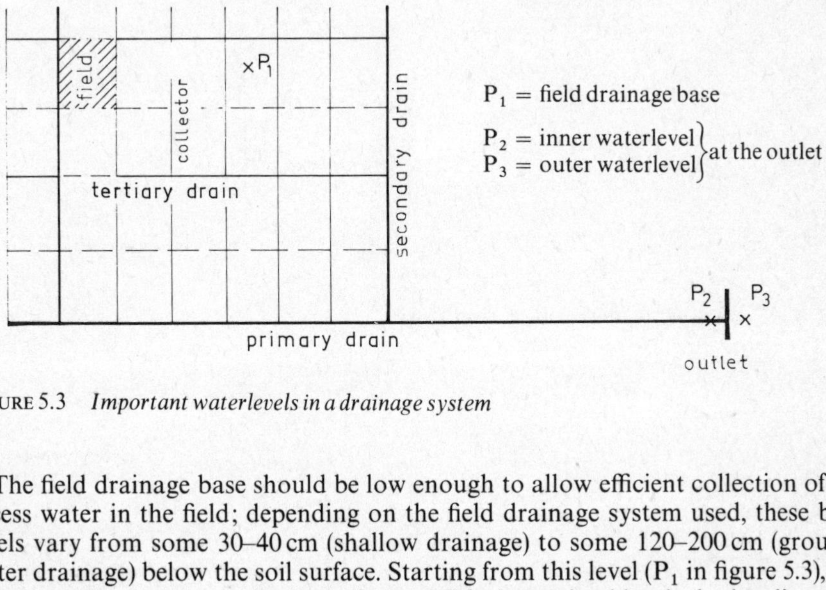

FIGURE 5.3 *Important waterlevels in a drainage system*

The field drainage base should be low enough to allow efficient collection of the excess water in the field; depending on the field drainage system used, these base levels vary from some 30–40 cm (shallow drainage) to some 120–200 cm (groundwater drainage) below the soil surface. Starting from this level (P_1 in figure 5.3), the inner waterlevel at the outlet (P_2 in figure 5.3) is determined by the hydraulic grade line adopted in the main canal system (see also figure 9.21):

$$P_2 = P_1 - H$$

where $H = h_1 + h_2 + \ldots\ldots =$ total headloss in the primary, secondary, etc, canals (including the structures).

When $P_2 > P_3$, a gravity outlet may be used, but pumping is required in situations where $P_2 < P_3$.

Often the outer waterlevel will vary in time. This is for example the case with outlets into the sea or into the lower reaches of rivers subjected to tidal influences. River levels also depend on the discharge which will often vary seasonally. In all these cases the frequency and duration of high outer levels should be studied, specifically for those periods in which the drainage discharge from the area occurs. These periods may or may not coincide with high outer water levels. On the basis of these investigations it may then be decided whether gravity or pump outlet is most feasible. Various combinations should also be investigated as well as the scope for storing water in the area during short periods when gravity discharge is not possible (see section 8.8).

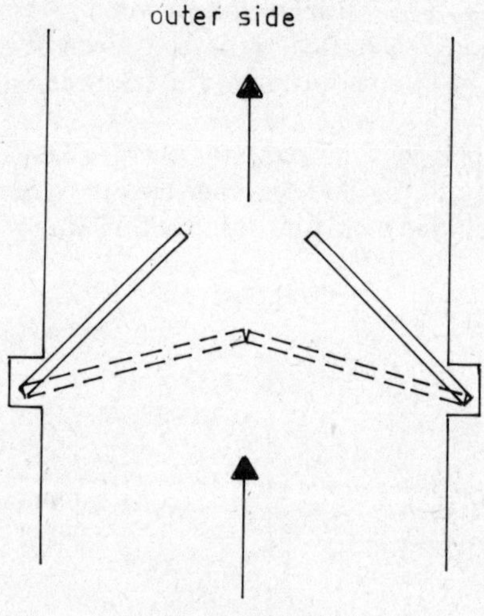

FIGURE 5.4 *Small outlet sluice with vertical doors*

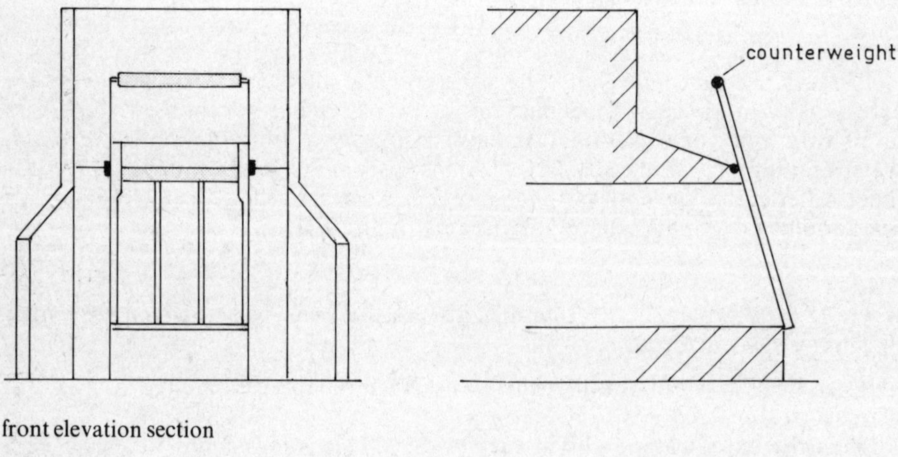

front elevation section

FIGURE 5.5 *Flap gate*

Outlet structures (figures 5.4, 5.5 and 5.6)

Various outlet structures may be used. Where the outer level is always well below the inner level, an open, free fall structure is suitable. Various types of gated structures are used to stop the intrusion of water due to high outer levels. The flap gate is popular as this allows water to flow out of the area when the outer waterlevel is below the inner level but closes automatically when the outer level rises above the inner level. Flap gates may be placed in a small outlet sluice or be incorporated into a culvert outlet passing through a dike (eg could be fitted in the culvert shown in figure 5.6). In the latter situation the gate is usually installed at the outer end of the culvert. The gate may hinge freely with a vertical rest position or rest slightly slanted outwards. The slanted design ensures the best closure (and as such is often preferred where salt instrusion is liable to present problems) but requires, of course, more pressure from the inside to open the gate.

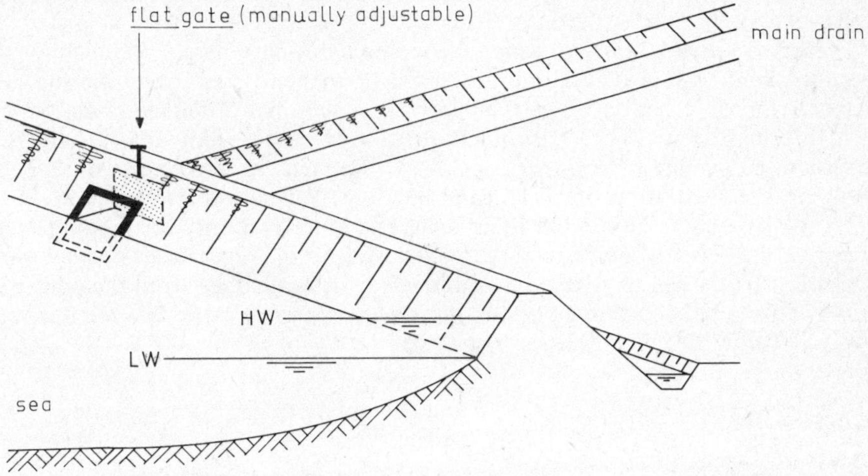

FIGURE 5.6 *Flat-gated culvert outlet through a dike*

Automatic gates are suitable for small tidal outlets. Where the outlet is on a river with a mainly seasonally fluctuating waterlevel, a manually operated gate would be preferable since the operation of the gate is not much of a burden and the infrequent operation of the gate makes automatic operation risky. Automatic gates may still be the best solution, however, for remote, inaccessible outlet points.

Large outlets may comprise single or multiple barrel, reinforced concrete box culverts or take the form of a sluice. In the case of tidal outlets, these structures may function automatically by providing the culverts with flap gates and the sluice with self-moving doors. Very large outlets may have specific operating rules or otherwise warrant non-automatic operation.

The (self priming) outlet syphon is a less common type of outlet structure. Here the excess water is syphoned through a pipe over the dike. Outlet syphons may be considered in tidal outlet situations as an alternative to the more conventional outlet culvert. The outlet syphon has the advantage that it is not necessary to make an opening in the dike while poor foundation conditions for the culvert may also make the syphon an attractive solution.

Pump outlets may vary from small sump-installations with a capacity of 20–50 $l.sec^{-1}$ to large stations occupied by several pump units, each with a capacity of some 10–20 $m^3.sec^{-1}$. Often pump outlet and gravity outlet are combined in one outlet structure, with the pumps mainly coming into operation during prolonged periods of high outer water levels. The main characteristics of drainage pumps are discussed in section 9.4.

5.2 POLDERS

Broadly defined, a polder is 'an area in which the open water level can be artificially controlled at a level which deviates from the prevailing regional open water level'. In the classical conception of a polder as developed in the Netherlands, polders maintain a *lower* waterlevel in the area. This is still the main aim in most cases where polders are established (although during summer or dry periods, the inner level may well be maintained above the outer level).

Polders are used in areas in which the regional drainage base is too high for the intended land use. This situation may exist permanently (swamps, sea and lake bottoms) or only temporarily (tidal lowlands, seasonally flooded river plains). Polder formation involves separating the area to be developed hydrologically from the surrounding area by constructing dams or weirs in the connecting canals or by endiking the area to protect it from flooding. When the outer waterlevel is permanently above the desired inner level, the latter can only be maintained by pumping the excess water out of the polder. Where the regional drainage base is controlled by the sea or river, excess water may be discharged from the polder by gravity during periods of low tide and during low river flow periods when the outer level falls below the inner level (see figure 5.7).

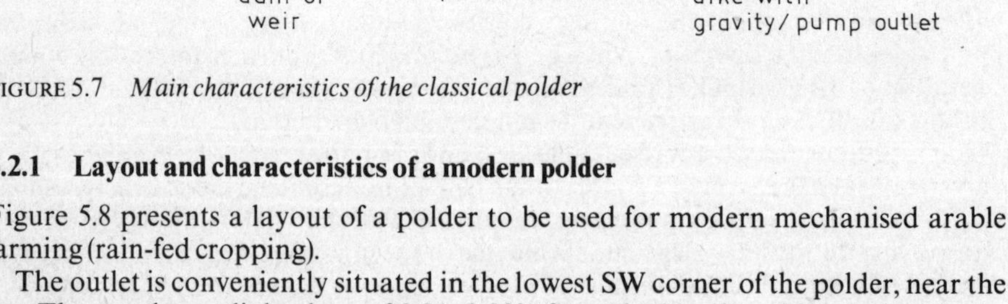

FIGURE 5.7 *Main characteristics of the classical polder*

5.2.1 Layout and characteristics of a modern polder

Figure 5.8 presents a layout of a polder to be used for modern mechanised arable farming (rain-fed cropping).

The outlet is conveniently situated in the lowest SW corner of the polder, near the sea. The area has a slight slope of 0·01–0·02% from the North and East towards the outlet.

Polderlevel

The target open waterlevel to be maintained in the polder is termed *the polderlevel.* It serves as a reference for the operation of the outlet. The polderlevel is established on

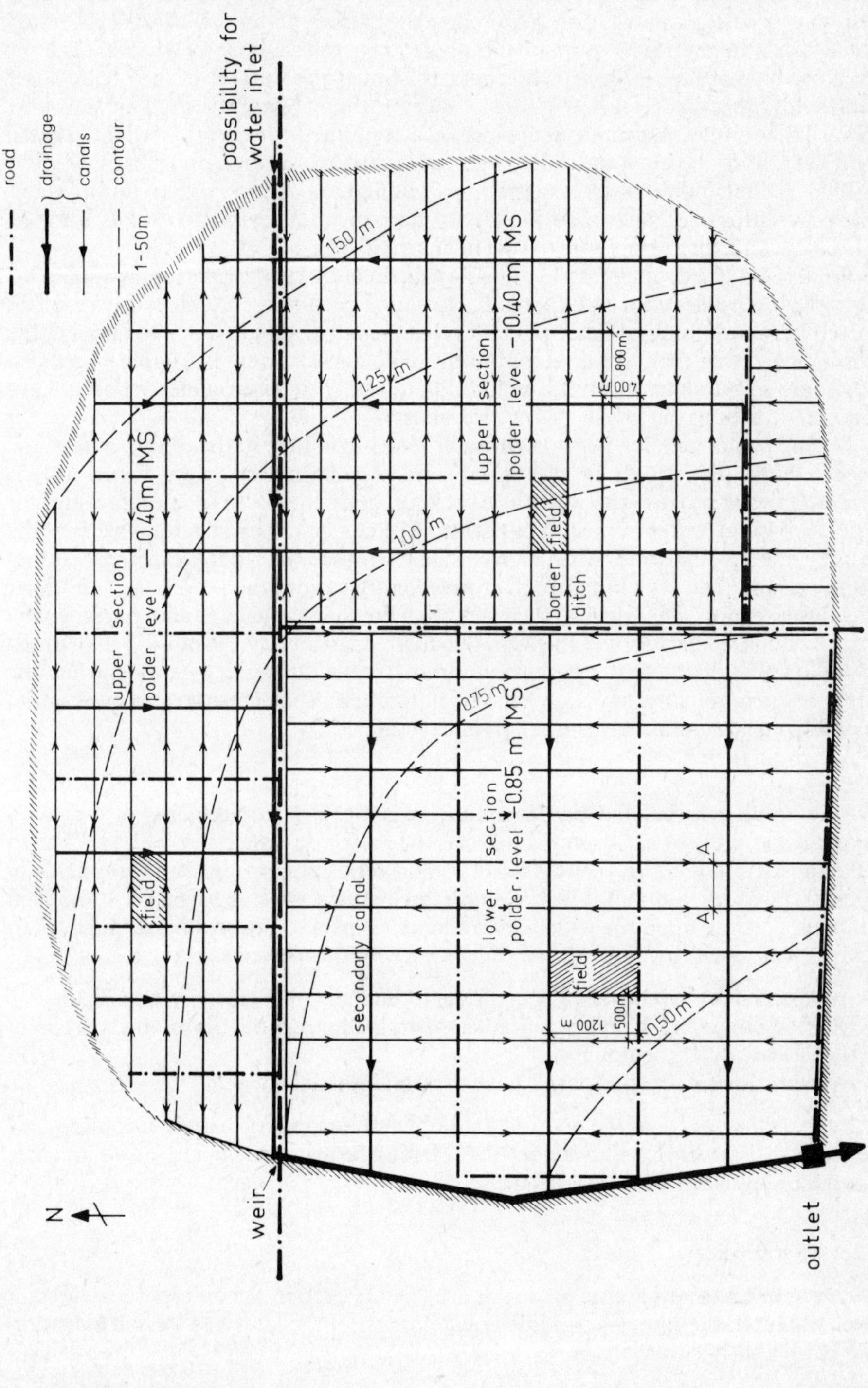

FIGURE 5.8 *Typical layout of a modern polder*

the basis of agricultural considerations (it should be deep enough to allow unobstructed discharge of field drains) although other considerations sometimes also play a role (eg navigation requirements). During periods of discharge, actual canal levels in the upper part of the polder will generally somewhat exceed the polderlevel (+10 to +20 cm) while near the outlet the waterlevel may be lowered well below the polderlevel (−30 to −40 cm), thus creating a gradient for flow towards the outlet. During the dry periods, continued evaporation will cause the waterlevel to drop below the polderlevel unless water is brought in from outside the polder. A high polderlevel is sometimes maintained in the polder during summer/dry seasons to encourage the lateral flow of water from the canals into the adjoining land, for example via the drain pipes (*sub-irrigation*).

Polders may be divided into *sections* having different polderlevels. This applies for example to a polder with an upper and a lower part which would probably be better served by maintaining a higher polderlevel in the upper part and a lower level in the lower part, rather than one level for the entire polder. Different polderlevels are also advisable where soil conditions within differ greatly or where different parts have different functions (eg urban v agricultural use).

During prolonged dry periods all water tends to collect in the lowest part of the polder whilst the canals in the upper part become (nearly) dry. Where this is undesirable (eg because the soil may overdrain or ditches need to contain water for cattle drinking), water in the upper part may be conserved by installing weirs in the canals. These will also retard the discharge from the upper part somewhat during flood periods. The weir in figure 5.8 conveniently divides the polder into an upper and lower section. The width and crest level of the weir should be such that when the design discharge passes over the weir, the upstream waterlevel is not too high (must still allow free discharge from the field drains) while during dry periods, when the canal levels in the upper section become horizontal at the crest level, a substantial water depth is maintained in all canals (see section 9.2.2).

Field layout

In the typically rather flat polder areas, the hydraulic requirements of the canal layout can usually be accommodated so as to harmonise rather well with the desired field and farm layout. In figure 5.8 canals and roads are arranged to form fields of 400–500 m width and 800–1200 m length, giving field sizes of 32–60 ha. If the field drainage is based upon the widely used singular pipe drainage system as depicted in figure 5.9, field widths should generally not exceed 600 m because:

- commercial drain pipe flushing equipment generally only reaches up to 300 m;
- with long lines and for normal gradients, drain depth at the beginning and end of the line will differ too much;
- with long lines, deep ditch levels have to be maintained.

Long fields save on the required number of secondary canals although this advantage is generally outweighed by various farm management disadvantages when the length exceeds 1000–1500 m.

5.2.2 River polders

Different drainage discharge problems are encountered in the different reaches of a river valley. In the upper to middle reaches, most of the land has a good elevation above the local drainage base and gradients are rather steep. In the upper reaches in particular, emphasis has often to be placed upon retarding rather than promoting the discharge. In the delta's, the land is usually rather flat and the natural drainage base is at a shallow depth or even above the land level (eg during high river discharges or high tide when the land is flooded). Here the classic polder situation exists.

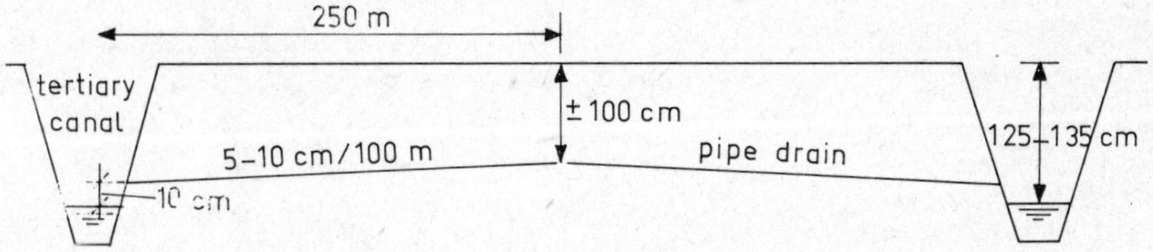

FIGURE 5.9 *Field cross-section (section A–A in figure 5.8)*

Drainage discharge problems in the middle to lower reaches of a river are illustrated in figure 5.10. Gradients along the river are usually of the order of 20–30 cm.km^{-1}. Natural raised levees are formed adjacent to the river and low lying basins occur further away. The basins are often semi-enclosed by the levees, the terrace land bordering the valley and higher ground on the valley floor (often levees related to a former river course), leaving only a narrow outlet at the downstream end. Tributaries entering the valley from the higher terrace land typically change direction in the basin area. With a slight gradient (less than the river gradient), they follow a course parallel to the river, centrally through the basin area, until sufficient head has been gained to break through the levee into the river. Often a number of tributaries combine. In the case of figure 5.10 tributary I flows into the main river some 50 km downstream of its point of entry into the valley.

This principle is used to provide gravity discharge for the polders shown in figure 5.10. The tributaries entering the valley from the side are diverted to join a neighbouring tributary. Within the plain they are normally endiked so as to convey high discharges safely (without flooding) to the main river. The polders discharge by gravity into the tributary when levels permit (flap gate), otherwise pumping is used. When floods on the tributary coincide with floods on the main river, the outfall drain is allowed to spill over near its outlet into a low lying reservoir area. Where a suitable reservoir area exists (low lying, low value land), this may be preferable to pumping (see also section 8.8).

5.3 ORGANISATIONAL AND OPERATIONAL ASPECTS

Disposal drainage systems typically have a communal function and the establishment of a proper legal and organisational framework is essential to prevent or settle disputes among the parties involved as these will generally be to the detriment of the drainage of the whole area. Almost invariably this leads to some degree of state involvement, the extent varying from country to country depending on the prevailing political system, and on the stage of development of the country.

Common arrangements in this respect are therefore described and discussed.

FIGURE 5.10 *Example of a river polder setting*

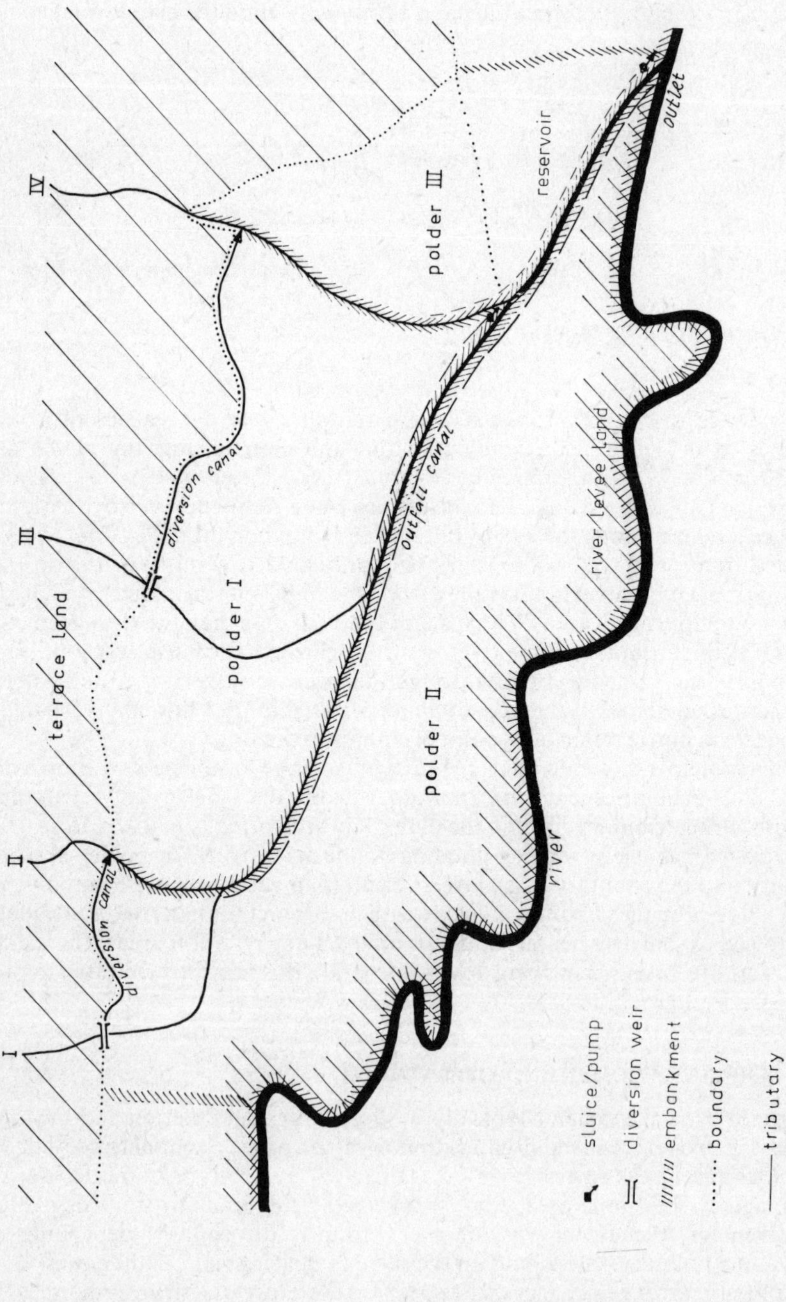

5.3.1 Legal frameworks

In most countries, landowners have certain rights and/or are subjected to certain restrictions in the disposal of the excess water from their land. Depending on whether the country's law system is based on the Roman civil law or on the English common law, the following basic rules apply:

Roman law: the upslope landowner is entitled to the free natural discharge of excess water from his land inherent to its elevated position.
English law: excess water is considered a 'common enemy'. Landowners may not burden other owners with drainage water over and above the amount which occurs naturally.

These rules deal adequately with the natural drainage discharge from the land but less so with the (man-made) improved drainage. Interpretations in the latter case have been made, generally giving the upslope owner a somewhat more favourable position under Roman than English law. Under both rules it follows that the owner who alters the natural drainage situation can be held responsible for the resulting damage. Clearly, landowners must make agreements and co-operate with neighbouring owners when they wish to improve the disposal of excess water from their land.

Eventually, all excess water will drain towards one of the natural streams, ranging from small to large, which comprise the natural disposal drainage systems of the region or country. Most countries have laws stating that all these natural streams are in principle under state ownership and control. Additional man-made drainage canals are almost always installed by a public drainage organisation; these organisations normally also retain legal ownership and control of these canals while the state frequently also delegates the management of the natural streams in the area to them.

5.3.2 Public drainage organisations

The communal organisation of the disposal of excess water from the farmers land has taken many forms in different countries. Interested parties are the landowners although cities, industries and other parties having an interest in an adequate disposal of the excess water from the area, may also participate in the organisation.

In it simplest form a group of farmers could co-operate on the basis of a legally recorded mutal agreement on the method of disposal of their excess water. More commonly, however, the organisation is established by a quorum-decision of the interested parties (simple or two-thirds majority, whatever the law requires for the establishment of this type of public organisation) or be set up by special act or government degree. The first applies for example to the Drainage District organisations in the USA (established by simple majority of the interested parties); the latter to the Water Management District organisations in the Netherlands (established by decree of the provincial government) and the Water Authorities in the UK (established by Act of Parliament). These organisations usually have wider tasks than simply drainage disposal. The Water Management Districts in The Netherlands are for example also responsible for water quality control, water supply, etc, although originally they were formed for flood control and drainage disposal purposes and these activities still remain their most important tasks.

These public drainage organisations are generally governed by a board, the members of which are elected by the general assembly of all interested parties. Some members of the board will often also be appointed by the government. The board may form an executive committee for the day-to-day management of the affairs of the organisation. For these tasks the board may avail itself of the services of a professional staff, headed by a general manager. Typically the staff consists of an administrative and an engineering division, subdivided as in the organisational chart in figure 5.11 (applicable to drainage basins of about 20 000–50 000 ha with mixed farming and predominantly medium sized farms). The functions, responsibilities and powers of the organisation would be laid down in the charter under which it was established, with the by-laws defining various aspects of the operational functioning of the organisation. For example, the by-laws would normally specify the maintenance obligations of the landowners. The charter and by-laws would also spell out how the expenses of the organisation were to be recovered from the participating parties. The participants are usually assessed on the basis of the benefits that each party gains from the work of the organisation. For landowners this would typically be a per hectare charge, varying with the classification of the land. As well as the operation and maintenance of existing main drainage systems, these organisations will often also undertake various improvement works with the charter/by-laws once again specifying how the relevant decisions should be made and the costs divided. Where there is no public drainage organisation, the impetus for its establishment frequently arises from the commonly felt desire of the parties that the drainage in the area should be improved and that an organisation through which this can be done, should be established.

In the developing countries, drainage organisations as described above are seldom in existence. Whatever natural and other drainage infrastructure exists is mostly installed and managed by the state or by large landowners/estates. Drainage systems within irrigation schemes generally come under the management of the overall scheme. Drainage improvement works are often initiated and implemented by the local, or more commonly by the central government (Ministry of Agriculture, Ministry of Irrigation, Ministry of Works). These organisations will normally also maintain the control of these works upon completion. Large scale drainage improvement works (often constituting part of, or being integrated with, a wider agricultural or rural development scheme) are often implemented and subsequently managed by so-called authorities, bodies with statutary powers created by the (central) government. The farmer's participation in these organisations and in the various decision-making processes is very limited.

5.3.3 Maintenance

The Chezy–Manning formula for open channel flow discussed in section 9.1 shows that a halving of the K_m roughness factor (as may well occur in a poorly maintained small canal compared to its well maintained state), halves the flow velocity. To maintain the same discharge four times as much gradient would be required. Poor maintenance as a rule reduces the discharge capacity of the system and raises the waterlevels in the canals. As the latter constitute the local drainage bases for the field drainage systems, the discharge from these systems is hindered preventing them from functioning as designed. Flooding may also occur, especially in the lower reaches of the disposal drainage system.

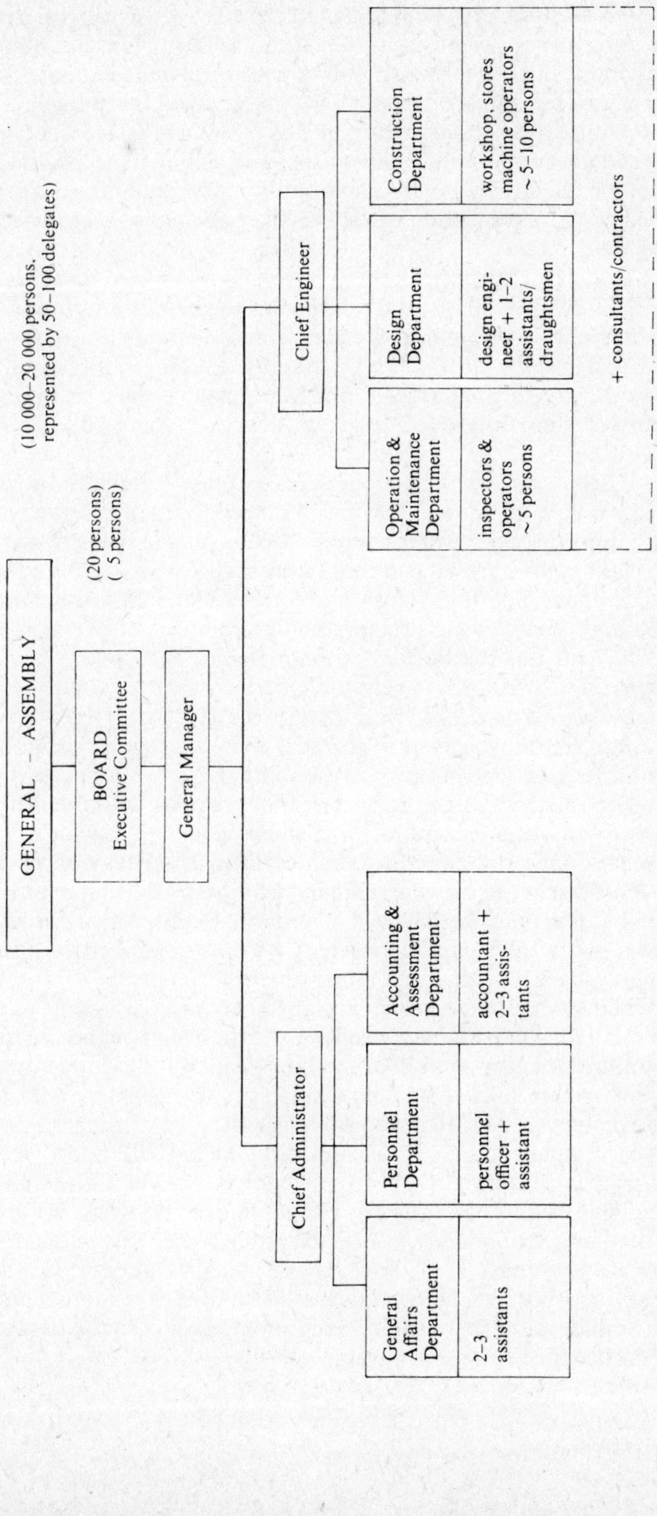

FIGURE 5.11 *Organisational chart for a medium sized public drainage organisation*

Good maintenance is thus clearly in the interest of the farmers. Where this is not appreciated, extension work may help to increase the farmer's motivation although some form of enforcement is usually required. This applies in particular to the lower order drainage canals which are usually maintained, on a fair and equal basis, by the landowners/farmers who stand to benefit. The drainage organisation should check that these maintenance obligations are fulfilled by making regular inspections (once or twice a year, duly announced in advance). The maintenance of the on-farm systems is of course the farmer's own responsibility. The main disposal canals are often maintained by the organisation itself (with their own personnel and equipment or using a contractor).

Maintenance requirements may to some extent be minimised and the maintenance work be facilitated by deliberately designing the canals and structures with these requirements in mind. Canal maintenance involves the clearance of silt and other deposits (and usually also some re-shaping of the cross-section) and the clearance of any flow retarding vegetation. Maintenance may be minimised by designing for non-eroding flow velocities, smooth curves, energy dissipation, bank protection, silt traps, etc. The growth of plants in canals is enhanced by light. Narrow, shaded, deep canals are less conducive to plant growth than wide open, shallow canals. Low flow velocities, allowing all suspended load to settle and light to penetrate deeply into the clear water, should also be avoided. There are however many different plant species growing in canals (bank vegetation, underwater plants, anchored/non-anchored floating plants and others), so many different growth conditions exist and designs which retard the development of one type may well promote another. Much of the burden of maintenance is therefore unavoidable; design can help to do the work effectively and efficiently.

Vegetation may have to be cleared once to three times a year. Heavy growth is to be expected during the rainy season in tropical climates, regular clearance being required as discharge requirements are correspondingly high. Discharge is of course much lower during the dry season; indeed drainage canals will often dry up and hardly any growth will occur (except for canals in irrigated areas which often carry some flow all year-round). In temperate climates, growth is heavy in spring and in early summer; most discharge, however, occurs in autumn and winter and so canals should be cleared well in time for the onset of autumn. Maintenance during summer should not however be neglected as canals may become so choked with vegetation that they are unable to even discharge summer rains adequately.

Silt clearance and re-shaping requirements are more variable, in time as well as to location. Stable canal sections may need very little maintenance. New canals normally remain unstable for some two to three years until a good sod becomes established. Plant growth may reduce flow velocities to the extent that fines that would otherwise remain suspended become deposited.

For the larger drainage canals it usually pays to provide good sized, easily accessible and trafficable berms alongside the canals on which the maintenance equipment (draglines, back-hoes, mowers, etc) can move and work. Boat mounted mowing and dredging equipment is also available, while the judicious use of herbicides may also be considered. Various maintenance alternatives should be weighed and be incorporated as an integral part of the planning and design processes. This applies to the technical designs and especially to the organisational aspects of the maintenance.

References and further reading

GREGORY, K J and D E WALLING 1973: *Drainage Basin, Form and Process*, Edward Arnold, London

ILRI 1964: 'The design of open watercourses', Bulletin 7, International Institute for Land Reclamation and Improvement, Wageningen

ILRI 1982: *Proceedings of the International Symposium 'Polders of the World'*, International Institute for Land Reclamation and Improvement, Wageningen

LUTHIN, J N 1966: *Drainage Engineering*, Wiley, New York

SCHROEDER, G 1968: *Landwirtschaftlicher Wasserbau*, Springer, Berlin (in German)

SOIL CONSERVATION SERVICE 1971: *Drainage of Agricultural Land*, National Engineering Handbook, section 16, US Department of Agriculture, Washington DC

SOIL CONSERVATION SERVICE 1977: 'Design of open channels' Technical Release no 25, US Department of Agriculture, Washington DC

THORN, R B *et al* 1959: *The Design of Land Drainage Works*, Butterworths, London

VOLKER, A 1981: 'Reclamation and polders', Lecture Notes, International Institute for Hydraulic and Environmental Engineering, Delft

PART III
PLANNING AND DESIGN OF DRAINAGE SYSTEMS

Chapter 6 Outline of principles and practices

Drainage planning involves the preparation of a plan for the solution of a drainage problem. The plan will generally consist of a number of measures to be taken and/or works to be constructed. In some cases the best solution to a drainage problem may well be a change in land use or in farm practices, making the agricultural use of the land less susceptible to excess water. In most cases, however, the core of a drainage plan consists of the construction of some new drainage works. The detailing of such works, being mostly in the domain of engineering, is commonly referred to as 'design'.

6.1 FIELD INVESTIGATIONS

To establish a basis for planning and design, a great deal of information on the project needs to be collected by means of field investigations. In particular, information is required to:

(a) diagnose the drainage problem in hand and to conceive possible solutions;
(b) prepare plans and designs.

Diagnosis of the problem

A correct diagnosis of the drainage problem is essential for the preparation of a sound drainage plan, and as such it is the first thing to be done. It involves establishing:

- *the nature and cause of the problem* (impeded percolation, high watertables, sources of excess water, bottlenecks in the collection/disposal of excess water, etc);
- *the harmful effects* (affected area, frequency and seasonal period of excess water, effects on crops and farm operations, etc).

A preliminary diagnosis can often be made on the basis of available information and field inspections, the latter preferably including actual observations of the occurrence of excess water and of the resulting damage.

Various types of maps are useful during this stage: geographical, hydrological, soils, administrative, etc, showing relief, characteristic elevations, the present (main) drainage systems, the main soil-patterns, land use, main infrastructural features and other relevant information. Aerial photo's are also particularly useful. The assessment of the outlet conditions and of 'foreign' water involvement, may require the investigation of a wider area than the project area.

A programme of further investigations can then be planned based upon the results of this preliminary diagnosis.

Plan preparation/elaboration

Once a satisfactory diagnosis of the problem has been made, plans for its solution can be conceived. Parallel with the planning and design process, further investigations are made to establish the environmental, technical, socio-economic and institutional conditions under which the planned measures and works are to be implemented and to operate. These provide the basis for the formulation of the boundary conditions and design criteria for the planning.

The required investigations are outlined in chapter 15.

6.2 Stages of project preparation

Investigations and planning for a drainage project normally proceed in stages. These form a sequence in which the intensity of the investigations and the degree of detail of the plans increase progressively whilst the overall level of generalisation decreases. Thus, by a process of successive approximations, increasingly more refined and optimised plans are made.

In general, the following four stages of project preparation may be distinguished. Depending on the project (available information, size, urgency, etc) certain stages may be omitted or combined (combination of stages 1) and 2) is quite common).

1 *Identification stage*
   - first formulation of the project, mostly on the basis of available information; hardly any analysis and/or appraisal.

2 *Reconnaissance/pre-feasibility stage*
   - information collected through reconnaissance type field investigations (broad and general);
   - preliminary diagnosis of the drainage problem;
   - rough outline of possible solution(s); delineation of the project area and its sub-divisions;
   - evidence that the proposed project is promising and desirable.

3 *Feasibility stage*
   - information collected through semi-detailed type of field investigations (mapping scale 1:10 000/50 000);
   - presentation of the proposed plans in sufficient detail to demonstrate convincingly that they are technically sound and to enable costs to be estimated within some 10% accuracy;
   - proof without doubt that the proposed project is the best available solution, is administratively workable and economically/financially viable.

4 *Final stage*
   - information collected through detailed type of field investigations;
   - elaboration of all plans to the extent that they can serve as working documents for the implementation (detailed plans and designs, construction drawings and specifications, etc).

The 'Feasibility Report', containing the results of the work done during this stage, represents the basic document upon which the decision to proceed with the project will be based. Most banks require such a report before they will consider a loan application. Once the go-ahead decision is made, the final stage can be entered to make the project ready for implementation.

Depending on the size and nature of the project, a feasibility study may be carried out by one drainage engineer or require a small multi-disciplinary team (composed for example of an engineer, an agronomist and an economist). This study may take, for a large complex project, almost a year to complete. The identification and reconnaissance/pre-feasibility studies often take the form of short missions by highly qualified and experienced drainage specialists. The final design stage usually takes the longest and would normally be carried out by a group of senior and junior engineers employed by the water management organisation or by a consulting engineering firm.

The required standards for the reconnaissance, semi-detailed and detailed type of field investigations are implicitely formulated by the objectives of the stage to which they apply. For some investigations, however, the required standards have been more explicitely formulated (eg for the soil mapping, see section 15.3).

6.3 PLANNING AND DESIGN; GENERAL APPROACH

In the preparation of a drainage plan the following three groups of variables have to be defined:

- *the physical characteristics of the drainage system*: the selection of suitable types of drains and structures; determination of alignments, spacings, depths, capacities; selection of materials and construction methods to be used; etc.
- *the land use characteristics*: suitable types of land use, farming systems, farming practices, etc.
- *the operation rules to be followed*: institutional organisation and procedures, maintenance, operation of pumps and structures, financial arrangements, etc.

Most of these variables are mutually dependant (drain type depends on land use; drain spacing depends on drain depth; etc) or are subject to outside constraints (dictating eg crops to be grown, location of the outlet, etc). The planner must endeavour to fix, within the limits imposed by the above constraints, the variables so as to optimise the objectives of the project (*optimal plan*). In practice, optimisation using straightforward planning methods such as systems analysis, can almost never be used due to the complex, still inadequately understood relationships between the variables themselves and between the variables and the objectives. Currently used drainage planning methods are essentially a loosely structured set of procedures by which the variables are determined partly on the basis of *a-priori* assumptions and *ad-hoc* decisions, partly on the basis of established relationships. These procedures may be tested and refined to the extent that, applied by experienced staff to routine situations, they result in reasonably optimised plans. No such guarantee, however, exists when applied to new areas and new situations.

Current methods all start with a formulation in technical terms of the desired excess water control, this being referred to as the *basic design criterion.* It is implicitely assumed that the drainage plan which best meets this criterion is the optimal plan. On the basis of this criterion, the physical characteristics of the drainage system can be worked out. Some land use characteristics and operation rules are considered beforehand and incorporated into the basic design criterion, whilst others are defined alongside in the process of defining the physical works. Various subsidiary design criteria are used as well, partly derived from the basic criterion, partly in addition to the latter. These are, implicitely or explicitly, discussed in the text where applicable.

6.4 BASIC DESIGN CRITERIA

Figure 6.1 illustrates the essential relationships to be considered in drainage planning. In this figure, the relationships between drainage and farming have been broken down into:

(a) The relationships between drainage and the occurrence of excess water on or in the land.
(b) The relationships between these occurrences and farming.

The relationships at (a) are fairly well established and having formulated the desired degree of excess water control (ie the basic design criterion), the required drainage system can be readily established. The formulation of the desired degree of excess water control on the basis of the relationships at (b) is more difficult since these are generally far less well understood. Current basic design criteria, therefore, are mostly based on experience rather than firm scientific knowledge.

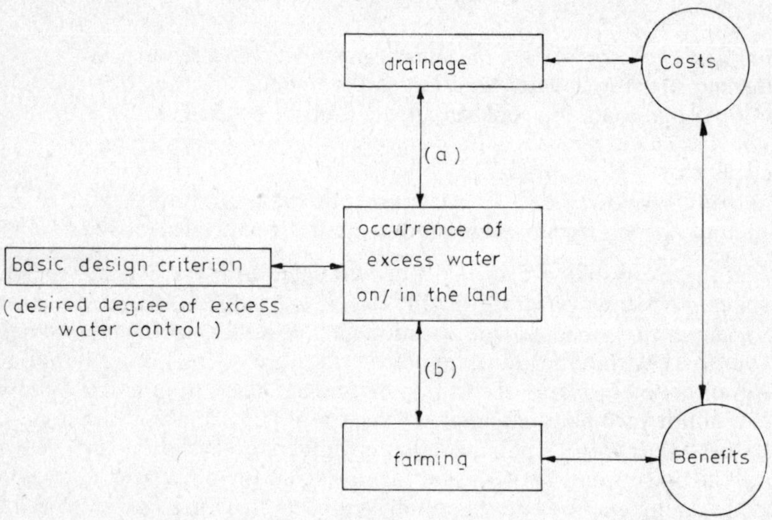

FIGURE 6.1 *Essential relationships to be considered in drainage planning*

The basic design criteria are formulated in different ways for groundwater drainage and for shallow drainage.

(a) *Groundwater drainage*:
The basic design criterion for groundwater drainage in humid regions prescribes the required watertable control during and after the occurrence of a specified high rainfall event (*design rainfall*). This of course applies where groundwater is recharged by rainfall. Elsewhere, recharge due to irrigation, seepage, etc, may need consideration. The required watertable control may be formulated in a steady state or a non-steady state form. Under steady state conditions, the recharge and discharge rate are assumed to be equal and steady (= constant in time) whilst the watertable level remains constant (*static watertable criterion*). Under non-steady state conditions, the criterion may specify how fast the watertable should fall after rising to an unacceptably high level (*falling watertable criterion*), or limits may be imposed on the frequency and duration of occurrence of unacceptably high water table levels (*fluctuating watertable criterion*).

(b) *Shallow drainage*:
The design criterion for shallow drainage prescribes within which span of time the excess water on the land, resulting from the design rainfall, must be evacuated. Other sources of excess water, such as snowmelt, irrigation losses, etc, may also be important, individually or in combination with rainfall, and should where relevant be incorporated in the design criterion.

These design criteria apply to field drainage systems. As main drainage is subservient to field drainage, criteria for the design of main drainage systems may in principle be derived from the field discharges they collect. Other requirements, however, also enter into the design of main systems and although the relevant criteria will always be closely related, they are seldom identical to those used for field drainage.

6.5 DESIGN RAINFALL

As the rainfall depth increases, its frequency of occurrence decreases. This holds true for all types of rainfall (hourly, daily, monthly, even annual rainfall). The nature of this relationship may be determined by noting how often different rainfall depths occurred during say the last 20–30 years. The results of such an analysis are commonly expressed as *frequency curves*, an example of which is presented in figure 6.2. The curve in figure 6.2 shows, for example, that a 48 hours rainfall higher than 40 mm is to be expected to occur on average once every year (1 × 1 year) while depths higher than 85 mm occur on average only once every five years (1 × 5 years).

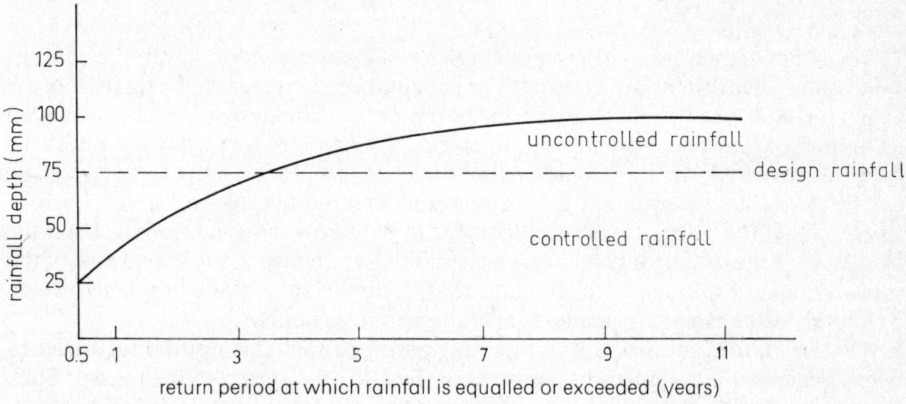

FIGURE 6.2 *Rainfall frequency curve for 48 hours rainfall periods*

The design rainfall is the most critical rainfall event that the drainage system should be able to cope with, ie it generates the most serious of all excess rainwater incidences which the system is designed to control. Even higher rainfalls may occur which are not fully controlled by the designed system and therefore these will cause damage (see figure 6.2). Such high rainfalls, however, only occur rarely and it is generally more economic to accept occasional damage than to construct a foolproof, expensive drainage system, capable of coping with the highest rainfalls.

Damage and benefits

A drainage system designed for a 48 hours rainfall of 75 mm will be able to cope with all 48 hours rainfalls up to 75 mm whilst rainfalls in excess of 75 mm will result in there being an uncontrolled element of rainfall which may cause damage. The relationship between the uncontrolled rainfall depth and the damage incurred is complex and is not well understood. Generally, however, the damage is proportional to (although usually not linearly) the depth of the uncontrolled rainfall, for example as in figure 6.3.

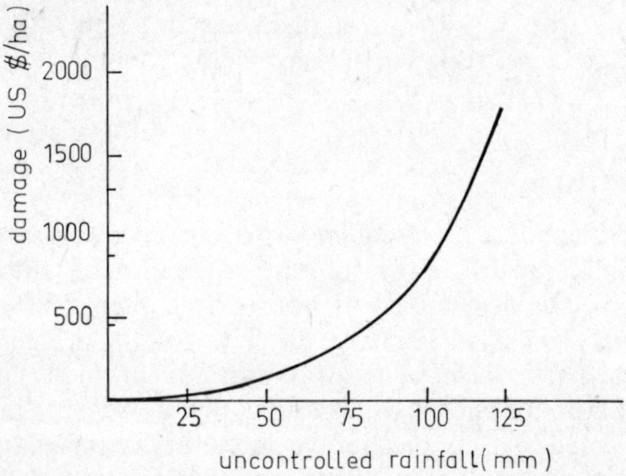

FIGURE 6.3 *Relationship between uncontrolled rainfall depth and damage incurred*

The benefits of a drainage system are derived from the reduction in the damage that would have otherwise occurred. The benefits may be increased by increasing the design rainfall but this involves an increase in cost. The design rainfall is thus an economic parameter and its selection essentially involves the optimisation of the expected benefits in relation to the costs.

The procedure for evaluating this optimum is set out in tables 6.1 and 6.2, whilst the results of the calculations are illustrated in figure 6.4. This figure shows that the total cost of the drainage system is composed of two elements, the actual cost of the installation and the cost of the damage that occurs in spite of the drainage system. The total costs decrease to a minimum at the optimum point A.

For agricultural drainage it is generally assumed that this optimum is reached when systems are designed on the basis of the 1 × 2 to 5 years rainfall event. Such systems may be expected to discharge safely and adequately all storms of the critical type occurring more frequently (these storms inflict little or no damage) but cannot fully control heavier storms (occurring less frequently). However, no hard and fast rules exist. In situations where uncontrolled rainfall can do considerable damage (eg in the case of a high value crop) and the cost of greater degree of control is not excessive, it may well pay to extend the control to less frequent events and base the design on, for example, the 1 × 10 year event. On the other hand a lower standard is often economically most feasible for drainage of land of low agricultural value. In the UK for example, the 1 × 2 years frequency storms are often taken as the design storm for the design of field drainage systems for grassland, the 1 × 5 years storms for arable land and the 1 × 10 years for horticultural land. Lower frequencies are

TABLE 6.1 *Analysis of frequency of occurrence of 48 hours rainfalls*

R 48 *hours rainfall* (*mm*)	*Return period of rainfall* ≥R[1] (*years*)	*Annual number of rainfalls* ≥R	*Rainfall interval* (*mm*)	*Annual number of rainfalls per interval*[2]
25	0·5	2·00	0–25	180·5
50	1·5	0·67	25–50	1·33
75	3·5	0·28	50–75	0·39
100	10·0	0·10	75–100	0·18
125	50·0	0·02	100–125	0·08
			> 124	0·02

[1] return periods based on figure 6.2

[2] there are 365/2 = 182·5 non-overlapping 48 hours rainfall periods per year. In two of these periods, the rainfall ≥25 mm, so there are 180·5 occurrences in the 0–25 mm interval. Similarly it follows that there are 2·00–0·67 = 1·33 occurrences in the 25–50 mm interval, etc.

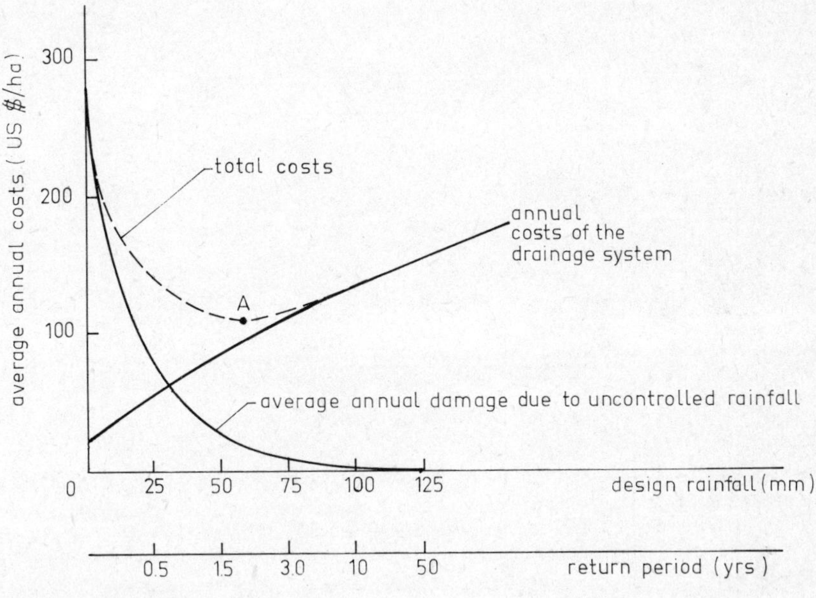

FIGURE 6.4 *Annual costs of drainage*

generally used for shallow field drainage systems as compared to groundwater drainage systems (1 × 2 to 5 years v. 1 × 1 to 2 years) and for main systems as compared to field systems (1 × 5 to 25 years v. 1 × 1 to 5 years), reflecting the differences in risk on damage for each of these systems.

Critical rainfall types

The most harmful type of rainfall varies with each drainage situation. For groundwater drainage, prolonged periods of moderately intensive rainfall, lasting

TABLE 6.2 *Analysis of average annual damage and marginal-average annual benefits due to drainage*

from table 6.1		*without drainage*		*with drainage*, taking as design rainfall							
				25 mm.day^{-1}		50 mm.day^{-1}		75 mm.day^{-1}		100 mm.day^{-1}	
rainfall interval (*mm*)	*annual number of occurrences*	*damage per storm* [1]	*average annual damage* [2]	*damage per storm* [1]	*average annual damage*	*damage per storm* [1 3]	*average annual damage*	*damage per storm* [1]	*average annual damage*	*damage per storm* [1]	*average annual damage*
0– 25	180·5	0	0	0	0	0	0	0	0	0	0
25– 50	1·33	40	53	0	0	0	0	0	0	0	0
50– 75	0·39	120	47	40	16	0	0	0	0	0	0
75–100	0·18	300	54	120	22	40	7	0	0	0	0
100–125	0·08	700	56	300	24	120	10	40	3	0	0
> 125	0·02	1500	30	700	14	300	6	120	2·4	40	0·8
Average annual costs of damage			240		76		23		5·4		0·8
Marginal average annual benefits[4]				164		53		17·6		4·6	

[1] the damage per storm US$/ha. based on figure 6.3.

[2] the annual damage due to 1·33 storms in interval 25–50 mm is 1·33 × 40 = $53/ha.

[3] with a design rainfall of 50 mm.day^{-1} the uncontrolled element of 48 hours rainfalls of 25, 50, 75, 100, 125 mm is 0, 0, 25, 50, 75 mm for which the damage per storm is US $ 0, 0, 40, 120 and 300 per/ha.

[4] equals the benefit (avoided damage) by increasing the design rainfall one step.

some three to five days and occurring during the rainy season, are usually most critical. Much of this rain can infiltrate into the soil and since during the rainy season soil profiles are at or near field capacity, these rains easily lead to saturated conditions in the soil and to high watertables. For shallow drainage, the more intensive shorter duration storms are usually more critical (12–48 hours storms).

For main systems fed predominantly by groundwater discharge, the critical rainfall types are the same or similar to those considered for the field drainage. This also applies to basins with a large proportion of shallow field drainage provided the land slopes remain minimal. In sloping basins, shallow field drainage and especially overland flow will respond directly to rainfall while in addition a concentration of the field drainage discharge is liable to occur. For these type of basins, storms with a duration equal to the time of concentration of the basin are most critical. Depending on the area and characteristics of the basin, critical durations vary from 0·5–6 hours (see chapter 8).

Of course the season of occurrence of the rainfall in relation to the farm calendar for the area is also of great importance. A high rainfall during the off-season would not normally be a good basis for design. Rather, design should be based on the rainfall occurring during periods when the crop is at a sensitive stage (planting and seedling stage) or during periods in which important farm operations have to be done (seedbed preparation, harvesting, etc). Interactions between rainfall frequency, critical rainfall type and critical season should also be considered.

6.6 DESIGN OF FIELD DRAINAGE SYSTEMS

The procedures for the design of the field drainage systems are briefly outlined below, further details being given in the following chapters 7 and 8.

Groundwater drainage systems: the following main variables have to be defined in the design (see also figure 7.1):

(a) type and layout of the system;
(b) discharge capacity of the system (q);
(c) watertable depths to be maintained in the field relative to the soil surface (H);
(d) the field drainage base depth (W), ie the installation depth of the pipe drains or the waterlevel to be maintained in the ditches;
(e) spacing of the field drains (L).

Various types and layouts of pipe drainage systems are discussed in chapter 3, together with the relevant selection criteria. The determination of the remaining variables are dealt with in chapter 7. Although the discussion focusses on pipe drainage, most of it applies equally well to ditch drainage.

Shallow drainage systems: here the main variables to be defined are:

(a) type and layout of the system;
(b) discharge capacity of the system (q);
(c) field drainage base depth (W), ie waterlevel to be maintained in the field ditches.

Aspects (a) and (c) are discussed in chapter 4. By comparing the pro's and con's of the relevant alternatives available, choices can be made. Very little design is required since the various parameters are either standard or are defined on the basis of experience.

The determination of the required discharge capacities are discussed in chapter 8 and having also defined the levels to be maintained, the dimensions of the drains and of the structures can readily be determined by means of standard hydraulic calculations.

6.7 Determination of design criteria

Design criteria are generally established partly on the basis of sound scientific theory and analysis and partly on the basis of experience. The scientific basis is presented where applicable in the following chapters 7, 8 and 9. The role of prior experience is reflected mainly in the use of (semi-)empirical formulae and engineering rules of thumb. Also, the design of a new system in an area usually relies heavily on experience obtained on similar work in the area or under comparable conditions elsewhere. All this should be judiciously applied, especially when the conditions to which the experiences apply are not well defined.

Often locally tested criteria will be unavailable, while results from test plots and/or pilot projects can often not be waited for (easily takes three to five years). In such cases estimated criteria may have to be used, at least for preliminary planning/early implementation. When a testing/monitoring programme is started early on in parallel with the progress of the work, preliminary adopted criteria may be verified and/or revised as work proceeds. However, by a judicious combination of theory, analysis and relevant experience, reasonable estimates can in fact be made in most cases.

The planner should of course be cost-conscious throughout the planning and design process. This applies especially to the selection of the design discharges which largely determine the required canal dimensions, pump capacities, etc, making it a main cost-determining factor. Other design parameters/criteria, however, also have a considerable influence on drainage costs, both on investments and on operational costs. With respect to construction, the local availability of skills and materials should always be an important consideration. It should also be stressed that design criteria should be established with a view to the future operation of the project. The future functioning of the project should be thoroughly analysed and defined prior to and alongside the design and the planner should make sure that the technical aspects of the drainage system and its envisaged operation are mutually compatible. Operational considerations should play an important and fully integrated role in the planning and design of a drainage project.

References and further reading

BURAS, N 1970: *Scientific Allocation of Water Resources*, American Elsevier, New York

HALL, WA and JA DRACUP 1970: *Water Resources System Engineering*, McGraw-Hill, New York

JAMES, LD and RR LEE 1971: *Economics of Water Resources Planning*, McGraw-Hill, New York

LINSLEY, RK and JB FRANZINI 1979: *Water Resources and Environmental Engineering*, McGraw-Hill, New York

MAAS, A *et al* 1962: *Design of Water Resource Systems*, Harvard University Press, Cambridge Mass

Chapter 7 Design of pipe drainage systems for groundwater control

Groundwater drainage by means of buried pipe systems is probably the most comprehensively studied subject in land drainage. As a result the relationships between the variables and parameters are well established. The remaining design problems are mostly due to the great variability in the geometry and in the hydraulic properties of the soil, and due to the many other interactions between the soil and the system (affecting the choice of materials, methods of installation, etc).

7.1 Drain spacing formulae

The basic design criterion for a pipe system for groundwater control specifies the recharge (q) that the system should be able to cope with while maintaining a desired watertable depth (H). Suitable values for W (= field drainage base depth) may be selected on the basis of considerations such as those discussed in section 7.6. This determines the watertable head (h) = W – H (see figure 7.1). Required drain spacings (L) may then be calculated using one of the drain spacing formulae to be discussed later (see figure 7.2).

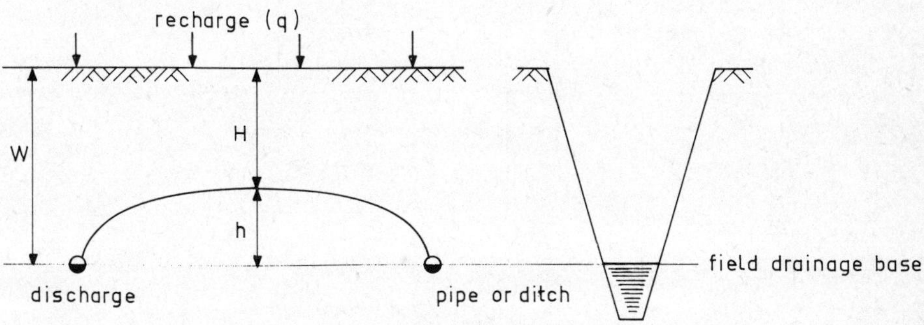

FIGURE 7.1 *The main variables in groundwater drainage design*

Drain spacing formulae may be categorised as either *steady state formulae* or *non-steady state formulae*. Steady state formulae are based upon the assumption that a steady constant flow occurs through the soil to the drains. Discharge equals recharge and the head (h) is also constant. In the non-steady state formulae all these parameters vary in time.

The steady state and the non-steady state formulae use respectively steady state and non-steady state basic design criteria as input.

In most cases design can be based on steady state conditions using one of the many available formulae, although the principles will only be discussed in relation to *Hooghoudts formula.* This formula has a wide applicability and a relatively simple structure. There are other formulae which may be superior in some cases but far worse in others, while some are quite unwieldy in use. In any event the accuracy with which drain spacing can be determined is limited by the accuracy of the soil parameters (due especially to the high variability of the hydraulic conductivity of the soil) rather than by the formula adopted.

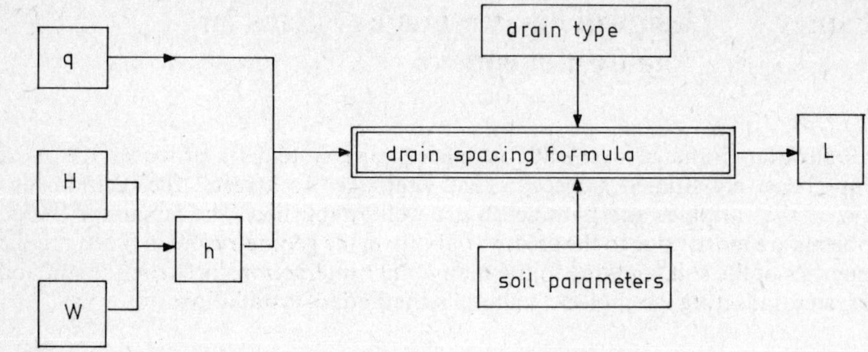

FIGURE 7.2 *Procedure for the determination of the drain spacing for parallel pipe drainage systems (for steady state conditions)*

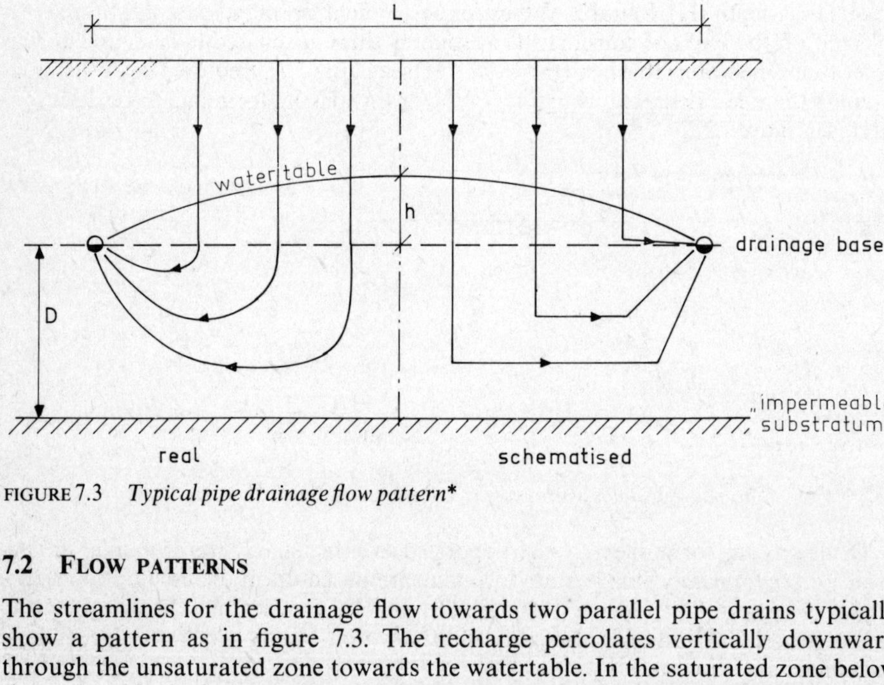

FIGURE 7.3 *Typical pipe drainage flow pattern**

7.2 FLOW PATTERNS

The streamlines for the drainage flow towards two parallel pipe drains typically show a pattern as in figure 7.3. The recharge percolates vertically downward through the unsaturated zone towards the watertable. In the saturated zone below the watertable the flow continues more or less in a *vertical*, downward direction, but soon turns into a lateral, *horizontal* flow towards the drains. Towards the end of its path the flow converges *radially* onto the drain. The extent of the three flow zones differs from case to case, depending especially upon the relative magnitude of h, L and D. When L is large in comparison to both h and D, the flow is predominantly horizontal.* An extensive radial flow sector is to be expected when L and D are of the

* Flow patterns are inevitably presented in a distorted form in diagrams like figure 7.3. In reality L is often 10 to 20 times larger than D and even 50–100 times larger than h. Due to the large difference in horizontal and vertical scale, the horizontal flow sector always becomes grossly under-represented in diagrams.

same order of magnitude. A distinct vertical flow zone only occurs when h is relatively large.

The horizontal flow may extend to depths down to $\frac{1}{4}$L below the drainage base. The radial flow zone is roughly confined to a circle with radius $\frac{1}{2}D\sqrt{2} \simeq 0{\cdot}7\,D$ around the drain.

The total headloss (h) may be visualised as being made up of the headloss due to vertical flow (h_v), horizontal flow (h_h), radial flow (h_r) and entry flow (h_e), or:

$$h = h_v + h_h + h_r + h_e \qquad \text{......(Ernst, 1962)}$$

The different component headlosses are indicated in figure 7.4, although in this case h_e is assumed to be zero ('ideal drain' case).

Vertical flow

The headloss due to a vertical flow of q (m.day^{-1}) through a soil layer of thickness D_v (m) and a hydraulic conductivity of K (m.day^{-1}) is according to Darcy's Law:

$$h_v = \frac{qD_v}{K} \qquad \text{......(eq. 7.1)}$$

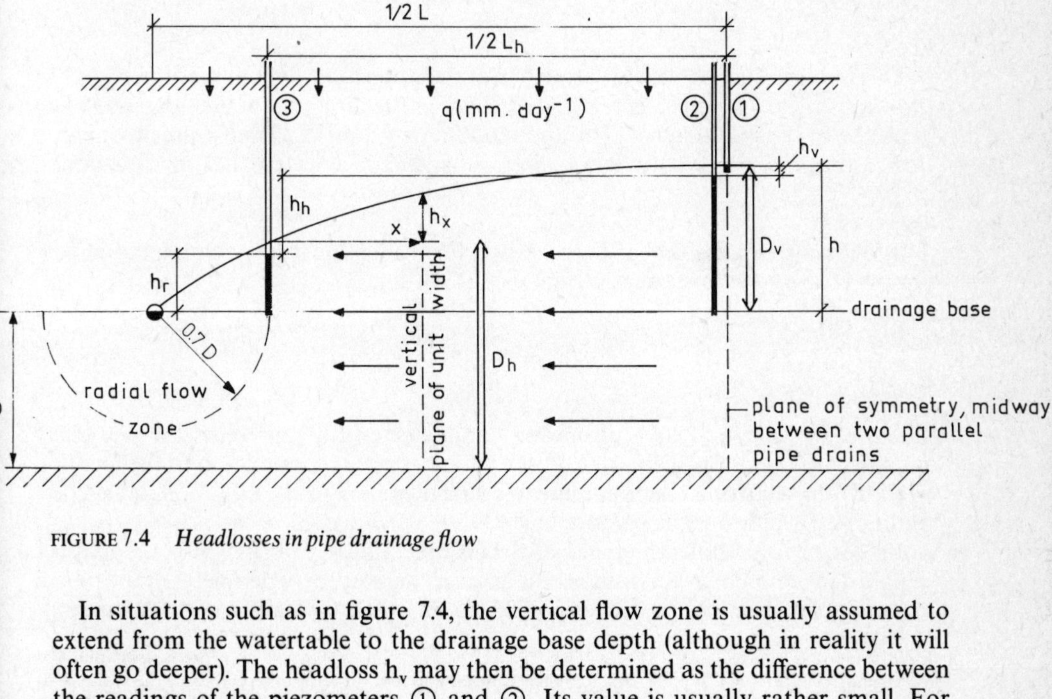

FIGURE 7.4 *Headlosses in pipe drainage flow*

In situations such as in figure 7.4, the vertical flow zone is usually assumed to extend from the watertable to the drainage base depth (although in reality it will often go deeper). The headloss h_v may then be determined as the difference between the readings of the piezometers ① and ②. Its value is usually rather small. For example for $q = 10$ mm.day^{-1}, $K = 0{\cdot}5$ m.day^{-1} and $D_v = 1{\cdot}0$ m, h_v is only 2 cm. The headloss due to vertical flow becomes significant only when it passes through very poorly permeable soil (heavy clay, impeding layer, etc).

Horizontal flow

The headloss over the horizontal flow zone equals the difference between the readings of the piezometers ② and ③ in figure 7.4. An analytical expression for this headloss may be derived by considering the (horizontal) flow Q_x through a vertical plane of unit width:

$$\text{(a)}\ Q_x = q(L_h/2 - x) \qquad \text{and} \qquad \text{(b)}\ Q_x = KD_h\frac{dh_x}{dx}^{*}$$

where D_h is the conceptual average thickness of the horizontal flow zone ($D_h \simeq D + \frac{1}{2}h$).

Equating these two expressions and solving the resulting differential equation for $x = 0; h_x = 0$ and $x = L_h/2; h_x = h_h$:

$$q\int_0^{L_h/2}(L_h/2 - x)dx = KD_h\int_0^{h_h} dh_x$$

$$h_h = \frac{qL_h^{\ 2}}{8KD_h} \qquad \text{......(eq. 7.2)**}$$

This implies a parabolic curvature *** of the watertable over the distance L^h. The gradient dh_x/dx becomes increasingly steeper in the direction of the flow as is to be expected in view of the fact that an increasing amount of water passes through a decreasing cross-sectional area (Q_x increases and h_x decreases in the flow direction).

Radial flow

The following expression has been derived for the headloss through the radial flow sector (EARNST 1962, see also WESSELING 1974):

$$h_r = q\frac{L}{\pi K}\ln\frac{aD_r}{u} \qquad \text{......(eq. 7.3)}$$

where aD_r is an indicative geometric parameter and u represents the wet entry perimeter of the drain. The value of aD_r varies, especially in relation to the location of the drain relative to the impermeable substratum. For the situation depicted in figure 7.4, its value may be taken equal to D.

Values of u for different types of drains are indicated in figure 7.5. For pipes installed in a trench, the u-value is best taken as the wet entry perimeter of the trench rather than of the pipe. For trenches of 20–25 cm width, typically u = 0·3–0·4 m.

* Darcy's Law applied using the so-called *Dupuit-Forchheimer assumptions* ie it is assumed that the streamlines in the vertical plane considered, are horizontal and that the flow velocity in the plane at all depths is proportional to the slope of the watertable (tangent).

**This formula may also be derived by noting that Q_x increases over the distance $L_h/2$ from $Q_x = 0$ for $x = L_h/2$ to $Q_x = \frac{1}{2}qL_h$ for $x = 0$, giving an average flow of $\frac{1}{4}qL_h$ through an average flow cross-section D_h. Applying Darcy's Law $Q = KiA$ gives $\frac{1}{4}qL_h = K\frac{h_h}{\frac{1}{2}L_h}D_h$ or $h_h = \frac{qL^2h}{8KD_h}$.

***Applies only for D_h = constant as is assumed in the integration underlying eg. 7.2. Integrating for $D_h = D + h_x$ yields the more precise equation of an ellipse for the curvature of the watertable.

The headloss h_r may be measured as the difference between the reading in piezometer ③ and the waterlevel in the drain (see figure 7.4).

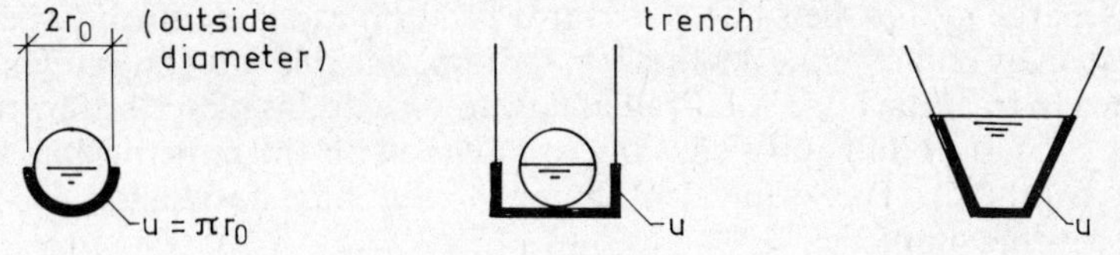

FIGURE 7.5 *The wet entry perimeter for different types of drain**

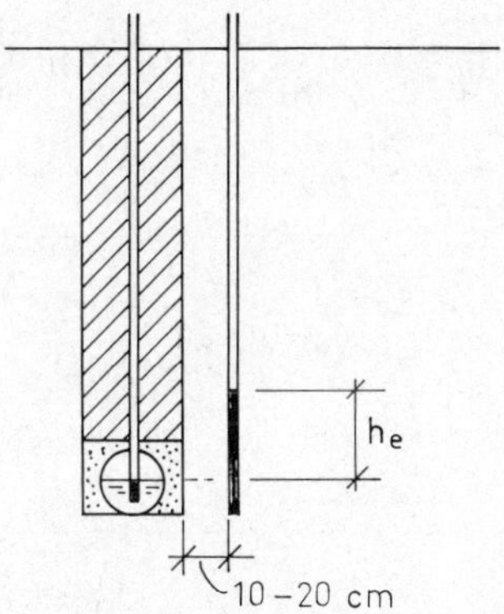

FIGURE 7.6 *Measurement of the headloss at entry*

Entry flow

The headloss h_e incurred in the flow through the pipe surround (envelope and/or backfill) to the openings in the pipe and then through these openings into the pipe, has been discussed earlier in section 3.1.3. Its value may be determined as the difference in readings between a piezometer placed by the side of the trench-wall (technically difficult and therefore usually placed in the undisturbed soil, just outside the trench) and a piezometer placed in the pipe itself (see figure 7.6).

In general the aim is to minimise the value of h_e (*ideal drain* case with $h_e \simeq 0$) so that the total head is available for flow through the soil. The open entry areas of drain pipes in common use, are sufficiently large to allow water to enter with negligible headloss during normal design discharges (this even applies to the drains which have the largest resistance factor α). The flow through the surround to the entry points may, however, constitute a bottleneck. To be an ideal drain, the hydraulic conductivity of the surround should be at least 10 times higher than that of the undisturbed soil outside the trench.

7.3 HOOGHOUDT FORMULA (steady state)

A steady state drain spacing formula for pipe drainage was developed in 1940 by the Netherlands drainage researcher HOOGHOUDT. In this formula only the headlosses

* For drain spacing calculations it is generally assumed that the pipes run half full although the actual water depths in the pipe may vary from full to nearly empty.

due to horizontal and radial flow to the pipe are considered (losses due to vertical flow usually being insignificant). Hooghoudt conceived that a parallel open ditch system with the ditches reaching to the impermeable substratum, could generate the same discharge (q) for the same watertable head (h) as an identically spaced pipe drain system by reducing the depth (D) to the impermeable substratum. This led him to the idea to treat the horizontal/radial flow to pipe drains (described by the rather complex eq. 7.4) as an equivalent flow to ditches with the impermeable base at a reduced depth (d). This equivalent flow is essentially horizontal and may be described by the simpler eq. 7.5.

– *real flow* (horizontal + radial)

$$h = h_h + h_r = \frac{qL_h{}^2}{8KD_h} + q\frac{L}{\pi K}\ln\frac{aD_r}{u} \qquad \text{......(eq. 7.4)}$$

– *equivalent flow* (horizontal)

$$h = h_{h*}(\text{equivalent}) = \frac{qL^2}{8KD_{h*}} \qquad \text{......(eq. 7.5)}$$

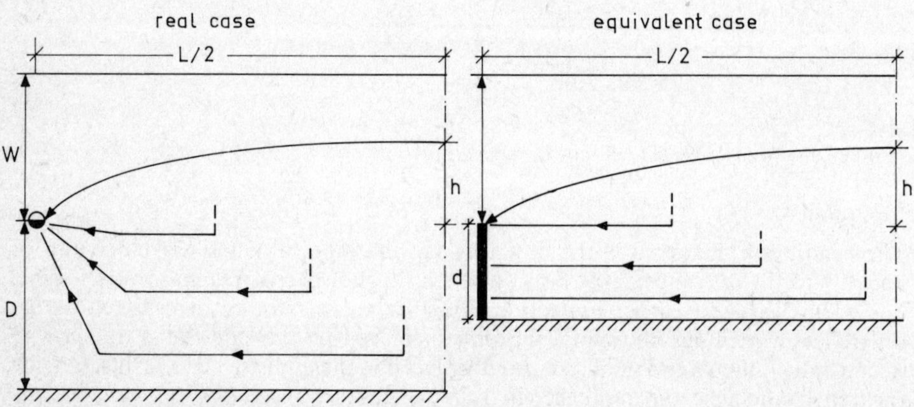

FIGURE 7.7 *Transformation underlying the Hooghoudt formula*

Since $d < D$, less cross-sectional flow area is available and consequently more head is lost in the horizontal flow in the equivalent case than in the real case, the difference just equalling the headloss over the radial flow zone in the real case.

The average thickness of the equivalent horizontal flow zone may be approximated as $D_{h*} = d + h/2$, which inserted in eq. 7.5 gives:

$$h = \frac{qL^2}{8K(d + h/2)} \quad \text{or} \quad q = \frac{8Kh(d + h/2)}{L^2} \qquad \text{......(eq. 7.6-a)}$$

$$q = \frac{8Kdh}{L^2} + \frac{4Kh^2}{L^2} \qquad \text{......(eq. 7.6-b)}$$

The equivalent horizontal flow takes place partly below the drainage base (average thickness of this flow zone being d and partly above the drainage base (thickness of this flow zone being h/2). These two flow components are respectively represented by the first and second terms in eq. 7.6-b. When the soil above drainage base has a different hydraulic conductivity (K_1) than below (K_2), this may be taken into account:

$$q = \frac{8K_2dh}{L^2} + \frac{4K_1h^2}{L^2} \qquad \text{......(eq. 7.7)}$$

This formula is commonly known as the *Hooghoudt spacing formula* for pipe drains (although also the formulae eq. 7.6-a and 7.6-b are often referred to as such). Hooghoudt also found that d = f(D,L,u) and prepared tables defining this functional relationship numerically for common sized pipe drains (table 7.1). Later an analytical expression was derived for this function, covering a wider range of drain types:

$$d = \frac{D}{\frac{8D}{\pi L}\ln\frac{D}{u} + 1} \;\; (\text{for } D > \tfrac{1}{4}L) \quad ; \quad d = \frac{\pi L}{8\ln\frac{L}{u}} \;\; (\text{for } D < \tfrac{1}{4}L) \qquad \text{......(eq. 7.8)}$$

In situations where there is no distinct impermeable substratum the depth D may be equal to the depth at which the K-value has decreased to 1/10 of the (average) K-value of the layer(s) above, provided no highly permeable layer occurs within 1–2 m below this depth.

7.3.1 Use of the Hooghoudt formula

The normal procedure for the determination of the drain spacing with the Hooghoudt formula involves the following steps (see also figure 7.2):

(a) formulation of the basic design criteria (q and H)
(b) establishment of the field drainage base W; h = W – H
(c) establishment of the soil parameters K (or K_1 and K_2) and D
(d) selection of drain type (pipe or ditch) and determination of u
(e) determination of the drain spacing (L) by solving the Hooghoudt formula (eq. 7.6-a/b or eq. 7.7).

The last step, solution of the Hooghoudt formula, may be done by trial and error or by graphical means.

Trial and error solution: since L depends on d and d depends on L, the Hooghoudt formula is not explicit in L, which can only be found by trial and error:

- assume a value for L and determine d from table 7.1 or eq. 7.8;
- solve the Hooghoudt formula for L and compare this value with the assumed value;
- modify the value of L and repeat until the calculated and assumed values are equal.

TABLE 7.1 *Equivalent depth 'd' for commonly used pipe drains* (r_o = 4–10 cm; $u \simeq 0{\cdot}3$ m)

D(m)	L (m)→ 5	7·5	10	15	20	25	30	35	40	45	50	75	80	85	90	100	150	200	250
0·50	0·45	0·50	0·50	0·50	0·50	0·50	0·50												
0·75	0·60	0·65	0·70	0·70	0·75	0·75	0·75	0·75	0·75	0·75	0·75								
1·00	0·65	0·75	0·80	0·85	0·90	0·90	0·95	0·95	0·95	0·95	0·95	0·95	0·95	0·95	1·00	1·00	1·00	1·00	1·00
1·25	0·70	0·80	0·90	1·00	1·05	1·10	1·10	1·15	1·15	1·15	1·15								
1·50		0·90	0·95	1·10	1.20	1·25	1·30	1·30	1·35	1·35	1·35								
1·75		0·90	1·00	1·20	1·30	1·40	1·45	1·50	1·50	1·55	1·55								
2·00			1·10	1·30	1·40	1·50	1·55	1·60	1·65	1·70	1·70	1·80	1·80	1·80	1·85	1·85	1·90	1·90	1·95
2·25			1·15	1·35	1·50	1·70	1·70	1·75	1·80	1·85	1·85								
2·50				1·40	1·55	1·70	1·80	1·85	1·95	2·00	2·00								
2·75				1·40	1·65	1·75	1·90	2·00	2·05	2·10	2·20								
3·00				1·45	1·65	1·85	1·95	2·10	2·15	2·25	2·30	2·50	2·50	2·55	2·55	2·60	2·70	2·80	2·85
3·25				1·50	1·70	1·90	2·05	2·15	2·25	2·35	2·40								
3·50				1·50	1·75	1·95	2·10	2·25	2·35	2·45	2·55								
3·75				1·50	1·80	1·95	2·15	2·30	2·45	2·55	2·65								
4·00					1·80	2·00	2·20	2·35	2·50	2·60	2·70	3·05	3·10	3·10	3·15	3·25	3·45	3·60	3·65
4·50					1·85	2·10	2·30	2·50	2·65	2·75	2·85								
5·00					1·90	2·15	2·40	2·60	2·75	2·90	3·00	3·50	3·55	3·60	3·65	3·80	4·10	4·30	4·45
5·50						2·20	2·45	2·65	2·85	3·00	3·15								
6·00							2·50	2·70	2·90	3·10	3·25	3·85	3·95	4·00	4·10	4·25	4·70	4·95	5·15
7·00							2·55	2·80	3·05	3·25	3·45	4·15	4·25	4·35	4·40	4·60	5·20	5·55	5·80
8·00							2·55	2·85	3·15	3·35	3·55	4·40	4·50	4·60	4·70	4·95	5·70	6·15	6·45
9·00								2·90	3·20	3·45	3·65	4·55	4·70	4·80	4·95	5·25	6·10	6·65	7·00
10·00									3·25	3·50	3·75	4·75	4·90	5·05	5·20	5·45	6·45	7·10	7·55
12·50												5·00	5·20	5·40	5·55	5·90	7·20	8·05	8·70
15·00												5·20	5·40	5·60	5·80	6·25	7·75	8·85	9·65
17·50												5·30	5·55	5·75	6·00	6·45	8·20	9·45	10·40
20·00													5·60	5·85	6·10	6·60	8·55	9·95	11·10
25·00													5·75	5·95	6·20	6·80	9·00	10·70	12·10
30·00																	9·25	11·30	12·90
35·00																	9·45	11·60	13·40
40·00																		11·80	13·80
45·00																		12·00	13·80
50·00																		12·10	14·30
60·00																			14·60
~	0·70	0·95	1·15	1·55	1·90	2·25	2·60	2·90	3·25	3·55	3·90	5·40	5·75	6·00	6·25	6·80	9·55	12·20	14·70

Roughly

d = constant for $D > \frac{1}{4}L$

Graphically: the nomograph in figure 7.8 covers most conditions normally encountered in pipe drainage design (h = 0·1–1·2 m, q = 1–10 mm.day^{-1}).

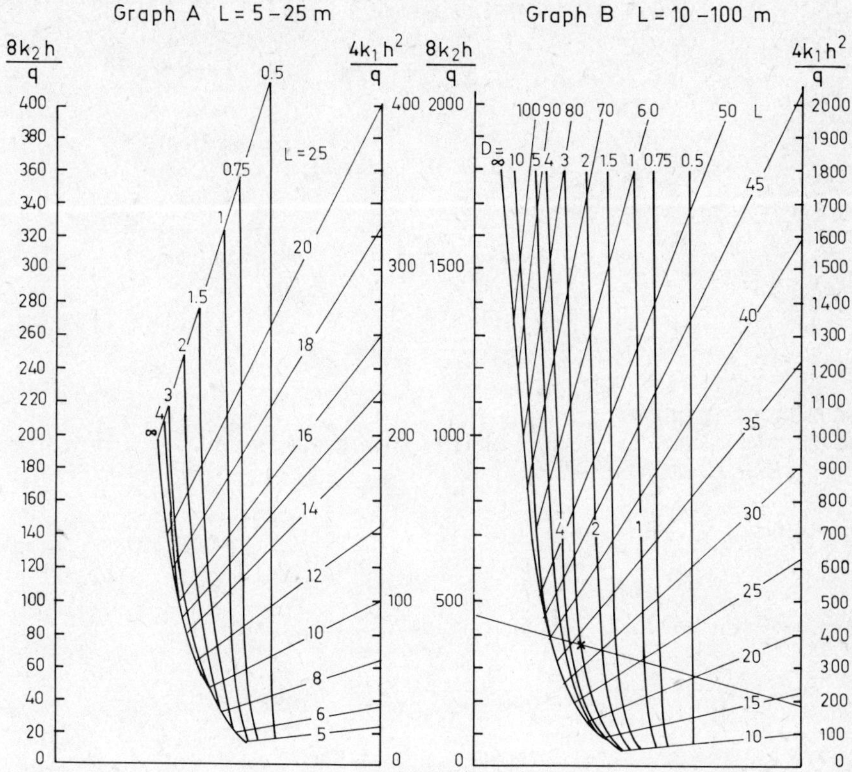

For pipe drains with r_o = 0·04–0·10 m, u = 0·30 m*

FIGURE 7.8 *Nomograph for the solution of the Hooghoudt drain spacing formula* (VAN BEERS 1965)

* For all practical purposes, the graph is valid for all pipes installed in 20–25 cm wide trenches (u ≃ 0·30 m).

Example: determine the required drain spacing for the basic design criteria q = 7 mm.day^{-1}, H = 0·6 m.

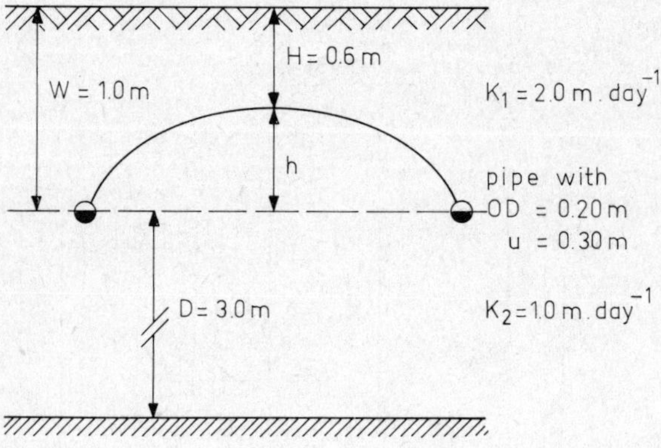

Trial and error solution:
For first trial assume L = 40 m. With D = 3·0 m, u = 0·3 m and L = 40 m, read from table 7.1 that d = 2·15 m

$$L^2 = \frac{4K_1h^2}{q} + \frac{8K_2dh}{q} = \frac{4 \times 2 \times 0{\cdot}4^2}{0{\cdot}007} + \frac{8 \times 1 \times 2{\cdot}15 \times 0{\cdot}4}{0{\cdot}007}; \quad L^2 = 1165; \quad L = 34\,\text{m}.$$

For second trial assume L = 32 m; from table 7.1 → d = 2·0 m.

$$L^2 = \frac{4 \times 2 \times 0{\cdot}4^2}{0{\cdot}007} + \frac{8 \times 1 \times 2{\cdot}0 \times 0{\cdot}4}{0{\cdot}007} = 1097; \quad L = 33\,\text{m}$$

Final solution: L = 33 m.

Graphical solution:

Calculation: $\frac{4h^2}{q} = 91{\cdot}5; \quad K_1 = 2{\cdot}0; \quad \frac{4K_1h^2}{q} = 183$

$\frac{8h}{q} = 457; \quad K_2 = 1{\cdot}0; \quad \frac{8K_2h}{q} = 457$

Figure 7.8: Connect 183 (on right hand scale) with 457 (on left hand scale); where the connecting line intersects with D = 3·0 m read L = 33 m (graph B).

7.3.2 Notes on the Hooghoudt formula

(1) The Hooghoudt formula shows that (all other variables remaining constant), the spacing L increases when:

- K increases (especially when K_2 increases; the value of K_1 has much less effect on L)

- q decreases ($L \sim q^{-\frac{1}{2}}$)
- D increases (has less influence when L is small than when L is large)
- h increases (implies increase of W or decrease of H).

(2) When the drainage flow above the drainage base may be neglected, the Hooghoudt formula reduces to:

$$L^2 = \frac{8Kdh}{q} \quad \textit{(simple Hooghoudt formula)} \qquad \ldots\ldots \text{(eq. 7.9)}$$

This formula may for example be used in situations where the watertable head (h) is small or where flow conditions below the drainage base are much more favourable than above it.

(3) A change in hydraulic conductivity at about drainage base depth is quite common in non-stratified soils. This depth normally occurs at 1·0–1·2 m corresponding to the depth of the most active part of the soil profile in terms of physical and chemical soil formation processes, soil biological life, soil moisture regime, root development, etc, all of which affect the pore geometry and distribution and thus the hydraulic conductivity of the soil. In stratified soils, however, the drainage base depth will rarely coincide with a boundary between two soil layers. Slight deviations, especially when the hydraulic conductivities of the two layers are similar, are of little concern. Otherwise composed K values based on horizontal flow should be used (see section 16.4). This is only acceptable when the K-values of the layers involved are similar in magnitude (differences of not more than a factor of 2 between the values).

Where the main flow zone includes contrastingly different layers, flow patterns may be quite different from those underlying Hooghoudt's formula, rendering the latter invalid. Drain spacings in these situations cannot be determined by standard drain spacing formulae but require special analysis. The (conceived) flow pattern may for example be schematized into vertical, horizontal and radial components, enabling the equations 7.1, 7.2 and 7.3 to be used to calculate the corresponding headlosses over the various flow zones. In this way an estimate of the required drain spacing may be established (see ILRI 1973).

(4) Where a significant vertical flow is to be expected and the relevant flow zone has a very low hydraulic conductivity, $(h - h_v)$, instead of h should be used in the Hooghoudt formula, with h_v determined according to eq. 7.1.

(5) The second part of the Hooghoudt formula ($q = \frac{4Kh^2}{L^2}$, see eq. 7.6-a), applying to the flow to the drains from above the drainage base should not be applied separately. When, as in figure 4.13, the drainage base coincides with the impermeable base, so that all flow occurs above the drainage base, the Fukuda formula is applicable (see ref. in chapter 4, van Hoorn 1974).

7.4 Non-steady state drainage formulae

Crop growth and soil workability are of course related to the actual watertable depth which fluctuates in response to the variable (rainfall) recharge. Quantitatively, little is known about the effect of fluctuating watertables on soils and crops and few soundly based non-steady state design criteria have as yet been established. However, soundly based and tested criteria do exist for steady state design procedures (partly due to the fact that these procedures were developed earlier in time and the underlying models and tests are so much simpler than in the non-steady case). Consequently, although it is recognised that the drainage process is in fact non-steady, design is normally based on steady state formulae and criteria. The latter criteria reflect (but are not necessarily equal to) average recharge (q) and average watertable depth (H) during the critical period(s). Systems designed according to these criteria control the high watertable occurrences to within acceptable limits as regards frequency and magnitude, indirectly taking into account the non-steady nature of the drainage process.

Steady state drainage criteria are not available or are not satisfactory in all cases. The first limitation applies to the many countries in which groundwater drainage has only recently been introduced or where there is no history of drainage research. The second limitation for example applies to climates in which the rainfall is generally distinctly non-steady in nature (coming in discrete storms rather than as prolonged periods with rather uniform, medium intensity rainfall). Steady state criteria are also unsatisfactory in situations where the soil hydrological conditions are such that watertables and drain-outflow respond directly and quickly to (rainfall) recharge, rather than the response being moderated. In all these cases the non-steady drainage formulae as discussed below, are of value.

7.4.1 **Falling watertable** (Glover–Dumm formula)

When $h < d \ll L$, the flow to the drains is essentially horizontal and it follows that (see figure 7.9):

(a) $$Q_x = KD_h \frac{\delta h_x}{\delta x}$$

(Darcy's Law applied under the Dupuit-Forchheimer conditions; compare with the derivation of eq. 7.2 in section 7.2.)

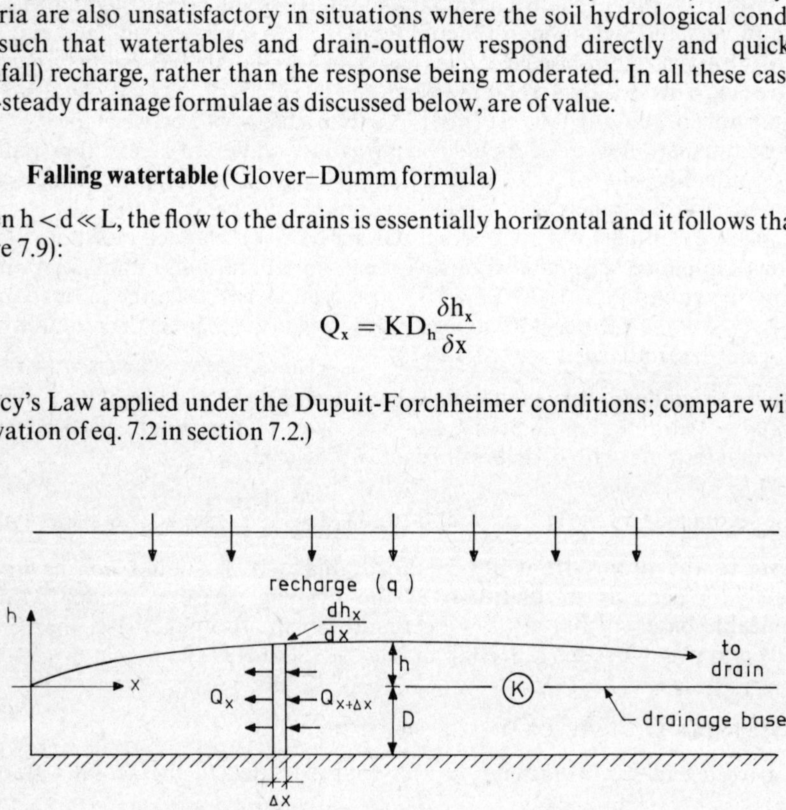

FIGURE 7.9 *Horizontal drainage flow to ditches*

(b) $$\mu\frac{\delta h_x}{\delta t}\Delta x = Q_{x+\Delta x} + q\,.\,\Delta x - Q_x$$

(waterbalance for element Δx below the watertable; μ = drainable pore space)

$$\frac{\delta Q_x}{\delta x} + q = \mu\frac{\delta h_x}{\delta t} \quad \text{(limit case with } \Delta x \rightarrow 0)$$

Combining these two equations:

$$\frac{\delta\left(KD_h\frac{\delta h_x}{\delta x}\right)}{\delta x} + q = \mu\frac{\delta h_x}{\delta t}$$

$$\frac{\delta h_x}{\delta t} = \frac{KD_h}{\mu}\cdot\frac{\delta^2 h_x}{\delta x^2} + \frac{q}{\mu} \qquad \text{......(eq. 7.10)}$$

Eq. 7.10 is a form of the so-called *Boussinesq equation* which describes the position of the watertable under non-steady recharge. Integration of eq. 7.10 for q = 0 (no recharge) and boundary conditions:

$$t = 0; \quad h_x = h_o \quad \text{for all x values}$$

(implies an intially horizontal watertable at a height h_o above the drainage base)

$$t > 0; \quad h = 0 \quad \text{at } x = 0$$
$$x = L$$

(implies an instantaneous lowering of the potential to zero in the drains)

yields (after some elaboration) the so-called *Glover–Dumm* drainage formula which describes the rate of fall of the watertable after the latter has risen instantaneously (at t = 0) to a height h_o above the drainage base (typical of the behaviour of shallow watertables in irrigated land which often rise sharply during a water application and then recede more slowly).

According to Glover–Dumm the mid-spacing watertable head h_t at t = t relates to the head h_o (at t = 0) as:

$$\frac{h_t}{h_o} = 1{\cdot}16\,e^{-\alpha t} \qquad \text{......(eq. 7.11)}$$

where $$\alpha = \frac{10\,Kd}{\mu L^{2}*} \qquad \text{......(eq. 7.12)}$$

where: t = time (days)
h_o = initial watertable head, at t = t_o (m)
h_t = watertable head at t = t (m)
α = reaction factor (days^{-1})
μ = drainable pore space ($m^3.m^{-3}$)

* The factor 10 is actually π^2 (9·87)

L = drain spacing (m)
d = equivalent depth to the impermeable substratum (m)
K = hydraulic conductivity (m.day^{-1})

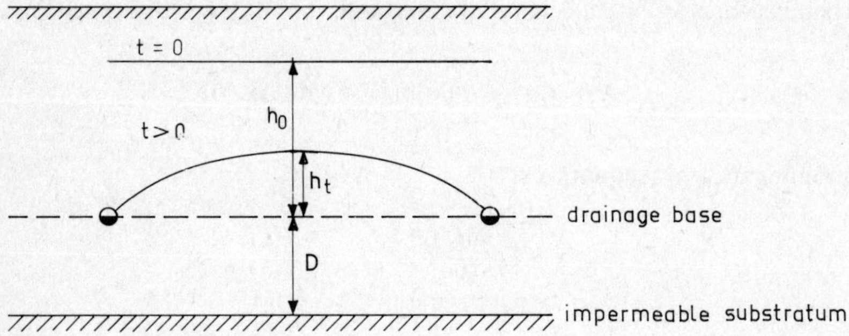

FIGURE 7.10 *Notations and boundary conditions used with Glover–Dumm non-steady state formula*

Combining eqs. 7.11 and 7.12 gives:

$$L^2 = \frac{10Kdt}{\mu}\left[\ln 1{\cdot}16\frac{h_o}{h_t}\right]^{-1} \qquad \text{(eq. 7.13)}$$

In these formulae it is assumed that the shape of the falling watertable conforms to a fourth degree parabola (explains the factor 1·16 in eq. 7.11). Radial resistances are taken into account by replacing the depth D to the impermeable substratum by Hooghoudt's equivalent depth d, making the Glover–Dumm formula applicable to pipe drainage.

A rather similar equation may be derived by following Bouwer and Schilfgaarde (1963) in assuming that in periods during which there is no recharge, the discharge (q) is related to the rate of fall of watertable $\left(\frac{dh}{dt}\right)$ as:

$$q = -C\mu\frac{dh}{dt} \qquad \text{......(eq. 7.14)}$$

The coefficient of proportionality is made up of μ (drainable pore space) and C (a correction factor accounting for the fact that during recession the watertable will not remain horizontal but falls more rapidly nearer to the drains; the value of C is of the order of 0·7–0·9 for h at mid-spacing). Combining eq. 7.14 with the simple Hooghoudt formula (eq. 7.9) gives:

$$\left.\begin{aligned} q &= -C\mu\frac{dh}{dt} \\ q &= \frac{8Kdh}{L^2} \end{aligned}\right\} \quad dt = \frac{-C\mu L^2}{8Kdh}dh$$

Integration for $t = 0$; $h = h_o$ and $t = t$; $h = h_t$ yields (after some re-arranging of terms and assuming $C = 0{\cdot}8$):

$$L^2 = \frac{10Kdt}{\mu}\left[\ln\frac{h_o}{h_t}\right]^{-1} \quad \text{(compare to eq. 7.13)}$$

Combining eq. 7.14 with the full Hooghoudt formula (eq. 7.6-a) yields similarly:

$$L^2 = \frac{10Kdt}{\mu}\left[\ln\frac{h_o(2d+h_t)}{h_t(2d+h_o)}\right]^{-1} \quad \text{(takes flow above the drainage base into account)}$$

The Glover–Dumm (and similar type) formulae may be used in design as illustrated by the following example.

Example

Given: situation as in the figure and the (non steady) basic design criteria:
$t = 0$; $H_o = 0\,\text{m}$ (watertable at soil surface)
$t = 4$; $H_4 = 0{\cdot}8\,\text{m}$

The basic design criteria thus state that the system should be able to lower the watertable from the soil surface to a depth of 0·8 m below the soil surface within a period of 4 days.

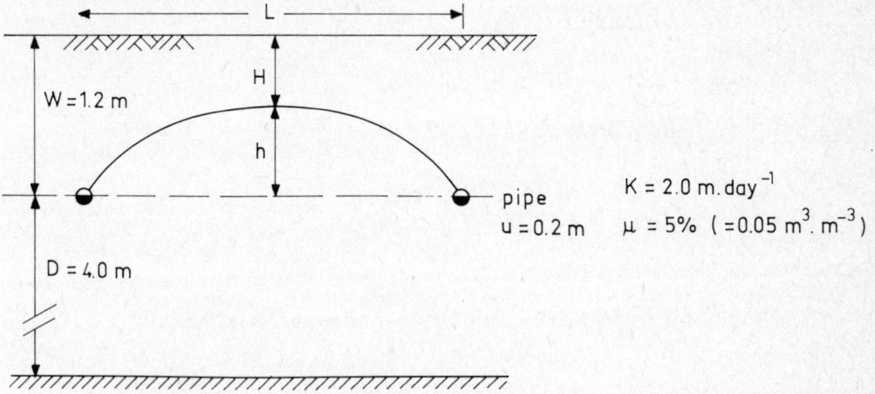

(a) *Direct solution*

$$\left.\begin{aligned} h_o &= W - H_o = 1{\cdot}2\,\text{m} \\ h_4 &= W - H_4 = 0{\cdot}4\,\text{m} \end{aligned}\right\} \frac{h_4}{h_o} = \frac{0{\cdot}4}{1{\cdot}2} = 0{\cdot}33$$

From eq. 7.11: $$\frac{h_t}{h_o} = \frac{h_4}{h_o} = 0{\cdot}33 \rightarrow \alpha t = -\ln\frac{0{\cdot}33}{1{\cdot}16} = 1{\cdot}24$$

and with $t = 4 \ldots . \alpha = \frac{1 \cdot 24}{4} = 0 \cdot 31$

In eq. 7.12: $\alpha = \frac{10Kd}{\mu L^2} = 0 \cdot 31 \quad \text{or} \quad L^2 = \frac{10Kd}{0 \cdot 31\,\mu}$

For $\mu = 0 \cdot 05$ and $K = 2 \cdot 0 \rightarrow L^2 = 1290 \times d$

Since L = f(d) and d = f(L), a trial and error procedure is followed to solve this equation for L:

- first trial: $L = 30\,m \ldots\ldots d = 2 \cdot 20\,m$ (from table 7.1)
  $L = \sqrt{1290 \times 2 \cdot 20} = 53 \cdot 3\,m$

- second trial: $L = 60\,m \ldots d = 2 \cdot 84$ (from table 7.1)
  $L = \sqrt{1290 \times 2 \cdot 84} = 60 \cdot 5\,m$

Solution is: $L = 60\,m$

(b) *Indirect solution*

The Glover–Dumm formula and the Hooghoudt formula can be interrelated, enabling the non-steady basic design criteria to be translated into steady-state criteria:

- Hooghoudt (simple, eq. 7.9): $L^2 = \frac{8Kdh}{q}$
- Glover–Dumm (from eq. 7.12): $L^2 = \frac{10Kd}{\alpha\mu}$

$$\frac{h}{q} = \frac{10}{8\alpha\mu}$$

For $\alpha = 0 \cdot 31$ and $\mu = 0 \cdot 05 \rightarrow \frac{h}{q} = 80 \cdot 65$

In Hooghoudt: $L^2 = 8Kd\frac{h}{q}$ with $K = 2$, gives $L^2 = 1290 \times d$

By trial and error, as above, it is found that $L = 60\,m$.

7.4.2 **Fluctuating watertable** (de Zeeuw and Hellinga formula)

Hooghoudt's simple formula (eq. 7.9) may be developed to show the non-steady response to periodic rainfall or irrigation. In this equation the drain discharge (q) is linearly related to the mid-spacing watertable head (h):

$$q = \frac{8Kd}{L^2} . h \qquad \ldots\ldots\text{(eq. 7.9)}$$

The variation of the drain discharge with time is thus also linearly related to the variation in time of the watertable head:

$$\frac{dq}{dt} = \frac{8Kd}{L^2} \cdot \frac{dh}{dt} \qquad \text{......(eq. 7.15)}$$

If the groundwater body is recharged by rainfall or by another source (R) and is depleted by drain discharge (q), it follows that the watertable will rise when $(R - q) > 0$ and fall when $(R - q) < 0$. By analogy with eq. 7.14, the watertable fluctuations may be described by:

$$\frac{dh}{dt} = \frac{(R-q)}{C\mu} \qquad \text{......(eq. 7.16)}$$

Combining eq. 7.15 and eq. 7.16 and taking $C = 0{\cdot}8$, gives:

$$\frac{dq}{dt} = \frac{10Kd}{\mu L^2}(R-q) = \alpha(R-q) \qquad \text{......(eq. 7.17)}$$

So the change in drain discharge $\left(\frac{dq}{dt}\right)$ is proportional to the excess recharge $(R - q)$, the constant of proportionality being α (*reaction factor*, see eq. 7.12).

Integration of eq. 7.17 between the limits $t = t$; $q = q_t$ and $t = t-1$; $q = q_{t-1}$ gives:

$$\int_{q_{t-1}}^{q} \frac{dq}{(R-q)} = \int_{t-1}^{t} \alpha dt$$

$$\frac{(R-q)_t}{(R-q)_{t-1}} = e^{-\alpha\Delta t}$$

which may be elaborated into:

$$q_t = q_{t-1}e^{-\alpha\Delta t} + R_{\Delta t}.(1-e^{-\alpha\Delta t}) \qquad \text{......(eq. 7.18)}$$

in which $R_{\Delta t}$, the mean value of R during the time interval $t-1$ to t, is assumed constant. Using the linear relationship between q and h of eq. 7.9 ($q = \frac{8Kd}{L^2}h = 0{\cdot}8\,\alpha\mu h$) it also follows that:

$$h_t = h_{t-1}e^{-\alpha\Delta t} + \frac{R_{\Delta t}}{0{,}8\mu\alpha}(1-e^{-\alpha\Delta t}) \qquad \text{......(eq. 7.19)}$$

Equations 7.18 and 7.19, first derived by DE ZEEUW and HELLINGA (1958), may be used to simulate drain discharge and watertable depth fluctuations on the basis of (historical) weather records, given the value of the reaction factor α. The non-steady processes are approximated by a series of intervals during which the recharge $R_{\Delta t}$ is assumed to be constant. The length of the time interval should be such that changes occurring during the interval may be described sufficiently accurately by the existing steady state formulae, using as input average rates during the interval. This requirement can generally be met by adopting intervals of length $\Delta t = 1$ day (expressing t in days makes the coefficient $\alpha\Delta t$ dimensionless since α has the dimension day^{-1}).

The first rains after a dry period are normally mostly retained within the soil. In general, the watertable will not rise above the drainage base until the overlying soil is at or above field capacity. Consequently, the best simulations are obtained when these conditions exist for example during the latter part of a rainy period, lasting several days, or in moderate climates during the winter when evaporation is low and watertables are continuously high. The low evaporation during such periods may be taken into account by a reduction in the rainfall. Soil moisture balance calculations as discussed in chapter 15 can help to establish when field capacity is reached and the watertable response to rainfall commences. It should also be noted that R refers to the infiltrated rainfall which can usually be taken to be equal to the full rainfall on the flat permeable soils in which groundwater drainage is typically used. An example of the use of equations 7.18 and 7.19 is detailed in table 7.2.

Reaction factor (α)

The reaction factor $\alpha = \dfrac{10Kd}{\mu L^2}$ is a direct index of the intensity with which the drain discharge rate responds to changes in the recharge. Values generally vary from $\alpha = 0{\cdot}1$–$0{\cdot}3$ for land with a slow response (low KD value, wide drain spacing, high drainable pore space) to $\alpha = 2{\cdot}0$–$5{\cdot}0$ for rapidly responsive land (high KD value, narrow drain spacing, low drainable pore space). It may be calculated according to eq. 7.12 (from known KD, L and μ values) but since the μ-value is especially difficult to determine, best estimates of α are obtained by observing the actual response of the system in the field. From equations 7.18 and 7.19, it follows that in periods during which there is no recharge ($R = 0$):

$$\alpha = \frac{\ln q_{t-1} - \ln q_t}{\Delta t} = 2{\cdot}30\,\frac{\log q_{t-1} - \log q_t}{\Delta t} \qquad \text{......(eq. 7.20a)}$$

$$\alpha = \frac{\ln h_{t-1} - \ln h_t}{\Delta t} = 2{\cdot}30\,\frac{\log h_{t-1} - \log h_t}{\Delta t} \qquad \text{......(eq. 7.20b)}$$

Observed q_t or h_t values may be plotted on log normal paper as in figure 7.11. If the system obeys the assumptions underlying the basic formulae equations 7.9, 7.15 and 7.16, the observed values more or less fit to a straight line with a slope equal to α. Observations are best made during periods of low evaporation shortly after the end of a few good rainy days when the recharge to the groundwater has ceased and the watertable starts receding (the recession sections of the watertable/drainflow vs. time graphs, see figure 16.8; the observations from day 6 to day 8 in table 7.2 may for example be used to estimate α).

TABLE 7.2 *Watertable head and drain outflow calculations on the basis of the de Zeeuw–Hellinga formulae*

Case: parallel pipe drainage system; $W = 1{\cdot}20\,m$
$L = 40\,m$
$Kd = 2{\cdot}5\,m^2.day^{-1}$
$\mu = 0{\cdot}05\,m^3.m^{-3}$

$$\alpha = \frac{10Kd}{\mu L^2} = \frac{10 \times 2{\cdot}5}{0{\cdot}05 \times 1600} = 0{\cdot}31; \quad \Delta t = 1{\cdot}0 \text{ day}$$

$$e^{-\alpha\Delta t} = e^{-0{\cdot}31} = 0{\cdot}73; \quad (1 - e^{-\alpha\Delta t}) = 0{\cdot}27; \quad 0{\cdot}8\,\mu\alpha = 0{\cdot}012$$

At the start of the rain (day 0), the watertable head is 0·10 m, or the mid-spacing watertable depth $H = W - h = 1{\cdot}20 - 0{\cdot}10 = 1{\cdot}10$ m below the soil surface. The corresponding drain outflow $q = (8Kdh)/L^2 = (8 \times 2{\cdot}5 \times 0{\cdot}10)/1600 = 0{\cdot}001\,m.day^{-1}$. The overlying soil profile can store 15 mm, bringing the soil up to field capacity. The evapotranspiration is low (1–2 $mm.day^{-1}$). All rainfall infiltrates and excess water in the profile readily percolates through to the groundwatertable.

Day	*Rainfall* (*P in m*)	*Evapotranspiration* (*E in m*)	*Recharge* (*R in m*)	*Watertable head*[1] (*h in m*)	*Drain outflow*[2] (*q in m.day*$^{-1}$)
0				0·10	0·001
1	0·020	0·001	0·004	0·16	0·002
2	0·020	0·001	0·019	0·54	0·007
3	0·010	0·002	0·008	0·57	0·007
4	0·025	0·001	0·024	0·96	0·012
5	0·005	0·002	0·003	0·77	0·010
6		0·002		0·56	0·007
7		0·002		0·41	0·005
8		0·002		0·30	0·004
9		0·002		0·22	0·003
10		0·002		0·16	0·002[3]

[1] $h_t = h_{t-1}e^{-\alpha\Delta t} + \frac{R_{\Delta t}}{0{\cdot}8\,\mu\alpha}(1 - e^{-\alpha\Delta t})$ ......(eq. 7.19)

for $t = 1$: $h_1 = (0{\cdot}10 \times 0{\cdot}73) + \left(\frac{0{\cdot}004}{0{\cdot}012} \times 0{\cdot}27\right) = 0{\cdot}073 + 0{\cdot}090 = 0{\cdot}16$

for $t = 2$: $h_2 = (0{\cdot}16 \times 0{\cdot}73) + \left(\frac{0{\cdot}019}{0{\cdot}012} \times 0{\cdot}27\right) = 0{\cdot}117 + 0{\cdot}427 = 0{\cdot}54$

for $t = 6$: $h_6 = 0{\cdot}77 \times 0{\cdot}73 = 0{\cdot}56$; for $t = 7$: $h_7 = 0{\cdot}56 \times 0{\cdot}73 = 0{\cdot}41$ etc

[2] $q_t = q_{t-1}e^{-\alpha\Delta t} + R_{\Delta t}(1 - e^{-\alpha\Delta t})$ ......(eq. 7.18)

for $t = 1$: $q_1 = (0{\cdot}001 \times 0{\cdot}73) + (0{\cdot}004 \times 0{\cdot}27) = 0{\cdot}002$
for $t = 2$: $q_2 = (0{\cdot}002 \times 0{\cdot}73) + (0{\cdot}019 \times 0{\cdot}27) = 0{\cdot}007$
for $t = 6$: $q_6 = 0{\cdot}010 \times 0{\cdot}73 = 0{\cdot}007$; for $t = 7$: $q_7 = 0{\cdot}007 \times 0{\cdot}73 = 0{\cdot}005$, etc

[3] evapotranspiration may be expected to start interacting

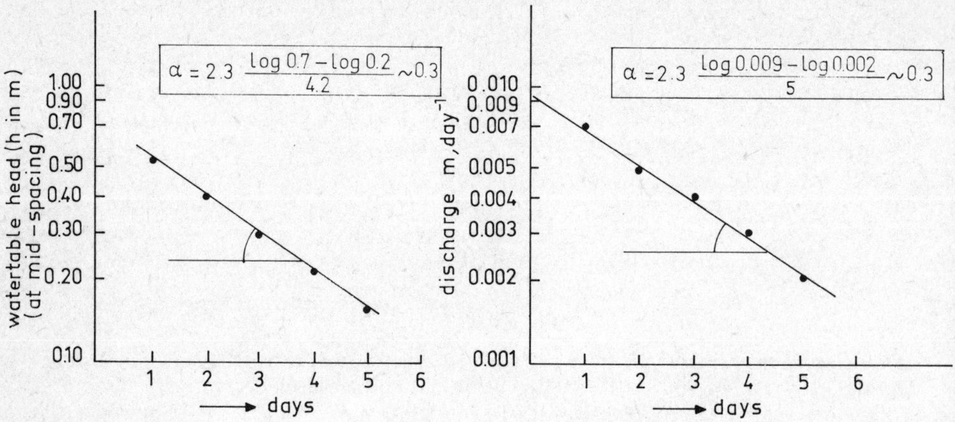

FIGURE 7.11 *Determination of the reaction factor (α) from observed watertable heads and drain discharges during recession*

7.5 Basic design criteria

The criteria to be used depend upon the objective(s) to be served by the drainage system. As stated in section 2.1 the overall objective is the control of the watertable in the soil in order to create favourable soil water conditions for crop growth and farm operations. Criteria for groundwater drainage for soil salinity control are treated separately in section 12.1.2.

Most of the criteria to be discussed are formulated in steady state terms although it will be shown that they can equally well be expressed in the non-steady state form.

A number of specific objectives may be distinguished:

- improvement of the rootzone aeration
- early soil workability after rain
- early warming up of the soil in spring
- prevention of soil structural deterioration
- promotion of useful biological, microbiological and biochemical processes (especially related to nitrogen availability to plants).

Criteria generally express the integrated requirements based on one or more of these objectives although the weight given to individual objectives varies from case to case. This is why different criteria are used for *off season* and *crop season* drainage, reflecting the different objectives of the drainage in these two seasons. The most critical season determines the criteria to be used.

7.5.1 Criteria for off-season drainage

These criteria apply to the majority of the groundwater drainage systems installed in the moderate, humid climates of NW Europe. The rainfall in this area is mostly well distributed throughout the year, totalling some 700–900 mm per annum. During the winter, P (precipitation) ≫ E (evapotranspiration) and watertables tend to be high (see figure 7.12-A). However, this is of little concern since there are very few farm operations which have to be carried out on the land during this period and crop growth is limited anyway by the prevailing low temperatures. Some control, however, is desirable to prevent continuous, severe waterlogging of the main rootzone as this has an adverse effect on the structure of the soil and on the nutrient availability in the following spring and summer, the main farming season (see also section 2.1). Excess water in the soil during the late winter/early spring also delays planting, retards the warming up of the soil and creates poor soil-water-air conditions during the critical germination and seedling growth periods.

During the summer $E > P$ and watertables tend to be low. They may rise occasionally well into the rootzone after heavy rain but even low intensity drainage, assisted by the considerable 'self drainage' provided by evapotranspiration, will generally suffice to lower the watertables in sufficient time to prevent crop damage. In the autumn as P once again overtakes E and watertables start rising, harvesting and other farm operations may be hindered by excess water conditions although these are generally somewhat less critical than in the spring.

To assure the desired excess water control during the critical winter and early spring period, groundwater drainage systems in NW Europe are commonly designed on the basis of criteria such as those detailed in table 7.3. For identical field drainage base depths, the parameter h/q is indicative of the drainage intensity of the system.* A system with a low h/q-value is able to cope with higher recharges and/or maintain lower watertables than a system with a high h/q-value. Under the prevailing climatic conditions in NW Europe, systems with h/q = 55 to 100 generally suffice. Such systems are able to control the watertable mostly below the topsoil and assure early return of the upper rootzone to FC (field capacity) in the spring, an essential condition for effective evaporative drying of the land. In this way drainage contributes to good seedbed preparation, timely planting, good germination and seedling growth.

TABLE 7.3 *Basic design criteria for groundwater drainage commonly employed in NW Europe* (ILRI 1973)

	q (*m.day*$^{-1}$)	*H* (*m*)	*h* = *W* − *H* (*m*)	*h/q* (*days*)	*Watertable rising up into the topsoil*
– tolerant and/or low value crops (applies to most grassland)	0·007	0·30–0·40	0·60–0·70	85–100	once or twice per year
– sensitive and/or high value crops	0·007	0·50–0·60	0·40–0·50	55– 70	once per year
– average conditions	0·007	0·50	0·50	70	

NB: valid for field drainage base depth W = 1·0 m and $\mu \sim 5\%$ (drainable pore space), see further section 7.5.4.

* The simple Hooghoudt formula eq. 9.7 $\left(L^2 = \frac{8Kdh}{q}\right)$ may be rewritten as $\frac{h}{q} = \frac{L^2}{8Kd}$ to show that for a given system h/q is a constant determined by the drain spacing L of the system and the Kd values of the soil. These parameters determine the degree of watertable control exerted by the system.

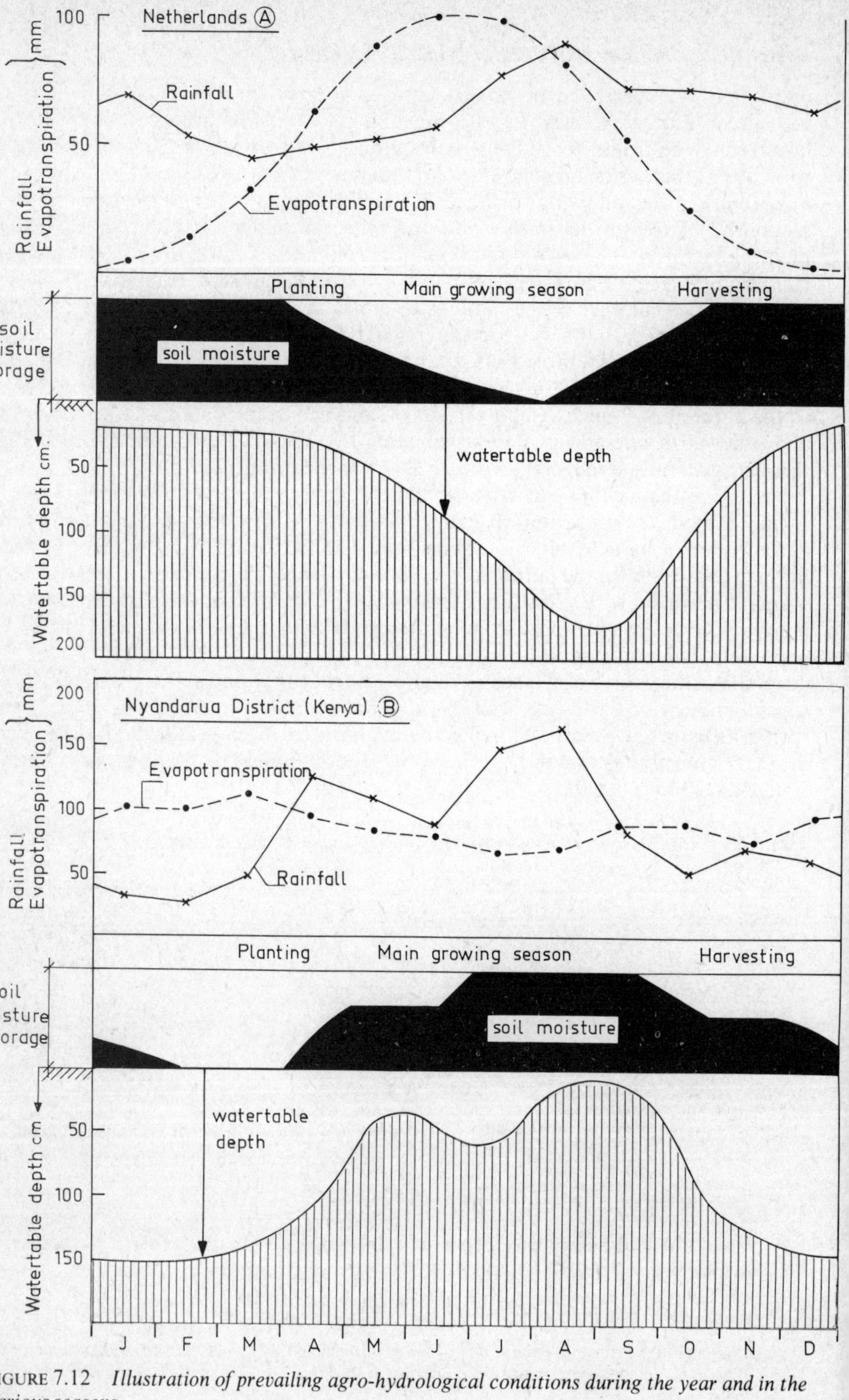

FIGURE 7.12 *Illustration of prevailing agro-hydrological conditions during the year and in the various seasons*

The criteria in table 7.3 may be adapted to other areas where the (groundwater) drainage is also primarily intended to control excess waterlogging in the winter and/or to assure an early return to FC in the spring. The same values for the watertable depth (H) would normally apply but the value of the design recharge/discharge (q) should be related to the prevailing rainfall conditions. For the value q = 7 mm.day^{-1} given in table 7.3 this relationship may be characterised as follows:

– 1 × 1 year winter rainfall during 5 days (design rainfall)...............	45–50 mm
– storage (under unfavourable conditions)	10–15 mm
– evapotranspiration..	– mm
– seepage minus natural drainage......................................	– mm
	35 mm

– required average discharge $= \dfrac{35\,\text{mm}}{5\,\text{days}} = 7\,\text{mm.day}^{-1}$.

Similar procedures (to suit local conditions) may be used to derive design discharges for areas with different rainfall conditions. Few situations warrant a system with a drainage intensity (h/q-value) of less than 50. Such a system, with an h/q = 50 and W = 100 cm, can cope with a recharge (q) of 10 mm.day^{-1} while maintaining the watertable 50 cm below the soil surface whilst it can cope with a maximum of 20 mm.day^{-1} when the watertable is at the soil surface (h = 100 cm). Daily rainfall in NW Europe seldom exceeds these rates, moreover the infiltration/percolation rate of the soil is often limiting. When rainfall rates are high, attention should be paid to improving surface/shallow drainage conditions rather than increasing the intensity of the groundwater drainage system.

7.5.2 Criteria for crop season drainage

In moderate climates, the main farming season normally coincides with the warm period (summer). In (sub)tropical climates crops are often grown all year round, including the rainy season. Indeed, cropping seasons are often specifically planned to coincide with the wet season to benefit from the rain (applies to most semi-arid/semi-humid climates, see figure 7.12B. The evapotranspiration during the warm period is considerable and contributes very significantly to controlling the occurrence of excess water and high watertables in the soil, both by removing water from the soil (which would otherwise have to be drained) and by depleting soil moisture during dry periods, thus creating storage for subsequent rainfall. Drainage requirements during warm growing seasons, therefore, are rather small, except for occasional heavy (summer) rains lasting several days which can create periods during which excess water occurs for sufficient length of time to cause harm.

Crop season drainage aims primarily at a rapid restoration of the rootzone aeration, following such a heavy rain. Other considerations for controlling excess soil water (early workability, warming-up, nitrification) are generally less critical. However, there are few well tested criteria for excess water/watertable control during this season. Values for H are often taken somewhat higher than for the off-season and could for example be taken equal to the depth of the main rootzone. Of course, steady state criteria depict a rather fictitious watertable situation which is not even an average situation and so no direct, straightforward relationship need exist between H and rootzone depth. The following depths serve as rough guidelines in selecting suitable values for H:

- shallow rooted and/or tolerant and/or low value crops........ H = 0·50 m
- deep rooted and/or sensitive and/or high value crops .... H = 0·75–1·00 m.

The corresponding value for q may be derived from the 1–2 year return period, 5 days rainfall. This rainfall should of course relate to the critical period (growing season, or even a specific period within this season). Fairly high reductions for storage and evapotranspiration may be allowed for during the main growing season.

Example

Criteria for crop season drainage of vegetables:

- H = 0·50 m (fairly sensitive and high value crops but mostly with a rather shallow root system)
- 1 × 1 year rainfall during 5 days .... 120 mm
  - storage (mostly in the soil) .... −30 mm
  - surface runoff (for the intense storms to be expected, assuming that a properly functioning surface drainage system has been installed) .. −20 mm
  - evapotranspiration (5 mm.day^{-1} on average)........................ −25 mm

45 mm

$$q = \frac{45\,\text{mm}}{5\,\text{days}} = 9\,\text{mm.day}^{-1}$$

An alternative (and often better) approach towards establishing crop season drainage criteria is to formulate them in *non-steady* terms, ie a certain rate of fall of the watertable (with corresponding rate of restoration of the rootzone aeration) is prescribed, following a heavy rainfall which is assumed to have caused the watertable to rise to the soil surface. It may eg be prescribed that within one to two days after the end of the rain the watertable should have fallen sufficiently so as to have restored the aeration of the upper rootzone, while within two to three days the entire main rootzone should be drained to field capacity. Relevant criteria are formulated in table 7.4.

TABLE 7.4 *Non-steady state criteria for crop season drainage*

	Tolerant and/or low value crops	*Sensitive and/or high value crops*
	t = 0 days H = 0·00 m	t = 0 days H = 0·00 m
Shallow Rootzone	t = 2 days H = 0·20 m t = 4 days H = 0·35 m	t = 1 day H = 0·20 m t = 2 days H = 0·35 m
Normal to deep Rootzone	t = 2 days H = 0·30 m t = 4 days H = 0·50 m	t = 1 day H = 0·30 m t = 2 days H = 0·50 m

These criteria may be used directly in the non-steady state Glover–Dumm formula to calculate L or they may be converted into steady state criteria and then used in the Hooghoudt formula (see section 7.4.1). It will be apparent that the drain spacings calculated on the basis of these criteria are rather similar to the spacings calculated on the basis of the corresponding criteria for off-season drainage given in table 7.3 (for $W = 1{\cdot}0\,m$ and $\mu = 5\%$).

The non-steady state approach is often most appropriate for tropical/sub-tropical conditions where the rain is too heavy for the watertable to be controlled below the topsoil during the actual storm and the main concern is to lower it rapidly once rain has ceased. Rapid restoration of the aeration of the upper rootzone is often all that is required. It ensures that the crop remains vigorous, capable of further 'self drainage' through its unimpaired transpirative capacity.

Tropical storms can be intense and prolonged and much of the rain which is unable to infiltrate into the soil needs to be discharged over the surface, requiring the provision of surface drainage in addition to the installation of a groundwater drainage system. The ditch systems (especially the system described in section 4.2.1 sub 2) combine these two functions and as such are well suited to tropical/sub-tropical conditions.

7.5.3 Drainage criteria determined by simulation

Non-steady state drainage formulae such as eq. 7.18 enable the watertable behaviour over a certain period to be simulated on the basis of the (infiltrated) rainfall and (actual) evapotranspiration data for that period (figure 7.13). Watertable hydrographs may thus be developed for, say, a 20 year period, using the historical daily weather data for a range of basic design criteria (q, H packages). From the results frequency tables, as in table 7.5 can be compiled on an annual or seasonal basis. For example, separate tables may be prepared for the off-season, for the crop season, or for a particular period posing its own particular drainage requirements, planting-early growth period, main growth period, harvest period, etc.

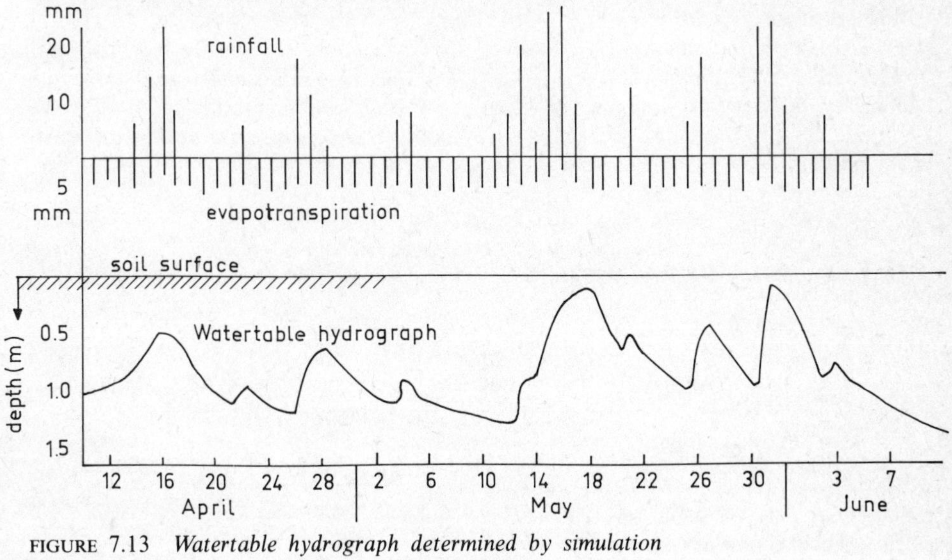

FIGURE 7.13 *Watertable hydrograph determined by simulation*

TABLE 7.5 *Average number of days per annum with watertable at or within depth indicated from the soil surface during the main growth period* (field drainage base depth W = 1·0 m)

Watertable depth below soil surface	$q = 7\,mm.day^{-1}$ H = 0·30 m	$q = 7\,mm.day^{-1}$ H = 0·50 m	$q = 10\,mm.day^{-1}$ H = 0·50 m
0·00 m	2 days	1 day	0 days
<0·20 m	4 days	2 days	1 day
<0·40 m	7 days	5 days	3 days
<0·60 m	12 days	7 days	5 days

For the case of table 7.5 with $q = 7\,mm.day^{-1}$ and H = 0·30 m, the watertable rises during the main growth period on average for two days per year to the surface, for four days per year to within 0·20 m from the surface, for seven days to within 0·40 m and for 12 days to within 0·60 m. If a crop is to be grown which can only tolerate the watertable in the rootzone for one to two days per year the more stringent criteria of $q = 7\,mm.day^{-1}$ and H = 0·50 m should be adopted.

7.5.4 The dependance of the criteria on the drainable pore space and on the drainage base depth

The steady state design criteria also depend to some extent upon μ (drainable pore space of the soil) and on W (the selected drainage base depth). This is due to the influence of these two parameters on the soil moisture storage. During rain-free periods watertables will fall, eventually to drainage base depth or even below. When W is large, watertables will on average have fallen deeper by the time the next rain arrives than when W is small. As a result, more rain can be stored in the soil without watertables rising to harmful levels thus reducing the quantity to be discharged. Consequently lower q-values should suffice when W is large.

Similar reasoning may be used to demonstrate that a lower q-value may be adopted when μ is large (>15%) since the rise of the watertable due to rainfall decreases as μ increases. The converse, however, does not apply for, although watertables rise rapidly when μ is low (<5%), they also fall quickly and so there is generally no need to increase q when μ becomes very small.

Some rough guidelines for taking into account the effects of μ and W on q are suggested in table 7.6. This table shows the variation in q relative to the standard

TABLE 7.6 *Guidelines for the adjustment of the design discharge (q) in groundwater drainage design for different values of the drainable pore space (μ) and drainage base depth (W)*

	Drainage base depth W			
	0·70 *m*	1·00 *m*	1·50 *m*	2·00 *m*
μ = 5%	1·5 × q	q*	0·8 × q	0·65 × q
μ = 10%	1·2 × q	0·8 × q	0·65 × q	0·5 × q
μ = 15%	q	0·65 × q	0·5 × q	0·5 × q

* Standard case with μ = 5% and W = 1·00 m

case of W = 1·0 m and $\mu = 5\%$ to which the criteria in table 7.3 apply. For a non-standard case such as W = 1·50 m and $\mu = 10\%$ the design discharge should be taken as $0{\cdot}65 \times q = 0{\cdot}65 \times 7 = 4{\cdot}6\,\mathrm{mm.day^{-1}}$. Since according to the Hooghoudt formula $q \propto -1/L^2$, the effects of these adjustments of q on L remain rather small (a 2 × smaller value of q only results in a 1·4 × wider drain spacing).

7.6 FIELD DRAINAGE BASE DEPTH

For a given value of watertable depth H the value of watertable head h increases with the value of drain depth W (see figure 7.1) and wider drain spacings may be used. This favourable effect of an increase of drain depth is reinforced by the fact that more excess water can on average be accommodated in the soil profile when W is large, allowing a lower value for q to be used which leads to a still larger drain spacing. Lowering the drainage base, however, raises the drainage costs since this requires deeper installation of pipe drains or deeper excavation of ditches and in addition the main system will generally need to be deepened. Up to a depth of 2·0–2·5 m these extra costs are normally more than compensated for by the savings resulting from the wider spacings. Further lowering of the field drainage base is, however, seldom profitable. Few machines are able to install drain pipe deeper than 2·5 m while the deep ditches pose other problems (land loss, caving in, maintenance, etc), resulting in pronounced increases in costs. Incremental benefits on the other hand tend to decrease at such low field drainage base depths, for a variety of reasons (see points below). Figure 7.14 illustrates these trends. As a general rule, field drainage should be made as deep as local conditions permit, up to a depth of 2·0–2·5 m.

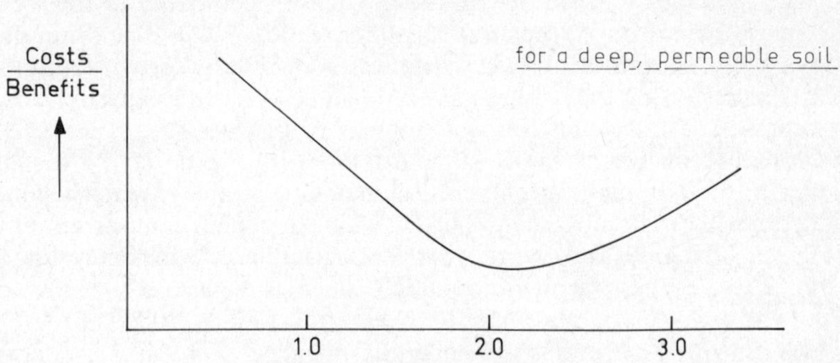

FIGURE 7.14 *Normal relationship between costs/benefits and the field drainage base depth of a groundwater drainage system*

The most important local conditions influencing the field drainage base depth selection, are:

(a) *The local/regional drainage base level*(s): the field drainage base cannot be lower than the local drainage base (= waterlevel maintained in the collector and lower order main system) while the latter cannot be lower than the regional drainage base (level in main outlet system). These local/regional levels are in their turn generally subjected to various other constraints. Where the natural drainage base is a constraint, polders with a pumped outlet may be considered although this of course involves extra costs. Other constraints may arise from established rights, the use of the drainage canals for navigation, etc.

(b) *Seepage*: a low local drainage base may constitute a 'sink', drawing excessive amounts of seepage water from the surrounding areas.

(c) *Soil conditions*: the stratification of the soil may favour or disfavour a certain depth. For example, pipe drains should not in general be installed in quick sand layers, in unconsolidated layers, in layers with a low hydraulic conductivity, etc. In situations where a rather poorly permeable upper soil profile is underlain (within the 2·0–2·5 m depth) by a more permeable substratum, advantage should be taken of the fact that drains are always much more effective when they are installed well into the more permeable layer. On the other hand, when a reasonably permeable upper profile is underlain by a very poorly permeable layer, little is to be gained by installing the drains into this latter layer.

Establishment of a deep drainage base in unconsolidated or unripened soils may be undesirable as it enhances subsidence and/or the formation of acid sulphate soils (see further chapter 14).

(d) *Drought risk*: a deep drainage base may 'over drain' the soil (see figure 2.1). Deeply drained soils retain somewhat less moisture in the rootzone at field capacity while also the supply of water to the roots by capillary rise from the groundwater will be less. Under certain climatic and hydrological conditions where the crops (partly) depend for their water requirements on soil moisture storage and on the supply from groundwater, a higher drainage base may need to be maintained during the dry season compared to the wet season, especially in drought sensitive soils (see section 5.2.1). Pipe drain depth will normally be based on the wet season situation. Under certain conditions 'over drainage' during the dry season may also be avoided by (partly) closing field drains.

(e) *Available machinery*: most standard type drain pipe laying machines for agricultural drainage in temperate climates are unable to reach beyond 1·50 m depth while their most economical working depth is usually around 1·0–1·2 m. Special machines are required for deeper installation which may not be locally available or can be made available only at high costs.

In NW Europe pipe drains are generally installed at 0·9–1·2 m depth, deeper installation generally not being feasible due to considerations in (a) to (e) above. This is also a common depth in other areas where the drainage is installed to control excess water during the wet season. In the case of pipe systems installed for salinity control, much deeper drainage bases are normally maintained (see chapter 12).

Drain pipes installed at shallow depths risk blockage by root penetration and a minimum depth of some 0·75 m is generally advisable. In cold climates drain pipes should be installed beyond the depth of frost penetration.

7.7 DETERMINATION OF THE REQUIRED DIAMETER OF A PIPE DRAIN

The hydraulic design of pipe drains is based on the standard pipe flow formulae which relate the discharge (Q) to the hydraulic gradient (i), the pipe diameter (d) and wall roughness. These formulae differ for smooth and corrugated pipes, and between pipes that simply transport a fixed discharge along their length (*uniform flow*) and those transporting a discharge which increases along the length of the pipe (*varied or non-uniform flow*). The latter condition applies to most field drains.

Uniform flow

The formulae for this type of flow are derived in the case of smooth pipes from the Darcy–Weisbach equation and for corrugated pipes from Chézy–Manning equation:

– smooth pipes: $Q = 50\,d^{2\cdot71}\,i^{0\cdot57}$ ......(eq. 7.21)

– corrugated pipes: $Q = 22\,d^{2\cdot67}\,i^{0\cdot50}$ ......(eq. 7.22)

in which Q = the discharge along the pipe, $m^3.sec^{-1}$
d = the pipes internal diameter, m
i = the hydraulic gradient, $m.m^{-1}$.

It is prudent to 'over design' the pipe to allow for subsequent partial silting and for mis-alignment during construction. This may be conveniently done by increasing the discharge Q by a factor varying from 1·33 for continuous plastics pipe installed in stable soil, to 2·0 for tile drains laid in siltation prone situations.

The hydraulic gradient is constant and it is assumed that for a pipe transporting its design discharge (Q_{design}), the hydraulic grade line will coincide with the top of pipe (in other words, it is assumed that the pipe runs full of water at all points without over-pressure, see figure 7.15-a).

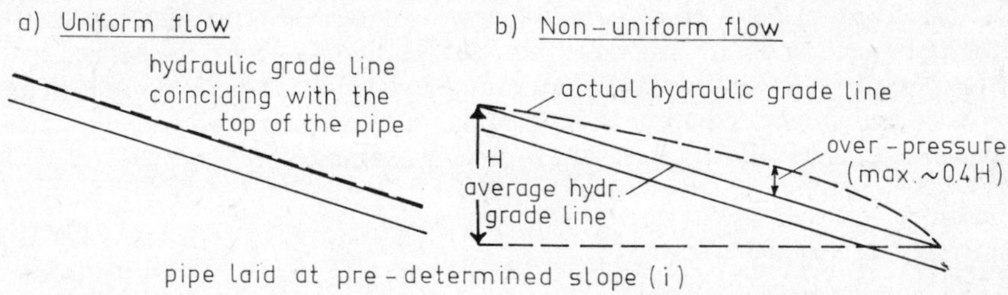

FIGURE 7.15 *Hydraulic grade lines under uniform and non-uniform pipe flow*

Example 1

- smooth concrete collector pipe
- slope of the pipe line 0·1%
- receiving water from 20 ha of land drained by field drains designed for q = 5 mm.day^{-1}
- an allowance to be incorporated for a subsequent 25% reduction in transport capacity due to siltation.

The discharge at the end of the line is:

$$Q = \frac{q \times A}{1000 \times 3600 \times 24} (m^3.sec^{-1}) \text{ where } A = \text{drained area } (m^2)$$

In this case $A = 200\,000\,m^2$; $q = 5\,mm.day^{-1}$, so $Q = 0{\cdot}0116\,m^3.sec^{-1}$. Allowing for situation $Q_{design} = \frac{1}{0{\cdot}75}(0{\cdot}0116) = 0{\cdot}0155\,m^3.sec^{-1}$. Eq. 7.21: $Q_{design} = 50d^{2\cdot71}\,i^{0\cdot57}$ or $d = \left[\frac{\cdot0155}{50(0{\cdot}001)^{\cdot57}}\right]^{\frac{1}{2\cdot71}} = 0{\cdot}217$ m, so select a pipe with ID (inside diameter) of 22 cm or the nearest larger size available.

Non-uniform flow

The formulae for non-uniform flow, derived from the same standard equations, are:

smooth pipes $\quad Q = 89\,d^{2\cdot71}\,\bar{i}^{0\cdot57}$ ......(eq. 7.23)

corrugated pipes $\quad Q = 38\,d^{2\cdot67}\,\bar{i}^{0\cdot50}$ ......(eq. 7.24)

in which Q and d are as previously described and $\bar{i}$ is the average slope of the pipe. The discharge in a field drain pipe increases linearly from zero at the upstream end to the maximum design value at the outlet. The slope of the hydraulic grade line also increases from zero at the upstream point of zero discharge to a maximum value at the outlet. For design it is assumed that the pipe runs full at all points and that the average hydraulic grade line coincides with the slope and position of the top of the pipe. As figure 7.15-b illustrates, this means that the water in the pipe is under a slight over-pressure for the whole of its length with the exception of the two ends.

Comparison of the equations for non-uniform and uniform flow indicates that a field drain of a given size and slope will transport a discharge which is approximately 75% larger than a similar pipe which simply conveys a fixed discharge along the length of the pipe.

The relationships between Q, i and d for non-uniform flow are also expressed in nomographs such as figure 7.16. Given the maximum discharge to be carried by the pipe (= discharge at outflow end), the average hydraulic gradient along its length, the type of drain pipe used, these nomographs enable the required pipe diameter to be determined. This is illustrated by the following example.

Example 2

- parallel system of pipe drains at 35 m spacing
- corrugated plastics drain pipe
- line length = 200 m
- $q = 5\,mm.day^{-1}$
- slope of the pipe line $i = 0{\cdot}10\%$

The discharge Q at the end of the line is calculated as:

$$Q = \frac{q \times A}{1000}(m^3.day^{-1}) \text{ where } A = \text{drained area } (m^2).$$

In this case $A = 200 \times 35 = 7\,000\,m^2$; $q = 5\,mm.day^{-1}$, so $Q = 35\,m^3.day^{-1}$ or $0{\cdot}0004\,m^3.sec^{-1}$. Taking a 50% safety factor to allow for siltation: $Q_{design} = 2 \times 35 = 70\,m^3.day^{-1}$ or $0{\cdot}0008\,m^3.sec^{-1}$. Entering the 'corrugated pipe, non-uniform flow' nomograph of figure 7.16 with this value for Q and with $i = 0{\cdot}10\%$, the required pipe diameter (ID) is read as 64 mm. This value is adjusted to the nearest higher commercially available size which in this case is ID = 72 mm (OD = 80 mm).

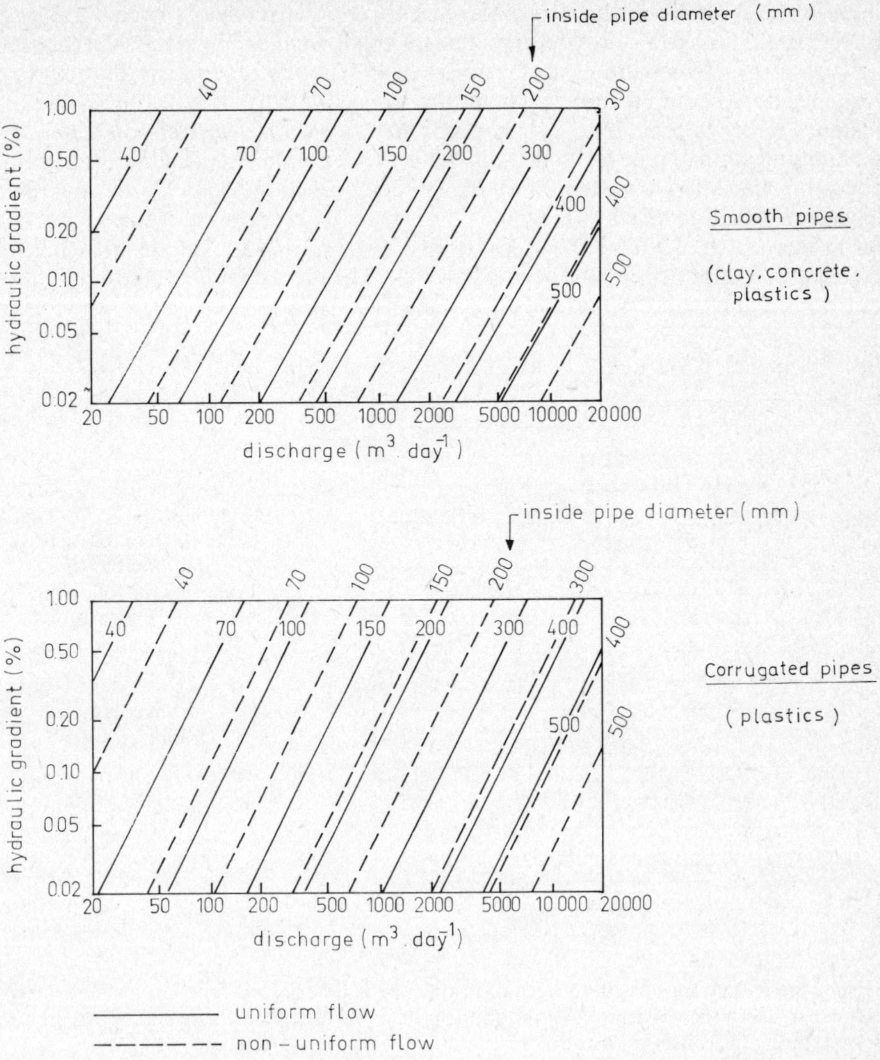

FIGURE 7.16 *Pipe diameter nomographs* (adapted from ILRI 1974)

Idem with eq. 7.24:

$$Q_{design} = 38\, d^{2\cdot 67}\, \bar{i}^{0\cdot 50}$$

$$d = \left[\frac{0\cdot 0008}{38(0\cdot 001)^{0\cdot 50}}\right]^{\frac{1}{2\cdot 67}} = 64\,\text{mm}.$$

Composition of a pipe out of sections of different diameter

For the larger collector type drain lines it is often economical to increase the pipe diameter in steps in the flow direction to adjust for the increasing discharge carried by the pipe. This practice is rarely justified for small diameter pipe drains because the savings in pipe costs are easily outweighed by the increase in installation costs (organisational complications). The design procedure entails selecting in the first instance either two or three possible pipe sizes, eg 10, 20 or 30 cm diameter. The maximum length of pipe for each diameter is then determined using the appropriate equation, the known q value and the predetermined spacing. These maximum lengths are then reduced by a factor* 0·85 when two pipe sizes are used and 0·75 when three or more sizes are used, to give the appropriate lengths at which the transition from one size to the next takes place. The procedure for determining these positions is illustrated by the following example.

Example 3

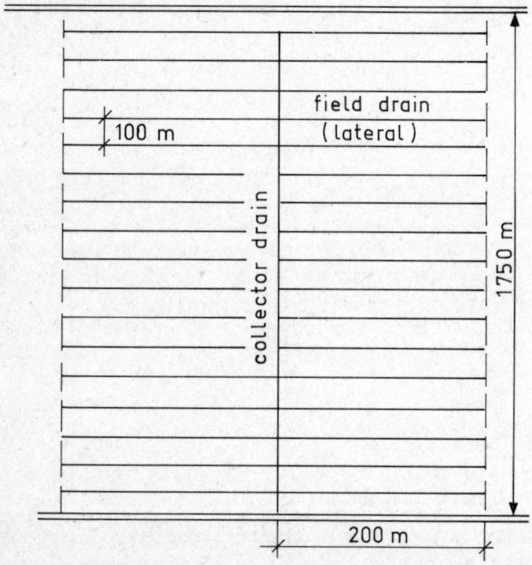

The collector drain is to be a concrete pipe of which the following diameters are available: ID = 10, 15 and 20 cm

$q = 2{\cdot}0\ \text{mm.day}^{-1}$ (typical drainage rate for salinity control)

$i = 0{\cdot}2\%$

Firstly, the maximum possible length of drain pipe is determined for each of the available diameters, using the equation for non-uniform flow (the discharge increases approximately linearly along the collector's length).

	Pipe diameter (ID in cm)			
	10	15	20	
Maximum discharge capacity of clean pipe: $Q_{design} = 89\, d^{2\cdot71}\, i^{0\cdot57}$	434	1 302	2 839	$m^3.day^{-1}$
Allowing for 25% reduction of capacity due to siltation $Q = 0{\cdot}75 \times Q_{design}$	326	977	2 129	$m^3.day^{-1}$
Drained area A = (1000 Q)/q	163 000	480 000	1 064 500	m^2
Maximum drain length = A/400	408	1 221	2 661	m

* These factors take into account the extra headloss incurred when part of the pipe line has a smaller diameter

For this case with the line being composed of pipes of three different diameters, the maximum single diameter length is 0·75 of the calculated maximum length. Applying this reduction factor, the collector is to be composed of:

- pipe with ID = 10 cm for section 0– 306 m (75% of 408 m)
- pipe with ID = 15 cm for section 306– 916 m (75% of 1221 m)
- pipe with ID = 20 cm for section 916–1750 m (75% of 2661 m, exceeds 1750 m).

7.8 CHECKING OF PIPE SYSTEMS

The overall functioning of a pipe drainage system may be checked by watertable observations, individual mal-functioning drains being identified by observing drain discharges.

The causes of mal-functioning of a pipe drain may be diagnosed and localised by measuring the flow resistances as evidenced by the head losses along the flow path by means of piezometers (see figure 7.17; also section 7.2).

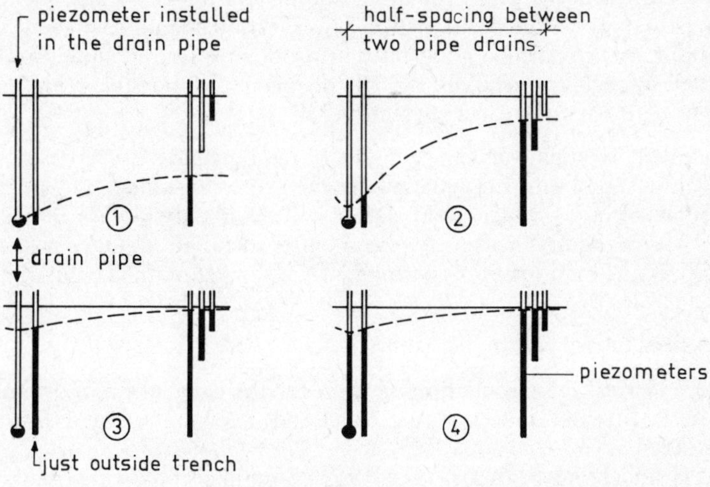

FIGURE 7.17 *Checking of the functioning of a pipe drain by means of piezometers*

Case 1: Impeded infiltration or percolation, usually evident by water ponded on the surface or by the occurrence of a perched watertable at some depth. Where the flow impedance is due to a hardpan layer, ripping may solve the problem. Often the cause of impeded infiltration is soil compaction, eg due to field traffic under wet conditions.

Case 2: High resistance in the groundwater flow towards the drain, usually due to under-design. Either the hydraulic conductivity was over-estimated, the impermeable layer occurs higher in the profile than assumed, the drainage coefficient was under-estimated or the drain spacing calculation formula used was not applicable. Horizontal flow and radial flow resistances may in principle be separated by placing an additional piezometer at about 0·7 D from the drain, although this is seldom necessary (they are normally correlated, when one is high so is the other). The appropriate remedial measure is installation of additional drains.

Case 3: High entry resistance, a common cause of mal-functioning where drains have been installed in a trench under wet conditions or by the trenchless method below the critical depth. Also, an envelope may have been omitted or may have become clogged. Not much can be done other than installing new drains. There is a high risk of drain failure when installation in a trench is done under wet conditions (high watertable, during rain, etc). Puddled soil around the pipe hinders the entry of water and such installation should as a matter of principle be avoided.

Case 4: obstructed pipe flow due to either:

- too high a waterlevel in the collector ditch (to be corrected by cleaning or deepening the ditch and/or by lowering the local drainage base);
- too small a pipe diameter (to be corrected by installing additional drain lines);
- mis-alignment of the pipes, eg due to poor installation or due to mechanical damage to the pipe (to be corrected by repairing the line; the trouble spot may be located during cleaning by measuring the length of flushing line inserted at the outlet to clear the blockage);
- clogging of the pipe by fines or by iron compounds (ochre) or blockage of the flow by root growth in the pipe. The latter should not normally be a problem when drains are deeply installed (see section 7.6). Siltation should be prevented as much as possible by providing a good filter in problem soils and should otherwise be alleviated by regular pipe cleaning. The problems and possible remedies of ochre clogging were discussed earlier in section 3.4.

The detailed investigations can generally be restricted to a few representative sites, from which extrapolations may be made (the same causal factors generally being instrumental across a wider area). Drain failure mostly occurs or will become apparent within the first year to two years after installation. Once past this initial period, pipes can function almost permanently when maintained regularly.

References and further reading

VAN BEERS, W F J 1965: 'Some nomographs for the calculation of drain spacings', Bulletin 8, International Institute for Land Reclamation and Improvement, Wageningen

BOUWER, H and J VAN SCHILFGAARDE 1962: 'Simplified prediction method for the fall of the watertable in drained land', Paper presented at the ASAE Winter Meeting, Chicago Illinois, December 1962

DIELEMAN, P J and B D TRAFFORD 1976: 'Drainage Testing', Irrigation and Drainage paper no 28, Food and Agriculture Organisation of the United Nations, Rome

DUMM, L D 1960: 'Validity and use of the transient flow concept in subsurface drainage', Paper presented at the ASAE Winter Meeting, Memphis Tennessee, December 1960

ERNST, L F 1962: *Grondwaterstroming in de verzadigde zone en hun berekeningen bij aanwezigheid van horizontale evenwijdige open leidingen*, Verslagen van Landbouwkundige Onderzoekingen *67* (15), the Hague (in Dutch)

HILER, E A 1976: 'Drainage requirements of crops', Proceedings of the Third National Drainage Symposium, American Society of Agricultural Engineers, St Joseph, Michigan

HOOGHOUDT, S B 1940: *Bijdrage tot de kennis van enige natuurkundige grootheden van de grond*, Verslagen van Landbouwkundige Onderzoekingen *46* (7): 515–707, the Hague (in Dutch)

ILRT 1974: Drainage Principles and Applications, Volume IV, publication 16, International Institute for Land Reclamation and Improvement, Wageningen

LUTHIN, J N 1957 (ed): *Drainage of Agricultural Lands*, Agronomy Monograph no 7, American Society of Agronomy, Madison, Wisc.

WESSELING, J 1973: 'Subsurface flow into drains'. In: *Drainage Principles and Applications*, volume II, publication 16, International Institute for Land Reclamation and Improvement, Wageningen

de ZEEUW, J W and F HELLINGA 1960: 'Neerslag en Afvoer', *Landbouwkundig Tijdschrift*, *70*: 405–421 (in Dutch)

Chapter 8 Design discharges

Design discharges (Q_{design}) are generally based on rainfall since this is almost always the most critical source of excess water with which a drainage system has to contend. For this reason other sources of excess water are not considered although the designer should remain aware that exceptions to this rule may occur locally (eg heavy seepage inflow into a polder or snowmelt).

Discharge generated by the intense storms normally taken for design, typically occurs as a wave in which the discharge in the drainage canals rises temporarily to a peak and then recedes. Hydrographs, showing the rate of discharge in the basin as a function of time, due to such storms have a peaked shape.

8.1 DISCHARGE TRANSFORMATION

Drainage is a series process whereby the rainfall subsequently passes through the field drainage system and the main canals of various order, to the outlet. In this process, the hydrograph is steadily transformed as schematically shown in figure 8.1. From rainfall to outlet discharge, the peak rate of discharge expressed in $mm.day^{-1}$ or in $l.sec^{-1}.ha^{-1}$, is reduced (although the actual discharges in $m^3.sec^{-1}$ are of course likely to increase in the direction of the outlet in line with the increase in basin area).

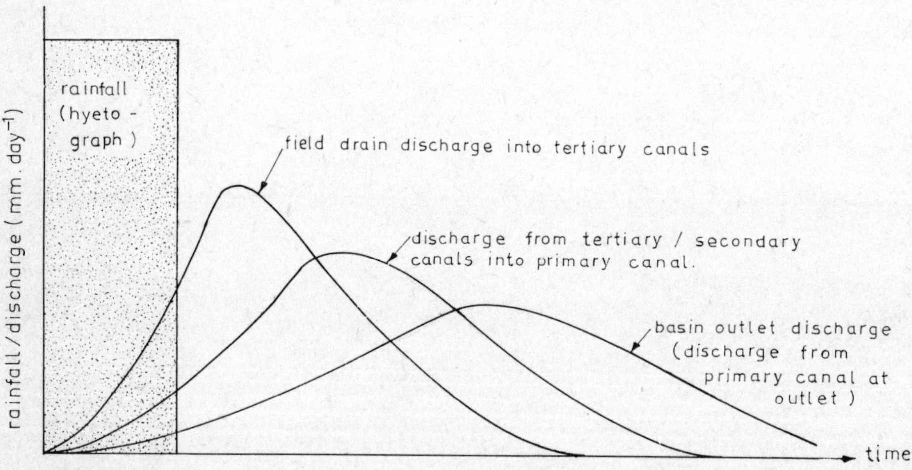

FIGURE 8.1 *Transformation of the hydrograph along the drainage path*

Effects of storage and flow resistance

Storage and flow resistance are the main factors causing transformation of the hydrograph. The head needed for the water to flow through the system is partly created by storing part of the inflow such that the 'upstream' levels rise above those 'downstream' (detention storage).

More head is needed (ie more water is stored), the higher the resistance in the system or the higher the discharge that has to pass through it. As the discharge decreases at the end of the discharge wave, the built-up storage gradually depletes, in line with the reduced head requirement. This is illustrated by figure 8.2 which shows that in the process the hydrograph is extended and flattened.

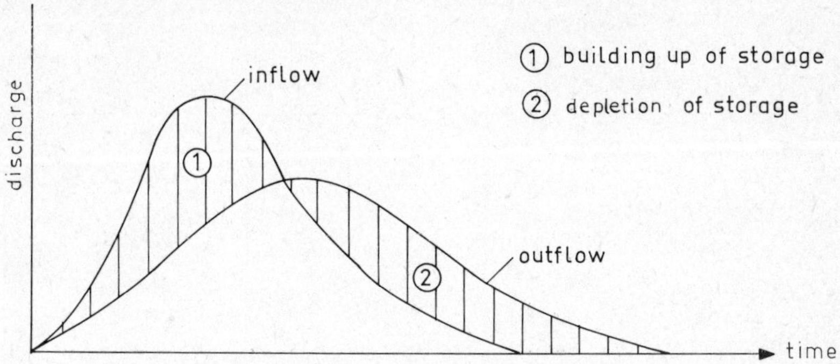

FIGURE 8.2 *Transformation of flow through a reservoir*

In general the greatest transformation occurs at the field drainage stage since in this stage, with the water moving over or through the land, the soil has potentially the largest storage capacity whilst it offers the highest resistance to flow of any stage. This accounts for the marked contrast between the hyetograph and the field drain discharge hydrograph. However, the extent and nature of the transformation within the field drainage process varies with the type of field drainage system used. Groundwater drainage normally results in more attenuated hydrographs than shallow flow. The hydrograph will be especially peaked in response to overland flow from sloping land, whereas on flat land it will be less peaked because more rain will be detained in depressions while the drainage, being mostly topsoil drainage/interflow, will also be slower.

Effects of time of travel

The field drains discharge into the collectors/tertiary canals at different points along the system whilst the collectors/tertiaries discharge at different points into the secondary system and so on into the primary canal. This spatial differentiation in inflow of lower order canals into higher order canals means that the flow in the latter canals is composed of flows having different times of travel. The effect this can have is illustrated for two identical tertiary canals, each draining a similar area of land, which discharge into a secondary canal at points distance L apart as shown in figure 8.3.

The hydrographs in the two tertiary canals will be more or less identical and synchronised. The discharge from tertiary ① will arrive at the junction point B with tertiary ② at time Δt after the discharge from tertiary ② has passed through this point. The travel time Δt between points A and B in the secondary canal equals L/v where v = the velocity of flow in this canal section. The combined hydrograph at point B may be developed by superposition of the hydrographs from the tertiary canals ① and ② taking due account of the time of travel, although it must be

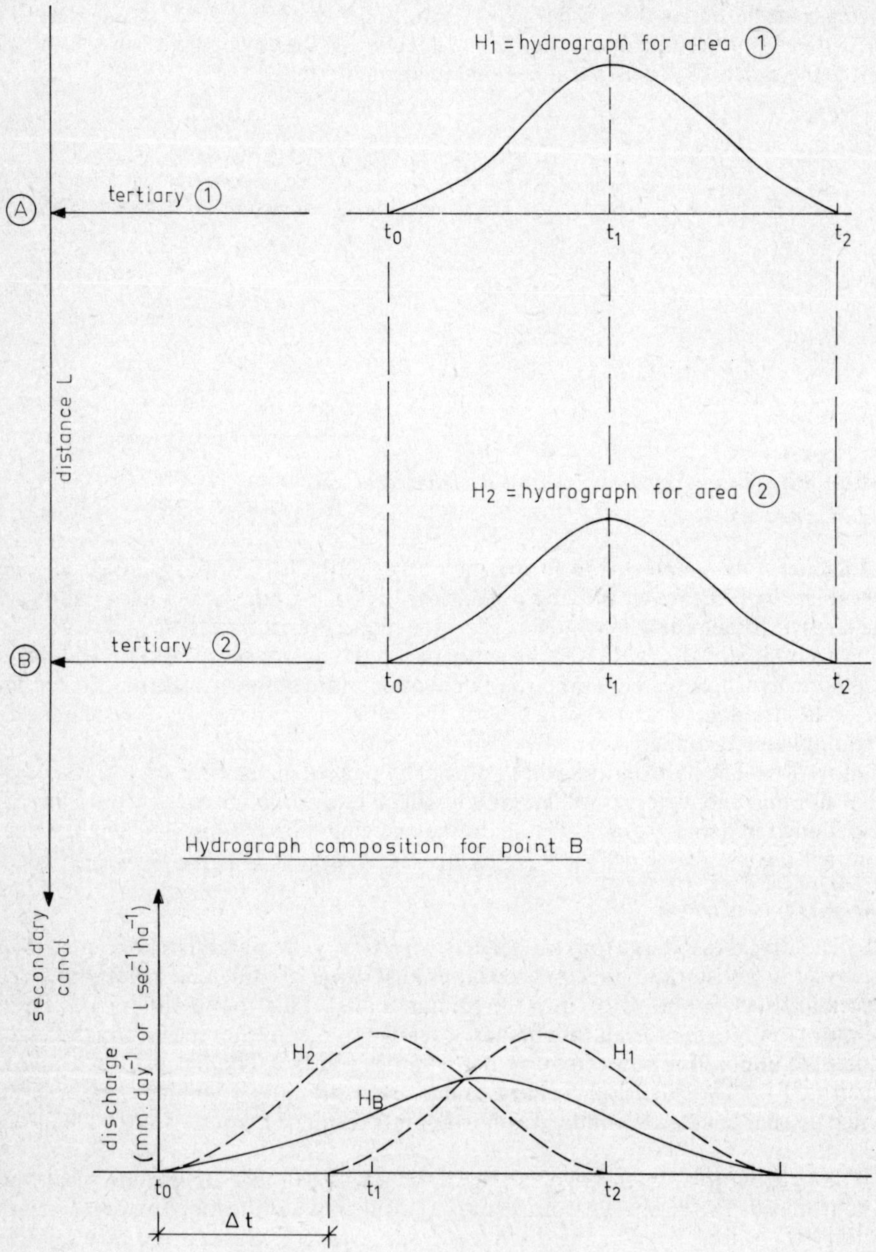

FIGURE 8.3 *Effect of time of travel on the hydrograph*

emphasised that this simple procedure ignores any transformation occurring in the discharge from tertiary ① as it moves between points A and B.*

The composed hydrograph at point B is quite different from the component hydrographs. The peak rate of discharge ($mm.day^{-1}$) will be equal to or less than the peak rates of the component hydrographs although the actual discharge ($m^3.s^{-1}$) will be higher due to the increase in the area of land involved. The peak rates will be equal when the peak rate of discharge ($mm.day^{-1}$) from tertiary canal ② occurs a time Δt later than the peak in tertiary canal ①, the two peaks arriving at point B simultaneously. However, it is more likely that the peaks will arrive at B at different times resulting in a reduced rate of discharge of the composed hydrograph, the more so for elongated basins as compared to fanshaped basins.

Basin v field discharge; drainage coefficient

Design discharges vary at different points along the drainage canal system, being generally calculated as:

$$Q = \frac{q.A}{1000} \qquad \text{......(eq. 8.1)}$$

where: Q = basin discharge (in $m^3.sec^{-1}$)
A = drainage basin area at the point considered (in ha)
q = the *drainage coefficient*, also termed the specific or unit design discharge, ie the design discharge per unit area of the basin considered (in $l.sec^{-1}.ha^{-1}$).

Basin discharges are formed by the combination of field drainage discharge although in general these cannot be simply summated because of the transformation occuring to the discharge in the canal system. There is also often a difference in the design frequency adopted for field drainage and for main drainage. Failure of the main system is generally of greater consequence and is more damaging than failure of the field system and so the main system is designed to cope with more extreme events than the field system, eg 1 × 15–25 year events compared to 1 × 1–2 year events.

All these influences mean that the drainage coefficient for main drainage normally differs from the design discharge for field drainage (also indicated with the symbol q in chapters 6 and 7) and that the drainage coefficient is not fully independent of the basin area A (as eq. 8.1 seems to suggest).

Flat v sloping land

As the slope of the land becomes steeper, an increasing proportion of the excess rain will discharge rapidly as overland flow or as another rapid type of shallow flow. As a result, hydrographs generally become more peaked with increasing land slope. This trend is often accentuated in the main system due to the reduction in travel time and the diminished canal storage of sloping basins. These features generally become

* The determination of the discharge of higher order canals on the basis of discharges of lower order canals, taking into account the effects of flow resistances, storage and travelling time, is generally referred to as *flood routing*. Formal flood routing procedures are seldom necessary in land drainage design work.

apparent when the land slope is >0·5% (*sloping land*) while they are normally insignificant when slopes are <0·2% (*flat land*). Other factors besides the degree of slope also influence the rapidity of the drainage discharge from the land (soil type, vegetative cover, cultivation method, etc). These factors largely determine whether the land with slopes from 0·2 to 0·5% should be classified as 'sloping' or as 'flat'.

Another difference between flat and sloping basins is that over-topping of the canals, leading to flooding of the adjacent land, is generally less harmful on flat than on sloping land. On flat land, the flood water will spread and retreat more calmly than on sloping land. On sloping land, flood water will travel further and faster and be generally more devastating due to erosion and siltation.

These differences underly the different approaches taken with respect to design discharges for flat and sloping drainage basins. In view of the potential flood hazards, design discharges for sloping basins are generally taken equal to *the peak discharge* generated by fairly extreme, short duration storms to ensure that flooding is a rare event (eg 1 × 25 year peak discharges lasting only 1 to 2 hours may be used in flood sensitive areas). Since hydrographs for these basins are in any event peaked, by implication the Q_{design} will also be quite high.

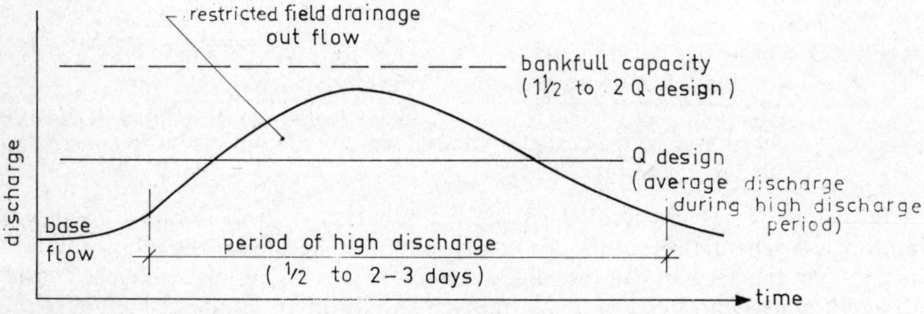

FIGURE 8.4 *Design discharge approach for flat basins*

For flat basins design discharges are much lower because they can be based on more frequent events (recurrence interval ≤5 years) and because they can be based on an *average discharge* during the period of high discharge considered. Figure 8.4 illustrates this last point. Periods of high discharge are likely to last from about half a day for basins with mainly shallow drainage to a period of two to three days for basins with mainly groundwater discharge. Taking Q_{design} equal to the average discharge implies that actual discharges will be higher than the Q_{design} during part of the discharge period. In view of the flat shape of the hydrograph, these discharges will not exceed the Q_{design} by much, so that in general they can be safely accommodated by the freeboard. During this period the drainage flow will also back up into the field drains, temporarily restricting the inflow into the main system.

Methods to determine design discharges

Three different methods are presented in the following sections, each having its own special area of application:

(a) *Statistical analysis* of observed discharges: this is the most straightforward method, generally producing the most reliable results. The required series of

Method	*Nature of field drainage flow*	*Information required*
1 *Statistical analysis* (section 8.2)	All types	Discharge or stage records
2 *Calibrated rainfall-discharge relationships:*		
– unit hydrograph method (section 8.3.1)	Predominantly shallow flow	Unit hydrograph
– reaction factor method (section 8.3.2)	predominantly groundwater drainage	α-value (basin)
3 *Uncalibrated rainfall-discharge relationships:*		
– flat basins		
• methods as in section 8.4.1	Groundwater drainage	Mostly readily available field and basin characteristics
• methods as in section 8.4.2	shallow drainage	
– sloping basins		
• rational formula (section 8.5.1)	Overland flow, including rapidly responding interflow	
• curve number method (section 8.5.2)		

discharges is, however, seldom available for agricultural basins (discussed in section 8.2)

(b) *Calibrated rainfall–discharge relationships:* here the starting point is the (design) rainfall which is converted into discharge on the basis of an established, calibrated rainfall–discharge relationship (see section 8.3)

(c) *Uncalibrated rainfall–discharge relationships:* procedure as for methods in (b) although in this case the rainfall–discharge relationships are not locally calibrated, therefore generally being somewhat less reliable (sections 8.4 and 8.5).

8.2 Statistical analysis of observed discharges

Discharges may be measured directly but are more usually derived indirectly from socalled *rating curves* which show the relationship between the discharge (Q) of a stream/canal and the stage level (h).* Rating curves may be established by plotting measured Q-values, representing the entire range of expected discharges, against corresponding stage levels. They may subsequently be used to determine the discharge passing through the stream/canal by simply reading the stage level (see example in figure 8.5).

In most countries, hydrological services maintain a number of official measuring sites (*gauging stations*) on the main streams/canals. Discharge data on the drainage canals within an agricultural basin are, however, seldom available and will need to be specifically collected for the project. The observation series will then as a rule be

* The stage level (h) is directly related to the water depth 'd' (see figure 8.5). According to the Chézy–Manning formula (eq. 9.3, section 9.1), Q = function (d) for a canal when the gradient I = constant, so also Q = function (h). Rating curves for trapezoidal canals are typically of the type depicted in figure 8.5 with Q increasing more than proportionally with increasing h.

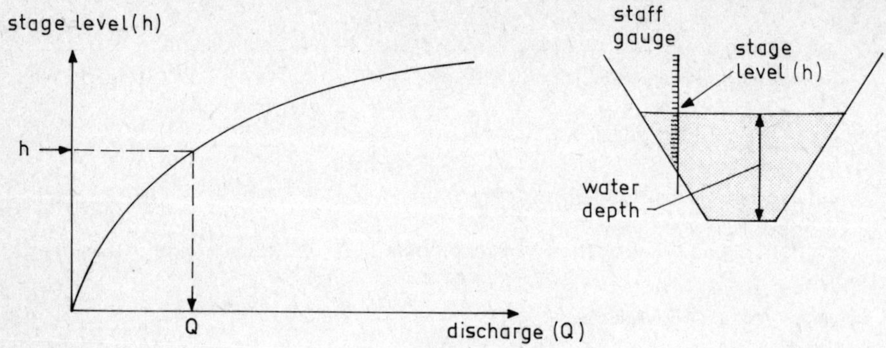

FIGURE 8.5 *Typical rating curve for a stream/drainage canal*

short (1–2 years only). Various methods exist to extend the series (see section 15.5), though generally at the expense of reliability.

Statistical analysis is applicable when a rather long series of discharge data is available (> 15–20 years). The observed annual maximum discharges are ranked in decreasing order as in table 8.1 (instead of annual maxima, the use of seasonal maxima per annum would be relevant if these are more critical).

TABLE 8.1 *Frequency analysis of drainage discharges*

Year	*Maximum discharge* ($m^3.sec^{-1}$)	*Rank number* *m*	*Rank number* *m*	*Discharge* ($m^3.sec^{-1}$)	*Probability* (*P*)	*Recurrence interval* (*T, years*)
1958	85·1	4	1	98·3	0·05	20
1959	50·1	17	2	90·2	0·10	10
1960	48·2	18	3	85·3	0·15	
1961	68·3	10	4	85·1	0·20	5
1962	60·4	13	5	80·7	0·25	4
1963	55·2	14	6	80·6	0·30	
1964	80·7	5	7	78·4	0·35	
1965	90·2	2	8	78·3	0·40	
1966	85·3	3	9	76·7	0·45	
1967	61·3	12	10	68·3	0·50	2
1968	98·3	1	11	61·5	0·55	
1969	78·4	7	12	61·3	0·60	
1970	80·6	6	13	60·4	0·65	
1971	36·7	19	14	55·2	0·70	
1972	50·2	15	15	50·2	0·75	
1973	61·5	11	16	50·2	0·80	
1974	50·2	16	17	50·1	0·85	
1975	78·3	8	18	48·2	0·90	
1976	76·7	9	19 (= N)	36·7	0·95	1

Note: Probability P = m/(N + 1): annual probability of occurrence or of exceedance of the discharge indicated
Recurrence interval T = (N + 1)/m (also called *return period*)

For example, basing canal design on the 1 × 5 year event, it follows from this table that $Q_{design} = 85{\cdot}1\ m^3.sec^{-1}$. Discharges outside the observed probability range may be estimated by extrapolation of the above series. This may be done graphically by plotting the Q-values on normal or other probability paper, or by plotting P against log Q, against $Q^{\frac{1}{2}}$ or otherwise, to see which plotting gives the best straight line, most suitable for extrapolation. Alternatively, the P and Q data may be processed on the basis of for example the Gumble or Pearson distribution functions which have at least a (weak) theoretical basis to claim validity for these type of discharges. However, in practice, extrapolation is seldom necessary since the infrequent events to be considered in this way (T > 20–25 years) are generally outside the scope of agricultural design discharges.

8.3 CALIBRATED RAINFALL–DISCHARGE RELATIONSHIPS*

Rainfall on the basin and the resulting drainage discharge are of course related. This relationship is, however, influenced and interacted upon by many factors limiting the applicability and reliability of the various established rainfall–discharge relationships. The reliability generally improves when they are calibrated for the type of rainfall and for the basin to which they are applied. This is the approach taken with the *Unit Hydrograph method* and with the *Reaction Factor method.* Application of these methods requires the prior recording of the passage of a discharge wave from the basin, generated under conditions comparable to the accepted design conditions. This of course limits the applicability of these methods. The methods discussed in sections 8.4 and 8.5 do not require such site discharge recordings but rely on more generally available basin characteristics, making these methods more universally applicable.

8.3.1 **Unit hydrograph method** (shallow drainage discharge)

The unit hydrograph theory is based on the premise that the hydrograph of a storm having generated aX mm discharge volume may be derived from the hydrograph of a storm of the same duration having generated X mm discharge volume, by simply multiplying all ordinates of the latters hydrograph by the factor aX/X = a (*principle of linearity*, see figure 8.6). This implies that the time base of the hydrograph is related to the storm duration (T)** but is independent of the discharge volume (V, mm) and the discharge intensity ($Q = V/T$, $mm.hr^{-1}$). This theory enables the hydrographs of all storms of duration T to be compiled, knowing the hydrograph of one of these storms. Hydrographs are most conveniently derived from *unit hydrographs* (UH), ie the hydrograph for a unit discharge volume of 1 mm 1 cm, etc (gives a convenient multiplication factor). This UH is a characteristic of the basin, reflecting the prevailing field drainage conditions as well as the discharge transformation taking place in the main system, up to the point under consideration (usually the outlet point of the entire basin or of a sub-basin). It follows that in each basin there are an infinite number of UHs, each being related to a particular storm duration.

* In modern hydrological terminology such relationships are commonly termed *models.*

**The duration of the recession of a hydrograph (falling limb period) is rather constant for a given basin and most of the variation in the time base is due to variation in the period of rise of the hydrograph which period is according to the Unit Hydrograph theory determined by the storm duration.

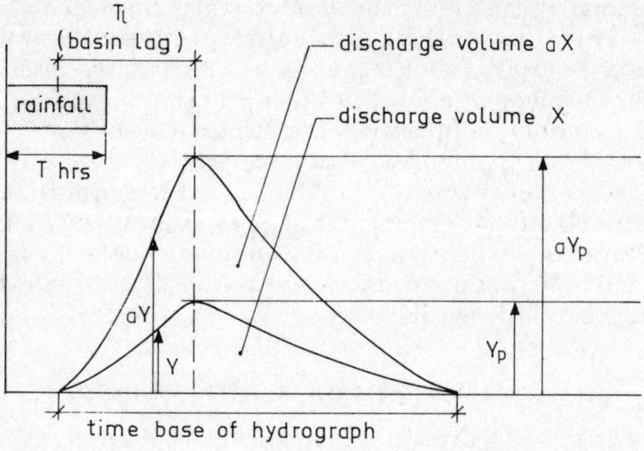

FIGURE 8.6 *Illustration of the principle of linearity in the unit hydrograph method*

Example

The use of the UH method to determine design discharges is illustrated by the following example. The 10 mm UH for the basin for a 3 hours storm is known and is illustrated in figure 8.7. The design storm to be considered is of 6 hours duration, composed of: 0–3 hours, 25 mm rain; 3–6 hours, 15 mm rain, totalling 40 mm. All rain is assumed to discharge as shallow flow (unfavourable conditions with negligible deep percolation and storage, commonly used as a basis for design). The hydrograph for the first 3 hours of the design storm is derived by multiplying the ordinates of the UH by the factor 25/10 = 2·5, while the hydrograph for the following 3 hours is derived by shifting the UH 3 hours later in time and multiplying its ordinates by the factor 15/10 = 1·5. The hydrograph for the entire design storm is obtained by superposition of these two hydrographs for the two (steady) storm intervals.

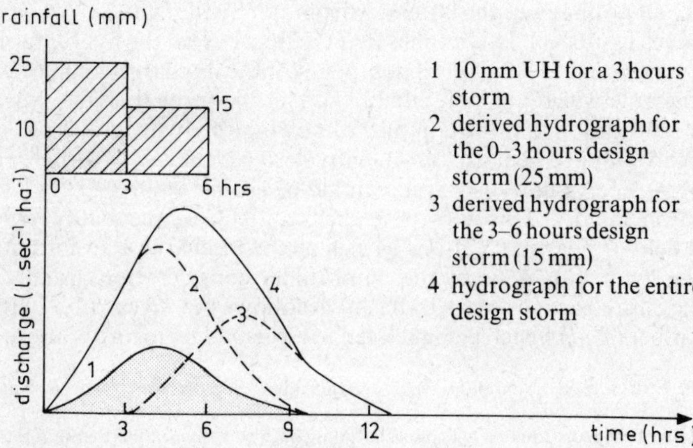

1 10 mm UH for a 3 hours storm
2 derived hydrograph for the 0–3 hours design storm (25 mm)
3 derived hydrograph for the 3–6 hours design storm (15 mm)
4 hydrograph for the entire design storm

FIGURE 8.7 *Application of the UH method*

The UH itself should ideally be established on the basis of measurements of the discharge from the basin considered, generated by a storm of the desired depth and duration. However, it would be very unlikely for such a storm to occur and the principle of linearity may be used to derive, for example, the UH 10 mm depth from an observed hydrograph of say 15 mm discharge, by simply multiplying its ordinates by 10/15. UHs may also be derived from more complex events in which the storm intensity varies and the duration does not equal the desired duration. In all cases, observed hydrographs should be adjusted to remove the groundwater drainage component (*base flow*). For details of this and of the more complex hydrograph analysis see the list of references at the end of this chapter.

Restrictions

The principle of linearity has proved to hold rather well under the following limiting conditions:

(a) the predominant type of field drainage should be shallow drainage flow (overland flow, interflow, etc). The principle may also be valid for the less rapidly responsive groundwater drainage but application is less straightforward as hydrographs for this type of drainage flow are generally less discrete due to their lengthy time bases and interactions with preceding or subsequent discharge events;

(b) the rainfall should cover the basin rather uniformly and have a rather steady intensity throughout the period to which the UH applies. The first condition means that the UH-method is less applicable to large basins which are unlikely to be uniformly covered by rain. On the other hand the requirement of steady rainfall intensity can be relaxed somewhat as the basin size increases since variations in intensity will be more evened-out (best UHs are those for storm durations $T \simeq \frac{1}{4}$ basin lag time as defined in figure 8.6; for medium sized basins bepween 10 000 and 50 000 ha, UHs for 2 to 6 hours storm durations are suitable).

(c) The design rainfall for which the hydrograph is to be determined should not be too widely different from the storm from which the UH was derived. The reasons for this are that the nature of the field drainage process and the time of travel in the main system are not fully independent of the rainfall intensity and the volume of discharge. For example, when the rainfall intensity increases a greater proportion of the rain will drain more rapidly as overland flow while the hydrograph recession will generally take longer. All in all the shape of the hydrograph will change, lessening the validity of the principle of linearity upon which the UH is based.

8.3.2 Reaction Factor method (*groundwater drainage discharge*)

This method developed in the Netherlands around 1960/65 (KRAIJENHOFF VAN DER LEUR, 1973, DE JAGER, 1965 and DE ZEEUW, 1973). It is applicable to basins in which almost all excess rain discharges as groundwater flow (as is generally the case in the Netherlands where mostly the soils are sufficiently permeable, rainfall is gentle and the land is flat). It was found that under these conditions the non-steady state formulae for groundwater field drainage could also be applied to determine discharges from basins in which the transformation of discharge in the main system is small compared to that in the field system.* Basin discharges may be described by (in analogy with eg 7.18):

$$Q_t = Q_{t-1}e^{-\alpha\Delta t} + R_{\Delta t}(1 - e^{\alpha\Delta t}) \qquad (\ldots \text{eq. } 8.2)$$

The reaction factor α in this case applies to the basin. Its value may be estimated from the recession limb of a representative basin discharge hydrograph in a similar manner to that outlined in section 7.4 (see eq. 7.20-a and figure 7.11). The most suitable hydrographs are those generated by storms whose total rainfall depth is close to that of the design storm and that have a recession limb which is unaffected by further rainfall for a 10 to 15 day period (methods do, however, exist to derive α-values from more complex hydrographs, see DE ZEEUW, 1973). Values of α for medium sized agricultural basins in the Netherlands are of the order of 0·3 to 0·7 day^{-1}. For basins with a rather slow response, reaction factors may be as low as 0·05–0·10 day^{-1} while for basins with a rapid response values may rise to 5–10 days^{-1}. High values often indicate that a considerable part of the rain is discharged as shallow flow, particularly overland flow, in which case eq. 7.18 is not strictly valid (although it seems to describe some types of shallow flow rather well). Reaction factors also generally increase as basins become smaller (approaching the values for field drainage mentioned in section 7.4.2).

Eq. 8.2 may be used with predetermined α values to simulate the basin discharges generated by various critical rainfall events (possible design storms), on the basis of which design discharges may be established (examples of application are presented by DE ZEEUW, 1973).

8.4 UNCALIBRATED RAINFALL–DISCHARGE RELATIONSHIPS FOR FLAT BASINS

In flat basins the main type of field drainage is either groundwater drainage or one of the flat land types of shallow drainage (topsoil drainage/interflow, mole drainage, etc, but not overland flow). The conditions under which one or other of these types predominate were described in chapters 3 and 4.

The determination of the basin discharge involves two steps:

1 conversion of rainfall into field drainage discharge. Various methods are used for this conversion, involving analytical as well as empirical relationships;
2 conversion of the field drainage discharges into basin discharges. In some circumstances it is argued that the transformation occurring in the main system is negligible and that field discharges may be directly equated to basin discharges. Otherwise, field drainage discharges are converted into basin discharges by means of simple flood routing.

8.4.1. Mainly groundwater drainage discharge**

Under these conditions most of transformation of the discharge occurs in the field drainage phase, the transformation in the main system being negligible by

* Note that this assumes the existence of a *linear relationship* between the discharge (Q_t) and the temporarily stored excess rain in the basin (detention storage) (S_t), see eq. 7.9 where the stored excess rain is expressed as a waterdepth h (= S/μ where μ = drainable pore space).

**Although much of the discussion is also relevant to groundwater drainage for salinity control, it is based on the humid climate type of groundwater drainage. Design discharges for salinity control are dealt with in section 12.3.

comparison. Field discharge may in this case be directly equated to basin discharges. As the time of travel from field drain to basin outlet for normal basins (say up to 25 000 ha) is seldom more than half a day and the peak rates of field drainage discharge are generally sustained throughout this period, the average basin discharge will be very similar to the average field drainage discharge.

As noted in table 7.3, groundwater drainage systems in humid climates in NW Europe are typically designed on the basis of the steady state criteria $q = 7\ \text{mm.day}^{-1}$ and $H = 50$ cm. The maximum possible discharge capacity of such a system (watertable at soil surface) is 2 to $2\frac{1}{2} \times q = 14$–$18\ \text{mm.day}^{-1}$ (frequency of occurrence about 1 × 1 to 2 years for short duration). The corresponding average basin discharge, allowing for the effects of storage and time of travel, would normally be of the order of $1\frac{1}{4}$ to $1\frac{1}{2} \times q = 9$–$11\ \text{mm.day}^{-1}$.

The non-steady state formulae eq. 7.18 enable the field drain discharge to be calculated for any rainfall event. As an example, drain discharges are calculated in table 8.2 for a 1 × 1 year, 3 day rainfall as normally found in NW Europe. The calculations are done for daily periods which is of the same order or even exceeds the travelling time in most medium sized basins. The reaction factors $\alpha = 0{\cdot}22$ and $\alpha = 0{\cdot}69$ represent respectively a slowly and a moderately rapidly responsive field system. With respect to antecedent rainfall it is assumed that previous rainfall has wetted the soil profile to field capacity and the field drains are discharging at a rate of $2\ \text{mm.day}^{-1}$ (tail end of recession curve). The results in table 8.2 are compared with those of an empirical rule according to which the daily rainfall discharges uniformly from the basin over a 3 day period starting with the onset of rain ($\frac{1}{3}$-$\frac{1}{3}$-$\frac{1}{3}$ rule, see table 8.2).

TABLE 8.2 *Groundwater drainage discharge calculations for the case of a 3 days, 1 × 1 year design rainfall*

Days	*Design rainfall* *(mm)*	*Discharge according to eq. 7.18* $\alpha = 0{\cdot}02$ *slow-response (mm)*	$\alpha = 0{\cdot}69$ *rapid-response (mm)*	*Discharge according to the rule of thumb** *day 1 rain (mm)*	*day 2 rain (mm)*	*day 3 rain (mm)*	*Total (mm)*
0		2	2				
1	20	5·6	11·0	7			7
2	10	6·5	10·5	7	4		11
3	8	6·8	9·3	6	3	3	12
4		5·4	4·7		3	3	6
5		4·3	2·3			2	2
6		3·4	1·2				
7		2·7	0·6				

* Until recently widely applied for the determination of outlet capacities for polders in the Netherlands (now replaced by the reaction factor method).

The result of the use of eq. 7.18 and of the rule of thumb do not quite agree. A slowly responsive field system gives lower discharge rates while a more rapidly responsive system gives similar though earlier peaks. The rule of thumb, however, applies to basin discharges and although it has been argued that groundwater drainage discharges and basin discharges are more or less identical, some transformation in the main system does in fact occur. In view of these considerations, the results of the rule of thumb are not too bad. Application of eq. 7.18 should generally be an improvement over the rule of thumb but best results are to be expected from the reaction-factor method as discussed in section 8.3.2 which is also based on eq. 7.18 but incorporates the basin effects on the field discharge (α derived from basin hydrographs).

8.4.2 Mainly shallow drainage discharge

Shallow drainage discharge on flat land mostly takes the form of interflow although considerable overland flow is also to be expected from field drainage systems especially designed to promote this type of flow (bedding, row drainage, etc, see chapter 4). The special discharge characteristics of mole drained land also have to be considered.

As stated, the hydrographs from shallow field drainage are much more peaked than those from groundwater drainage, and the variation in inflow into the main system during the typical half-day time of travel through this system is too great to permit design discharges for the main system to be established by simple averaging. Instead, two full steps should be applied, ie

(i) determination of the field drainage discharge from design rainfall;
(ii) conversion of these discharges into basin discharges (taking into account the effects of differences in travel time but ignoring the effects of storage in the main system).

(i) *Field discharges*

The generation of the shallow drainage discharge due to rainfall cannot be based on any flow formulae as can be done for groundwater drainage, as shallow drainage flows such as interflow, overland flow, mole drainage flow, etc, occurring singly or in combination, have not as yet been found amenable to any type of physical-

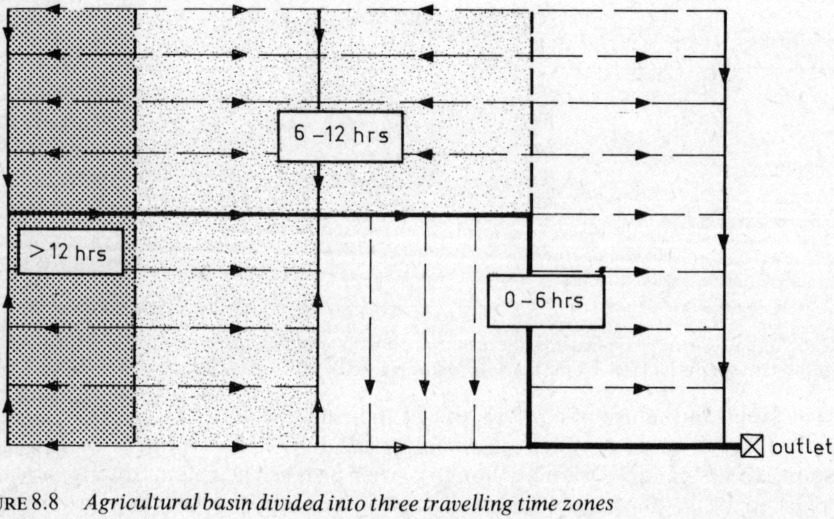

FIGURE 8.8 *Agricultural basin divided into three travelling time zones*

TABLE 8.3 *Determination of the shallow field drainage discharge from 12 hour, 1 × 5 year design rainfall*

Rainfall period (mm)	*Rainfall depth (mm)*	*Discharge from rainfall during periods (mm)*	*(mm)*	*Total discharge (mm)*
0–6 hr	32	1 (15 mm storage)*		1
6–12 hr	18	16	9	25
12–18 hr			9	9
	50	17	18	35

* Estimated conservatively low as is prudent for design conditions (note, however, that the combination of a 1 × 5 year storm with minimal storage involves conditional probability, the probability of the combined event being less than the 1 × 5 year storm probability).

mathematical analysis. As a rough approximation it may be assumed that half of the rainfall occurring during a 6-hour period will be discharged from the field during that same period while the remainder will be discharged in the following 6-hour period. ($\frac{1}{2}$-$\frac{1}{2}$ rule, compared to $\frac{1}{3}$-$\frac{1}{3}$-$\frac{1}{3}$ rule used in table 8.2.) In the example in table 8.3, this rule is used for a 12 hour, 1 × 5 year design rainfall of 50 mm.

(ii) *Basin discharges*

The basin should be divided into zones of equal time of travel as shown in figure 8.8. A division into zones at 6 hour intervals will normally suffice. In the case of figure 8.8, the basin is divided into three zones:

time of travel 0–6 hours: 40% of the basin
6–12 hours: 40% of the basin
> 12 hours: 20% of the basin

100%

TABLE 8.4 *Outlet discharge composition on the basis of field drainage discharge, taking into account times of travel in the basin*

Time of travel	*Basin area*	*Discharge during time interval, mm* 0–6 *hrs*	6–12 *hrs*	12–18 *hrs*	18–24 *hrs*	24–30 *hrs*
0		1	25	9		
0– 6 hrs	40%	0·4	10·0	3·6		
6–12 hrs	40%		0·4	10·0	3·6	
12–18 hrs	20%			0·2	5·0	1·8
total mm. 6 hrs^{-1}		0·4	10·4	13·8	8·6	1·8
mm. 24 hrs^{-1}		1·6	41·6	55·2	34·4	7·2
$1.sec.^{-1}.ha^{-1}$		0·2	4·8	6·4	4·0	0·8

Taking this into account, the discharge hydrograph for the outlet point may be composed from a field drainage hydrograph (derived in table 8.3) as illustrated in table 8.4. In this example, the maximum discharge at the outlet occurs during the 12 to 18 hour period. Compared with the rainfall peak of 32 mm occurring during the 0–6 hour period, the peak is shifted by some 12 hours and is reduced by some 18 mm.

Variant for bunded land

Bunded land as for example used in wet rice cultivation (*sawahs* as in figure 8.9) can store a lot of rain on the fields. Generally excess water discharges from these fields by spilling (directly over the bunds or over a low spot or weir in the bunds) into the drainage system. Figure 8.9 illustrates a suitable graphical procedure to determine the drainage coefficient under these conditions. The design rainfall is depicted by the rainfall duration line for the chosen frequency, in this case 1×5 years. In this example the field water level at the start of the rain is assumed to coincide with the level of the spill weir, so that discharge begins immediately. As the (assumed) uniform discharge rate (line 2) is initially lower than the rainfall rate (line 1), water accumulates on the field until the permitted excess depth is reached (75 mm is normal for modern rice varieties). Thereafter the discharge rate exceeds the rainfall rate and reduces the depth of ponding. Graphically this means that the discharge line should be tangential to the rainfall line, starting from the 75 mm point on the vertical (rainfall) axis. Excess depth should generally not last more than 3 to 5 days (the time tolerance varying with the growth stage of the rice crop; when the tangent line fails to comply with this duration limit, a steeper discharge line should be chosen).

8.4.3 Further guidelines

The rainfall–discharge relationships discussed in the preceding sections enable a historical rainfall series, covering for example some 15 to 25 years to be converted into discharges. These can then be statistically analysed as in table 8.1 to arrive at a suitable design discharge. The conversion of rainfall into discharge can be readily programmed into a computer. The rainfall data may also be statistically analysed beforehand to establish a design storm or a series of design storms, which are then converted into discharges by hand as in tables 8.2, 8.3 and 8.4 (or by means of the reaction factor method discussed in section 8.3 when the α-value of the basin is known).

In the example of groundwater drainage in table 8.2, the 1×1 year 3-day rainfall taken as the design rainfall is quite normal. The basin discharge may be taken equal to the average discharge during a 2 to 3 day period, depending on how much canal storage can be permitted in the basin. This leads to drainage coefficients of the order of 10 mm.day^{-1} for areas of mainly groundwater drainage throughout NW Europe. Groundwater drainage coefficients of this magnitude apply to many other areas where the rainfall may be higher although the average high groundwater drainage discharge during 2 to 3 day periods seldom exceeds this value.

The discharge from a mole drainage system often rises rapidly to a peak which may be of the order of 20 to 30 mm.day^{-1}. This high discharge rate is only sustained for short periods of time before declining rapidly. In general, the discharge hydrograph of a mole system generated by a critical 12 to 24 hour event, is entirely contained within a 24 to 36 hour period so that for example in the UK the design

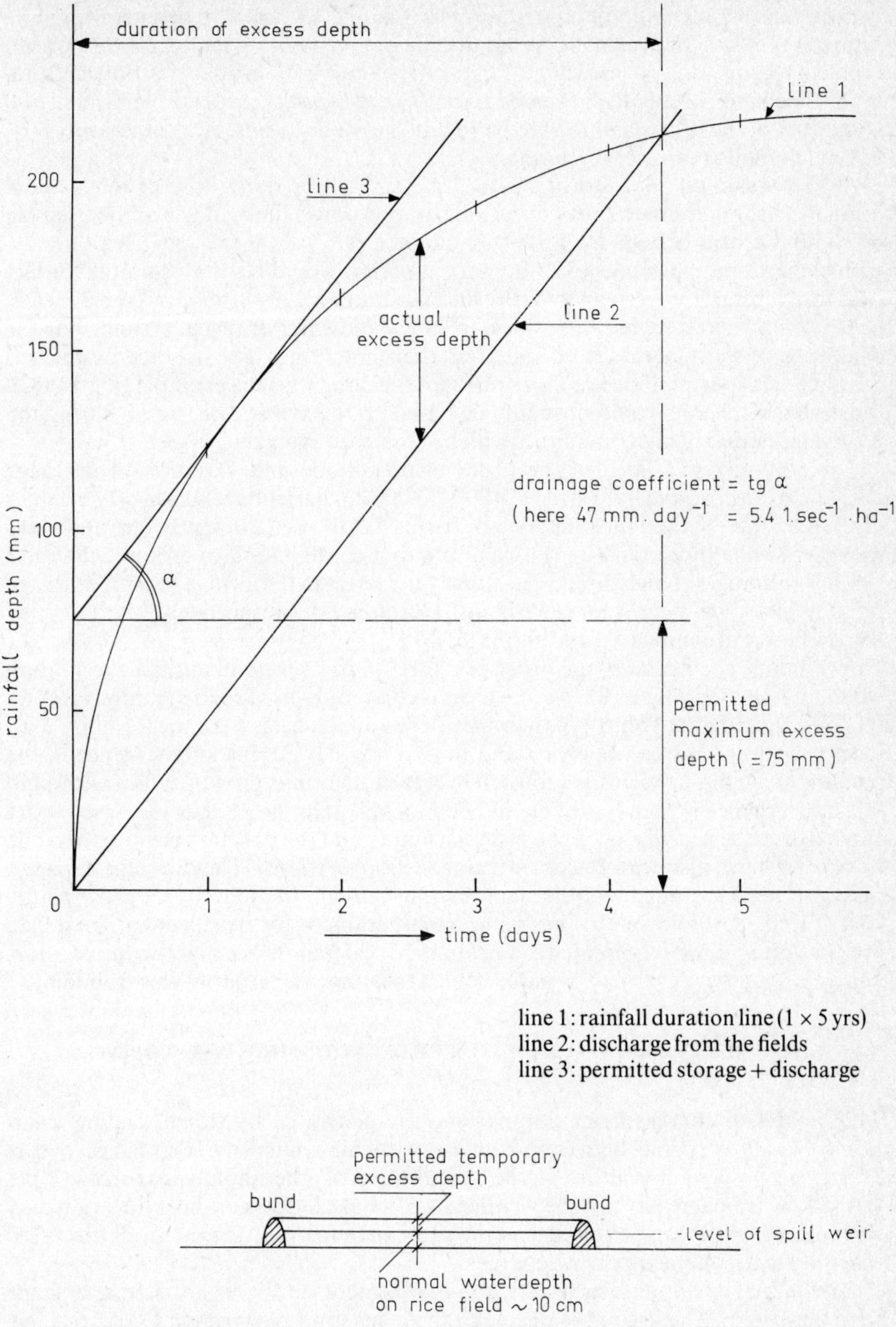

FIGURE 8.9 *Determination of the drainage coefficient for flooded rice fields*

capacity of the intercepting pipe drains (see figure 4.12) tends to be based upon 24 hours, 1 × 1 to 2 year rainfall, to be discharged within the same 24 hour period (applies to grassland; somewhat stricter criteria apply to arable and horticultural land). The land is generally assumed to be at field capacity prior to the rainfall and some 20% of the rain is assumed to be retained in depressions, etc, so that only 80% of total rainfall appears as discharge.

A 1 × 5 year, 12 hour storm as used in table 8.3 is quite normal for shallow drainage although other storm durations and response times may also be used (eg use 6 hour storm periods for a well maintained flat bedding system, clean furrow with some slope, etc, while a 12 hour storm period would be more suitable for less responsive interflow systems). With shallow drainage, drain discharge can reach peaks several times higher than those from groundwater drainage systems but the peaks are of short duration (usually less than one hour). The average discharges during 6–12 hour periods are more relevant for design. In the example in table 8.4, the design discharge would normally be based on the average discharge during the 3 × 6 hour period of high discharges which in this case averages 5.1 $l.sec^{-1}.ha^{-1}$.

Table 8.5 gives an idea of the order of magnitude and variation of drainage coefficients for flat, medium sized (10 000–25 000 ha) agricultural basins. The values given for the Netherlands apply to basins with predominantly groundwater drainage, while those for Yugoslavia apply to a combination of groundwater and shallow drainage (at high discharges mostly the latter in the form of topsoil drainage by interflow). The figures for Sudan and Tanzania refer respectively to exclusively furrow flow and composed overland/inter flow).

A drainage coefficient of the order of 5 $l.sec^{-1}$, ha^{-1} is quite normal for modern wet land rice cropping in SE Asia. Modern commonly used high yielding varieties (HYV) require better control of the water depth on the fields than the formerly used local varieties. They can mostly stand one to two days of full submergence in the seedling stage but are sensitive to such events in their mid-growth stages (although full submergence is then less likely to occur as the plant height has increased; part submergence, especially when the water is muddy and the duration is more than one to two days can also lead to considerable yield reductions). The drainage capacity may also be determined by other factors than rainfall. In modern rice growing the fields need periodically to be completely drained for fertilizer or pesticide application and the requirement to drain a 10 to 15 cm water layer within a short period (about two days) may be more critical than the discharge of excess rainfall.

8.5 UNCALIBRATED RAINFALL-DISCHARGE RELATIONSHIPS FOR SLOPING BASINS

The highest discharges from sloping land are generated by storms lasting a few hours, which combine high rain volume with high intensity. Discharge occurs mostly as overland flow although the contribution of other rapidly responsive types of shallow drainage may also be significant. High peak flows of short duration may be expected which can be very devasting if left uncontrolled. The methods discussed therefore focus on these peak discharges.

Discharges in sloping basins are more dependent on the basins size and shape than flat basins. The use of the drainage coefficient and the formula $Q = q \times A$ (eq. 8.1) are therefore less relevant to sloping basins than to flat basins.

Many methods have been developed to determine design discharge for sloping

TABLE 8.5 *Typical basin drainage coefficients*

	1 × 5 yr 24 to 48 hrs rainfall (mm)	*Slope of the land* (%)	*Drainage coefficient* ($l.sec^{-1}ha^{-1}$)
Netherlands (grassland/mixed cropping; rainfed)	31–44	0·05–0·20	1·0–1·3
Yugoslavia (grassland/mixed cropping; rainfed)	52–64	0·25–0·50	3·5
Sudan (cotton, furrow irrigation)	64–90	0·05–0·10	4·0
Japan (flooded rice)	80–115	0·05–0·20	5·0
Tanzania (sprinkler irrigated sugar cane)	145–165	0·20	7·0

basins. However, most are highly empirical in nature and consequently are of limited, local applicability. Two of the more soundly based methods, the *rational formula* method and the *curve number* method, which are both widely used, generally with satisfactory results, will be discussed.

8.5.1 Rational formula method

Discharge starts as soon as the rain has replenished the surface retension. Assuming that this occurs at the same time throughout the basin, the order in which the discharges from the various locations in the basin pass through the outlet (point B in figure 8.10) will differ due to differences in their time of travel. The longest time will apply to water travelling from the most hydraulically remote point (point A in figure

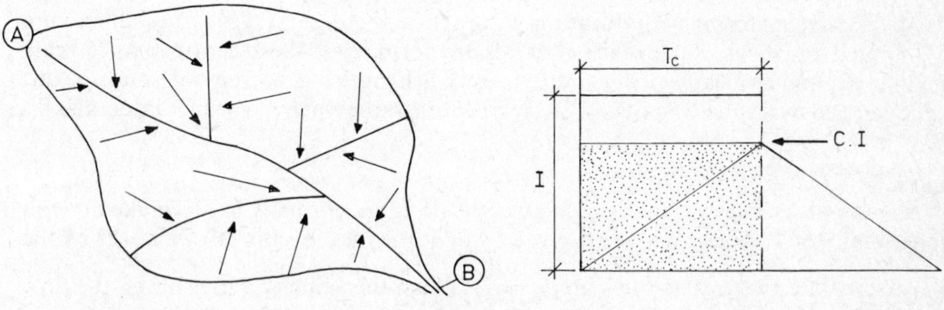

FIGURE 8.10 *Concentration of discharge in a basin*

8.10). This time is called the *time of concentration* (T_c) and it is an important hydrological characteristic of the basin. The basic principle underlying the rational formula is that the highest discharge from the basin occurs in response to a storm with a duration equal to the time of concentration, the reasoning being:

(a) when the storm duration $< T_c$, only part of the basin contributes to the total discharge at the outlet at any one time for the discharge from the area close to the outlet B will have passed through B by the time the discharge from point A arrives

(b) when the storm duration $> T_c$, there is a time at which the entire basin contributes to the discharge through B although the discharge rates will normally be less than those generated by a storm of duration T_c since rainfall intensities decrease with increasing storm duration (see figure 6.2).

This reasoning only applies to rapidly responsive types of field drainage (overland flow, rapid interflow) where very little transformation occurs in the field drainage system (the field drainage hydrograph resembles the hyetograph).

The peak discharge may be estimated by the following formula:*

$$Q_p = \frac{C.I.A}{360} \qquad \text{......(eq. 8.3)}$$

where: Q_p = peak discharge ($m^3.sec^{-1}$)
I = 'uniform' intensity of the design storm of duration T_c ($mm.hr^{-1}$)
C = discharge coefficient (proportion of rainfall discharging rapidly as shallow flow)
A = basin area (ha).

The formula was originally developed to estimate peak discharges from small urban basins, generally having a large proportion of impervious area (for which C approaches 1·0). Its extension to agricultural areas is most appropriate for basins not exceeding 100–200 ha in area (for large basins with T_c values of several hours, the assumed steady, uniform intensity of the design storm is less realistic and considerable transformation of the discharge will occur in the main system).

Design storm

As explained, storms with duration T_c generate the highest discharges and so are adopted for design. Recurrence intervals are normally of the order of 10 to 25 years (this method typically being used to establish outlet capacities of small basins controlled by a culvert, spillway, etc, warranting rather high recurrence intervals).

Time of concentration (T_c)

As observed values are rarely available, the designer normally has to make do with estimates of T_c. These may be derived by dividing the travelling distance (L) by the velocity of flow (v), according to $T_c = \frac{L}{v}$. When the velocity varies along the flow path, T_c may be calculated as $T_c = \Sigma 1 \frac{L_i}{v_w}$ where L_i and v_i respectively represent the travelling distance and the velocity of flow in the individual reaches. Distinctions may for example be made into an overland flow reach, an upper canal flow reach with predominantly small canals and moderate to fairly steep gradients, and a lower canal flow reach with large canals and rather flat gradients. The velocity of overland

* The term rational formula was coined at the time of its development to distinguish this formula from the then widely used empirical formulae.

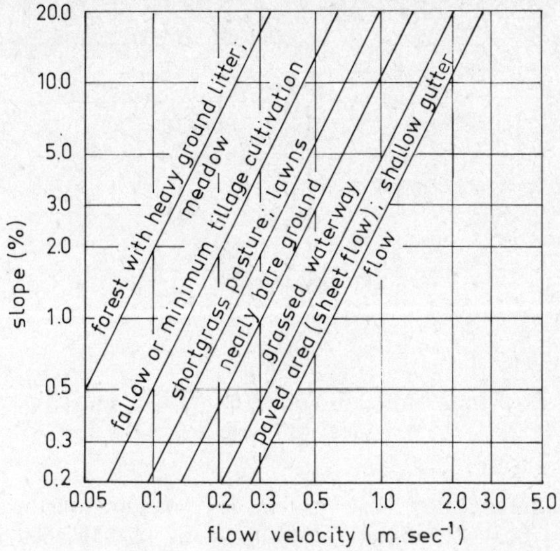

FIGURE 8.11 *Overland flow velocities for different land surfaces* (*Wanielista* 1978)

flow may be estimated using figure 8.11 while the velocity of flow in the drainage canals may be estimated using the Chézy-Manning formula (eq. 9.3).

Several formulae have also been developed relating T_c to the relevant basin characteristics such as the basin area, basin shape, slope, soil conditions, etc. The most accepted formula was developed by Kirpich, 1940 for small ($A < 50$ ha) agricultural basins:

$$T_c = \frac{L^{1 \cdot 15}}{3080\, H^{0 \cdot 38}} \quad \text{(min)} \qquad \ldots\ldots \text{(eq. 8.4)}$$

where: T_c = time of concentration (hours)
L = maximum travelling distance in the basin (m)
H = difference in elevation over the above distance (m)

Discharge coefficient (*C*)

This coefficient indicates the proportion of the design rainfall that actually discharges rapidly from the basin and which contributes to the peak discharge (see figure 8.10). Its value is of course directly dependant on the infiltration/percolation characteristics of the soil and on the retention/detention characteristics of the land/basin, as indicated in table 8.6 (in this table differences in retention/detention conditions in a basin are incorporated as differences in slope and in land use). The values also vary somewhat with the rainfall intensity and the storm duration, all in all making the determination of C the weakest part of the formula.

Example: application of the rational formula to determine the Q_{design} for the small agricultural basin illustrated with soil of medium infiltrability and (during the critical season) 40% arable and 60% pasture land.

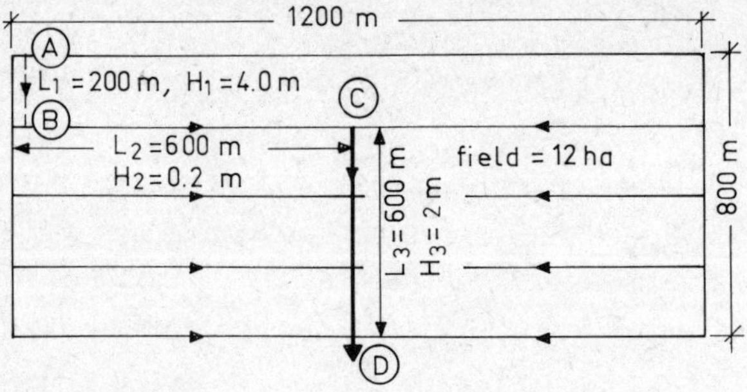

Time of concentration (T_c):

- for reach A–B: overland flow, $L_1 = 200$ m, H_1 (fall) $= 4$ m (slope is 2%); from figure 8.11 select $v = 0{\cdot}3\,m.sec^{-1}$ (pasture) — $T_c = 0{\cdot}19$ hr
- for reach B–C: channel flow, $L_2 = 600$ m, $H_2 = 0{\cdot}2$ m; use the Chézy-Manning formula with $I = 0{\cdot}2/600 = 0{\cdot}0003$, $K_m = 20$, b = 1m, d = 1m, side slope 1:1 ($v = 0{\cdot}22\,m.sec^{-1}$) — $T_c = 0{\cdot}74$ hr
- for reach C–D: channel flow, $L_3 = 600$ m, $H_3 = 2$ m; use the Chézy–Manning formula with $I = 2/600 = 0{\cdot}003$ $K_m = 20$, b = 2m, d = 1m, side slope 1 (vertical): 2 (horizontal) ($v = 0{\cdot}84\,m.sec^{-1}$) — $T_c = 0{\cdot}2$ hr

total from A to D — $T_c = 1{\cdot}13$ hrs

Design storm: 1 × 10 yrs, 1·13 hrs rainfall = 34 mm rainfall; $I = 34\,mm/1{\cdot}13\,hrs = 30\,mm.hr^{-1}$.

Discharge coefficient: $C = (40\% \times 0{\cdot}5) + (60\% \times 0{\cdot}3) = 0{\cdot}38$, using the guidelines as in Table 8.6.

Rational formula (eq. 8.3):

$$Q_{design} = Q_p = \frac{C.I.A}{360} = \frac{0{\cdot}38 \times 30 \times 96}{360} = 3{\cdot}04\,m^3.sec^{-1}$$

8.5.2 Curve number method*

This method is applicable to larger basins than the rational formula (up to several thousand ha) and it has the advantage that the complete hydrograph, including the peak discharge generated by the design storm, may be determined.

The method involves the following steps:

(a) Conversion of rainfall into discharge (mainly surface runoff) using the curve-number graph (CN-graph)

* Developed by the US Soil Conservation Service (1972)

TABLE 8.6 *Guidelines for the determination of the discharge coefficient C in the rational formula**

	Infiltrability of the soil		
	High	*Medium*	*Low*
Arable land			
slope < 5%	C = 0·30	C = 0·50	C = 0·60
5–10%	0·40	0·60	0·70
10–30%	0·50	0·70	0·80
Pasture			
slope < 5%	C = 0·10	C = 0·30	C = 0·40
5–10%	0·15	0·35	0·55
10–30%	0·20	0·40	0·60
Forest			
slope < 5%	C = 0·10	C = 0·30	C = 0·40
5–10%	0·25	0·35	0·50
10–30%	0·30	0·50	0·60

* US Soil Conservation Service 1972. Further information is presented in Annex 8.1

(b) Development of these discharges into a basin hydrograph using the SCS/USDA dimensionless unit hydrograph (Soil Conservation Service, US Department of Agriculture)
(c) Determination of the drainage coefficient (q) as the peak of the hydrograph and $Q_{design} = q \times A$, where A = basin area

Design storm

The frequency of occurrence of the design storm is normally taken to be 10–25 years for agricultural land and 30 to 100 years for urbanised basins. The hydrograph is normally developed from a standard 6 hour storm (provided $T_c < 6$ hours as is usually the case for small basins; when $T_c > 6$ hours a storm of duration equal to T_c should be used). In situations where only daily (24 hours) rainfalls are available, the 6 hour rainfall may be estimated on the basis of established 6 hours/24 hours rainfall ratios (often of the order of 0·5–0·7, see section 15.1).

The CN method also requires that the rainfall distribution within the 6 hours period be known. This information will seldom be available in which case a distribution such as in table 8.7 may be assumed to apply (the S-type distribution is rather unfavourable in the sense that it results in conservatively high discharges).

Hydrological conditions in the basin

The conditions are characterised by the following two parameters, *viz* the 'CN-value' and the 'time to peak'.

(i) *The CN-value* is estimated from table 8.10. The values read from this table are then moderated on the basis of the antecedent moisture conditions (AMC) as classified in table 8.8. The CN-values given in table 8.10 apply to AMC-class II,

TABLE 8.7 *S-type distribution curve for a 6 hour storm (WMO 1974)*

Time (hours)		*% of storm occurred*
0·0	(beginning of storm)	0
0·5		2
1·0		8
1·5	....	15
2·0		22
2·5		60
3·0		70
3·5	....	78
4·0		84
4·5		88
5·0		92
5·5		96
6·0	(end of storm)	100

TABLE 8.8 *Classification of the antecedent moisture conditions (AMC) in the basin*

AMC-class	*Total rainfall in the 5 days preceding the design storm*	
	Dormant season	*Growing season*
I	< 12·5 mm	< 35 mm
II	12·5–27·5 mm	35–52·5 mm
III	> 27·5 mm	> 52·5 mm

TABLE 8.9 *Corresponding CN-values for different AMC-classes (to be used for conversion)*

AMC-class	*Corresponding CN-values*										
I	100	78	63	51	40	31	22	15	9	4	0
II	100	90	80	70	60	50	40	30	20	10	0
III	100	96	91	85	78	70	60	50	37	22	0

the corresponding values for AMC-classes I and III being given in table 8.9. When conditions in the basin vary a composite, weighted CN-value may be determined.

Example: basin covered by small grains, on straight rows, in good hydrologic conditions with low infiltration rates, during the dormant season with 40 mm of rain recorded in the preceding 5 days

- from table 8.10: CN = 83, for AMC-class II
- from table 8.8: AMC-class III (40 mm rain)
- from table 8.9: corresponding value for AMC-class III, CN = 92·5.

(ii) *The time to peak* (T_p) = time lapse between the beginning of the storm and the occurrence of the peak flow (see figure 8.12). The T_p-value is a hydrological

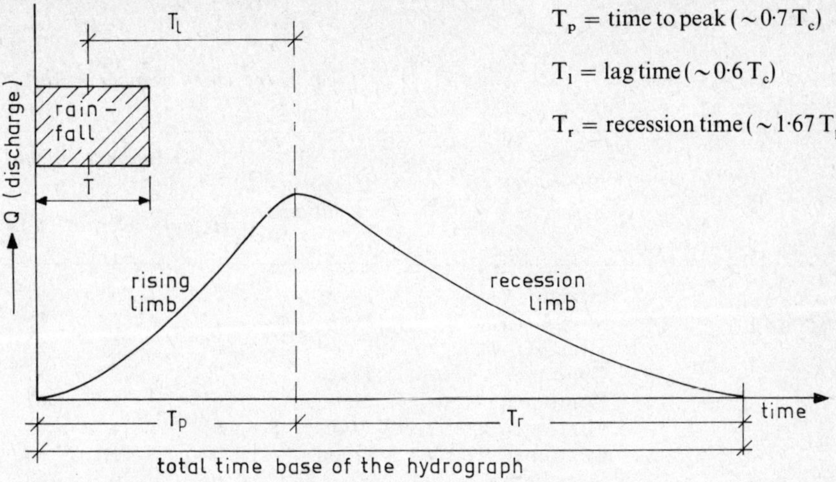

FIGURE 8.12 *Illustration of the various time parameters in a hydrograph*

characteristic of the basin. It should preferably be determined on the basis of rainfall and stream flow records but it may also be estimated as $T_p = 0{\cdot}7 \times T_c$ (time of concentration, to be estimated as outlined in section 8.5.1).*

Calculation procedure (see table 8.11)

The application of the curve-number method is illustrated step by step for the following case:

- design storm 60 mm falling in 6 hours distributed as in table 8.7 (1 × 10 year storm)
- time to peak T_p = 2 hours (observed)
- CN-value = 87

Runoff calculations (lines 1 to 5 in table 8.11)

- line 1: time (T) since beginning of the storm; the time interval (ΔT) should not exceed 0·25 T_p, therefore in this case ΔT = 0·5 hours
- line 2: design storm, cumulative in time (R_t)
- line 3: R_t from line 2 converted into runoff (Q_t) using figure 8.13 (CN = 87)
- line 4: depth of runoff (ΔQ) per time interval, calculated as $\Delta Q = Q_{t+1} - Q_t$
- line 5: peak discharge (q_p) generated by each ΔQ, calculated using the formula**: $q_p = 2{\cdot}08 \dfrac{\Delta Q}{T_p}$ where q_p is in $l.sec^{-1}.ha^{-1}$, ΔQ in mm, T_p in hours so that in this case $q_p = 1{\cdot}04 \times \Delta Q$

* The relationship $T_p = 0{\cdot}7 \times T_c$ is established empirically. Note that the rational formula assumes that $T_p = T_c$ (see figure 8.10)

**Numerical integration of the unit hydrograph of figure 8.14, using the given tabular data, gives $\Delta Q = 1{\cdot}34\, q_p\, t_p$ with ΔQ in mm, q_p in $mm.day^{-1}$ and T_p in hours. With q_p in $l.sec^{-1}.ha^{-1}$ this gives $q_p = 2{\cdot}08\, \Delta Q/T_p$

TABLE 8.10 *CN-values for AMC class II to be used in the curve number method (Soil Conservation Service 1972)*

Land use or cover	*Treatment or practice*[1]	*Hydrologic conditions*[2]	*Infiltration rate*			
			High	*Moderate*	*Low*	*Very low*
Fallow	Straight rows	Poor	77	86	91	94
Row crops	Straight rows	Poor	72	81	88	91
	Straight rows	Good	67	78	85	89
	Contoured	Poor	70	79	81	88
	Contoured	Good	65	75	82	86
	Contoured & terraced	Poor	66	74	80	82
	Contoured & terraced	Good	62	71	78	81
Small grains	Straight rows	Poor	65	76	84	88
	Straight rows	Good	63	75	83	87
	Contoured	Poor	63	74	82	85
	Contoured	Good	61	73	81	84
	Contoured & terraced	Poor	61	72	79	82
	Contoured & terraced	Good	59	70	78	81
Close-seeded	Straight rows	Poor	66	77	85	89
legumes or	Straight rows	Good	58	72	81	85
rotational	Contoured	Poor	64	75	83	85
meadow	Contoured	Good	55	69	78	83
	Contoured & terraced	Poor	63	73	80	83
	Contoured & terraced	Good	51	67	76	80
Pasture, range		Poor	68	79	86	89
		Fair	49	69	79	84
		Good	39	61	74	80
	Contoured	Poor	47	67	81	88
	Contoured	Fair	25	59	75	83
	Contoured	Good	6	35	70	79
Meadow (permanent)		Good	30	58	71	78
Woodland (farm woodlots)		Poor	45	66	77	83
		Fair	36	60	73	79
		Good	25	55	70	77
Commercial areas (85% impervious)			89	92	94	95
Industrial areas (72% impervious)			81	88	91	93
Residential, $\frac{1}{8}$ acre lots (65% impervious)			77	85	90	92
Residential, $\frac{1}{2}$ acre lots (25% impervious)			54	70	80	85
Roads with curbs and storm sewers			98	98	98	98
Roads, gravel or paved without storm sewers			74	84	90	92
Roads, dirt			72	82	87	89

[1] Straight rows = rows running with the slope of the land
[2] Primarily depends on density of the vegetation
For grassland: poor = heavily grazed, poor stand
good = dense cover, ungrazed
For row crops: poor = early growth stages, poor stand
good = closed cover, good stand.

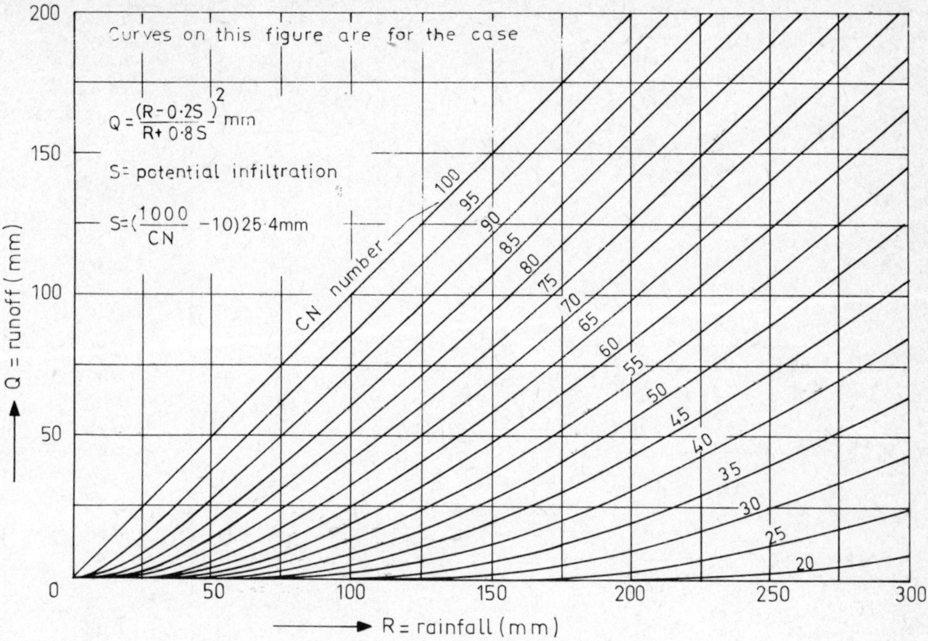

FIGURE 8.13 *Curve number graph for the conversion of rainfall into runoff*

Hydrograph construction (lines 6 to 37 in table 8.11)

- column 1 : time scale (T)
- column 2 : T from column (1) divided by T_p
- column 3 : q/q_p corresponding to each T/T_p value according to figure 8.14
- column 4–12: hydrographs for each time interval producing runoff; the discharge rate (q) is calculated as $q = (q/q_p).q_p$, so the q-values in column (4) are arrived at by multiplying the q/q_p values in column (3) by 1·04, the q-values in column (5) by multiplication by a factor 11·44, etc.

  Each hydrograph is scaled in on the time scale (column 1) by noticing that the peak discharge (q_p) occurs T_p hours after the beginning of the rain (= beginning of the time interval), so for the time interval running from T = 1·5 to T = 2·0 the peak discharge occurs at T = 1·5 + 2·0 = 3·5 hours.
- column 13 : the hydrograph of the design storm, arrived at by superposition (horizontal addition of the q-values in columns 4 to 12).

The peak discharge of the hydrograph of the design storm is taken as the drainage coefficient, so in this case q = 19·7 l.sec^{-1}, ha^{-1}. The $Q_{design} = q \times A \times 10^{-3}$ with Q in m^3.sec^{-1} and A in ha.

8.6 COMPOSED DISCHARGES

Most main systems will receive groundwater as well as water from shallow drainage but their contributions are likely to be quite different in magnitude and occur at

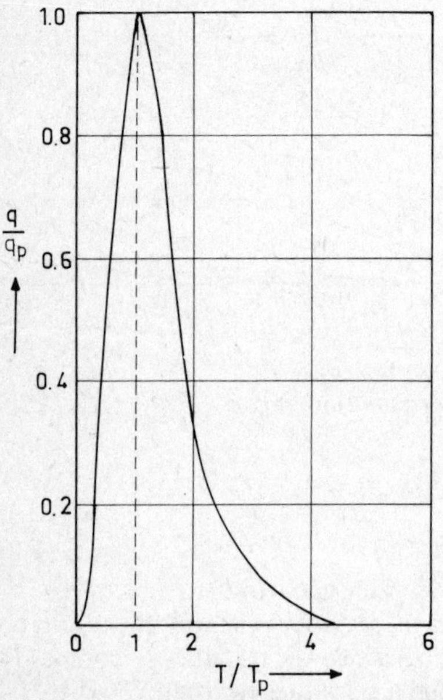

(1) *Time ratio* T/T_p	(2) *Discharge ratio* q/q_p
0	0
0·1	0·015
0·2	0·075
0·3	0·16
0·4	0·28
0·5	0·43
0·6	0·60
0·7	0·77
0·8	0·89
0·9	0·97
1·0	1·00
1·1	0·98
1·2	0·92
1·3	0·84
1·4	0·75
1·5	0·66
1·6	0·56
1·8	0·42
2·0	0·32
2·2	0·24
2·4	0·18
2·6	0·13
2·8	0·098
3·0	0·075
3·5	0·036
4·0	0·018
4·5	0·009
5·0	0·004
Infinity	0

FIGURE 8.14 *The SCS/USDA dimensionless unit hydrograph*

different times. The design discharge should be based on the highest discharge. In sloping basins, the discharge due to shallow drainage (especially the overland flow) is almost always several times higher than the concurrently occurring groundwater drainage discharge (the peak of which comes later) so that the contribution of the latter may well be neglected in establishing design discharges. For flat basins in humid climates this may be different. In these areas the concurrently occurring discharges due to shallow drainage (mostly interflow) and groundwater drainage differ less and an approach taking their joint contribution into account may have to be adopted.

The problem is further complicated by differences in discharge between the different parts of the basin for the same type of drainage flow. In the Netherlands, for example, basins are often divided into two parts, one with rapidly responsive

TABLE 8.11 *Calculations based on the Curve-Number method*

	Design storm													
(1)	T (hours)	0	0·5	1·0	1·5	2·0	2·5	3·0	3·5	4·0	4·5	5·0	5·5	6·0
(2)	R_t (mm)	0	1	5	9	13	36	42	47	50	53	55	58	60
	Run off													
(3)	Q_t (mm)	0	0	0	0	1	12	17	20	22	24	26	28	30
(4)	ΔQ (mm)					1	11	5	3	2	2	2	2	2
(5)	q_p (l.sec^{-1}.ha^{-1})					1·04	11·44	5·20	3·12	2·08	2·08	2·08	2·08	2·08

	(1)	(2) Dimensionless T/T_p	(3) hydrograph q/q_p	(4)	(5)	(6)	(7)	(8)	(9)	(10)	(11)	(12)	(13) *Hydrograph of the design storm*
				Hydrographs per time unit (q in l.sec^{-1}.ha^{-1})									
(6)	T = 0 hrs	0	0										q = 0·00 l.sec^{-1}.ha^{-1}
(7)	0·5	0·25	0·12										0·00
(8)	1·0	0·50	0·43										0·00
(9)	1·5	0·75	0·83										0·00
(10)	2·0	1·00	1·00	·12									0·12
(11)	2·5	1·25	0·88	·45	1·37								1·82
(12)	3·0	1·50	0·66	·86	4·92	·62							6·40
(13)	3·5	1·75	0·45	1·04	9·50	2·24	·37						13·15
(14)	4·0	2·00	0·32	·92	11·44	4·32	1·34	·25					18·27
(15)	4·5	2·25	0·22	·69	10·07	5·20	2·59	·89	·25				19·69
(16)	5·0	2·50	0·15	·47	7·55	4·58	3·12	1·73	·89	·25			18·59
(17)	5·5	2·75	0·11	·33	5·15	3·43	2·75	2·08	1·73	·89	·25		16·61
(18)	6·0	3·00	0·08	·23	3·66	2·34	2·06	1·83	2·08	1·73	·89	·25	15·07
(19)	6·5	3·25	0·05	·16	2·52	1·66	1·40	1·37	1·83	2·08	1·73	·89	13·64
(20)	7·0	3·50	0·04	·11	1·72	1·14	1·00	·94	1·37	1·83	2·08	1·73	11·92
(21)	7·5	3·75	0·03	·08	1·26	·78	·69	·67	·94	1·37	1·83	2·08	9·70
(22)	8·0	4·00	0·02	·05	·92	·57	·47	·46	·67	·94	1·37	1·83	7·28
(23)	8·5	4·25	0·01	·04	·57	·42	·34	·31	·46	·67	·94	1·37	5·12
(24)	9·0	4·50	0·01	·03	·45	·26	·25	·23	·31	·46	·67	·94	3·60
(25)	9·5	4·75	0·01	·02	·34	·21	·16	·17	·23	·31	·46	·67	2·57
(26)	10·0	5·00	0·00	·01	·23	·16	·12	·10	·17	·23	·31	·46	1·79
(27)	10·5	5·25		·01	·11	·10	·09	·08	·10	·17	·23	·31	1·20
(28)	11·0	5·50		·01	·11	·05	·06	·06	·08	·10	·17	·23	0·87
(29)	11·5	5·75			·11	·05	·03	·04	·06	·08	·10	·17	0·64
(30)	12·0	6·00				·05	·03	·02	·04	·06	·08	·10	0·38
(31)							·03	·02	·02	·04	·06	·08	0·25
(32)								·02	·02	·02	·04	·06	0·16
(33)									·02	·02	·02	·04	0·10
(34)										·02	·02	·02	0·06
(35)											·02	·02	0·04
(36)												·02	0·02
(37)													0·00

groundwater drainage and one with slowly responsive groundwater drainage. The rapidly responsive part typically comprises the lower areas of the basin where the groundwater levels are always at shallow depth (within 1–2 m from the surface). The slowly responsive part comprises the naturally better drained areas of land at a higher elevation in which groundwater levels fall to considerable depths during dry periods. Much of the rainfall on these areas of high land will be stored in the soil above the watertable while the groundwater recharge will be delayed. Moreover, as the groundwater drains are likely to be widely spaced (little need to improve the natural drainage), the outflow will be even more gradual and delayed, similar in character to base flow. In contrast, in the lower part the drain outflow will increase soon after the start of the rain, reaching its peak within one to two days, and then recede rather rapidly. Design discharges for basins should obviously be based mostly on the discharge from the rapidly responsive part of the basin, the contribution of the other part being taken into account as a base flow only or even being neglected.

There is very little authoritative guidance as to how composed design discharges should be determined. Land conditions may be analysed to form some idea of the relative contributions of the different types of drainage discharge, noting such characteristics as the slope of the land, infiltrability, possibilities for retention/storage of water on the surface and in the soil, groundwater depths, KD-values and other characteristics of the substrata, etc. For example, in the case described for the Netherlands, it was found that the two parts could be delineated on the basis of the (available) information on the annual groundwater regime. Special consideration should also be given to the discharge from concentrations of impervious surfaces within the basin (built-up areas, glasshouse complexes, etc). The runoff from such surfaces is virtually 100% of rainfall and occurs rapidly, increasing the peak discharge of the relevant main drains very considerably (see also section 8.8).

8.7 Area reduction factors

As discussed, the drainage coefficient (q) in the general formula Q = q.A (eq. 8.1 where Q = basin discharge and A basin area), should not in general be expected to be independant of the area A. This is partly due to the discharge transformation occurring in the main system, the effect of which increases as the size of the basin increases, generally resulting in lower values for q with increasing area A (see figure 8.1). Another factor is that the rainfall distribution in large basins is unlikely to be uniform (especially not for the short duration, convective type of storms forming the design basis for shallow drainage which are typically of a highly scattered, localised nature). Also, with increasing times of concentration in larger basins, the average rainfall intensity decreases. The overall effect is that the average basin rainfall depth and intensity decrease with increasing basin area and a similar trend should apply to the drainage coefficients.

Sloping basins

The formula Q = q.A and the underlying concept of the drainage coefficient are most relevant to flat basins. In sloping basins the influence of the area is taken into account partly implicitely in the different empirical relationships and factors (emphasising once again that these should only be used for the range of basin areas for which they were established) and partly explicitely in the time of concentration which takes into account the decrease of rainfall intensity with increasing storm duration (although not the effects of the scattered occurrence of storms). These comments mean that the effects of area have, with the exception of the scattered nature of the short duration storms used for design, been accounted for in the development of the formulae. This variation of rainfall depth across the area is really only of relevance for the curve number method when used for large basins. It can be taken into account either by using a known average basin rainfall based upon sufficient point measurements or by using ratios such as in table 8.12 between the point rainfall and the average basin rainfall.

TABLE 8.12 *Ratios between the average basin rainfall and the point rainfall for basins of different sizes (for 6 hour storms, based on observations in the USA)*

Basin area	*Basin rainfall/point rainfall*
100 ha	1·00
5 000 ha	0·98
10 000 ha	0·94
25 000 ha	0·99
50 000 ha	0·85

Flat basins

For flat basins these area reduction effects are often taken into account by using the empirical formula:

$$Q = q_o . A^{\alpha} \qquad \text{......(eq. 8.5)}$$

where: Q = discharge capaity ($m^3.sec^{-1}$)
q_o = drainage coefficient ($m^3.sec^{-1}.km^{-2}$)
A = basin area (km^2; $A > 1\ km^2$)
α = area reduction factor (usually of the order of 0·80–0·95).

This formula shows that the reduction is applied to the area A rather than to the drainage coefficient (not physically sensible but mathematically of course equally effective). The amount by which the area needs to be reduced depends of course on the nature of the drainage coefficient q_o. When this represents shallow field drainage discharge (as is the case with values determined by the method outlined in section 8.4.2 (i), a large reduction is applicable to include the effects of both the discharge transformation in the main system and the rainfall depth variations. When q_o has been determined as in section 8.4.2 (ii), most of the discharge transformation effects have already been taken into account in adjustments to the time of travel although a further correction due to the rainfall variations might still be in order. The nature of q_o in the formula is not always clear although it may generally be assumed to apply to field drainage discharge.

A well known example of this type of formula is the *Cypress Creek formula*, reading $Q = q_o A^{5/6}$ with Q in cusec, A in sq miles and q_o in cusec per sq miles. This formula was derived in the USA for basins <5 000 ha, slopes <0·5% and with a mixed shallow/groundwater drainage discharge (at high discharges mostly the former type). The value of q_o may be estimated as: $q_o = 16{\cdot}4 + 14{\cdot}75\ R_e$ where R_e is the excess rain to be discharged (in inches per hour). For normal farm crops, the 48 hours, 1 × 2 to 5 years design storm is used with the excess to be discharged within the 48 hours storm duration period (high value crops take the 1 × 10 year, 24 hours storm, to be discharged in 24 to 36 hours). The excess rainfall may be determined by means of the CN-graph (figure 8.13). For a 100 mm (4 in.) design storm and CN = 90, the excess is 75 mm = 3 in., so $R_e = 3/48 = 0{\cdot}06\,\text{in.hr}^{-1}$, $q_o = 17{\cdot}3$ cusecs/sq mile = $1{\cdot}9\,\text{l.sec}^{-1}.\text{ha}^{-1}$ for A = 1 sq mile = 259 ha.

No area reduction factors are used in the Netherlands with its mostly groundwater drainage and low intensity frontal type of rainfall, for areas up to 10 000 ha while for areas of the order of 100 000 ha a reduction of only some 10% is used.

8.8 Discharge reduction through storage

Flood storage involves the temporary impounding of drainage discharge in order to reduce downstream peak flows. This might provide a more economical solution to the prevention of flooding than the alternative of providing for a large disposal capacity in the canals and structures, especially when this is liable to involve pumping.

Flood storage may be particularly suitable for agricultural basins faced with a growth in urban areas. The increased proportion of impervious surface and the installation of gutters and storm water sewers in these areas, increases the surface runoff while reducing the time of concentration, leading to high urban discharges into the main system (see figure 8.15).

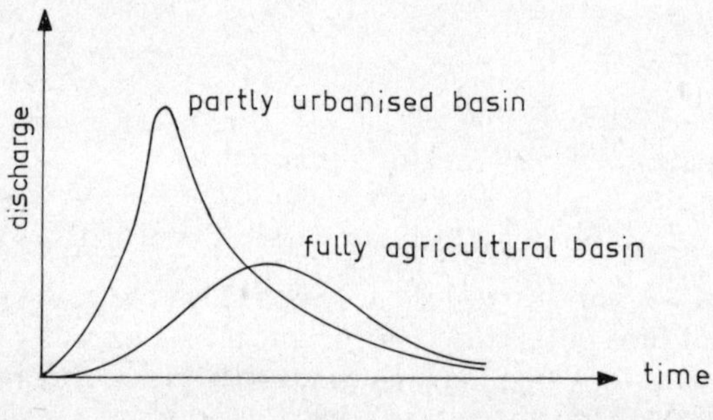

FIGURE 8.15 *Characteristic hydrograph from an agricultural and a partly urbanised basin*

Flood storage in a drainage basin may take different forms. It may be localised in the form of *retention reservoirs* or be spread out over (part of) the basin as *canal storage*. Canal storage may involve either allowing temporarily higher water levels in the canals or the provision of larger canal cross-sections than needed for transport alone. Canal storage involves backed-up canal flow and is generally only applicable in flat basins (typically applied in polders to reduce pumping capacities).

8.8.1 Retention reservoirs

These reservoirs may be distinguished into the off-stream type and the on-stream type. The characteristic differences between the two types is illustrated in figure 8.16, showing that the choice between them depends mostly on the prevailing local conditions.

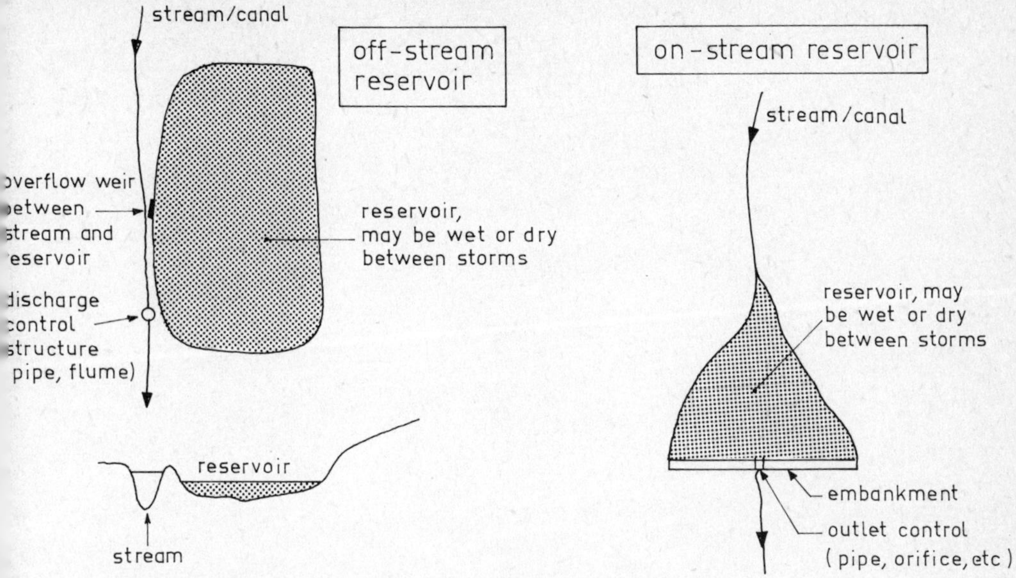

FIGURE 8.16 *Types of retention reservoirs* (see HALL and HOCKIN 1980)

The off-stream reservoir is suitable for flat basins. It operates by diverting all the discharge above a permitted value into the reservoir. The discharge which is allowed to pass downstream is based on a sensible downstream canal capacity and is controlled by an orifice, pipe or flume, the latter having the advantage that it will allow debris to pass down the stream. The stored water can normally drain back to the main stream, once the storm has passed, via a pipe with a non-return flap valve at its end.

The on-stream reservoir is usually constructed in valley situations by building an embankment across the valley. The stream flow is controlled by the outlet from the reservoir and the inflow hydrograph is attenuated by the storage provided by the reservoir.

Both types of reservoir may contain water at all times to provide shallow lakes or they may be allowed to dry out between flood seasons to provide grassed areas.

Evaluation of flood risks

A flood risk analysis is carried out to establish the extent and frequency of flooding with and without the reservoir. The level of sophistication of such an analysis should be related to the importance and consequences of the basins development. In this discussion only the principles are illustrated, using the relatively simple procedures applicable to small basins (upstream basin <200 ha).

In the rare case that a suitable series of discharge records exist, they may be analysed statistically, enabling the effect of recent growth in urban areas to be established and extrapolations to be made. In the more common case with no (or no long term) discharge records available, a discharge series may be generated from a rainfall series, using one of the methods discussed in sections 8.3, 8.4 or 8.5, distinguishing between the situation before and after. The unit hydrograph, the

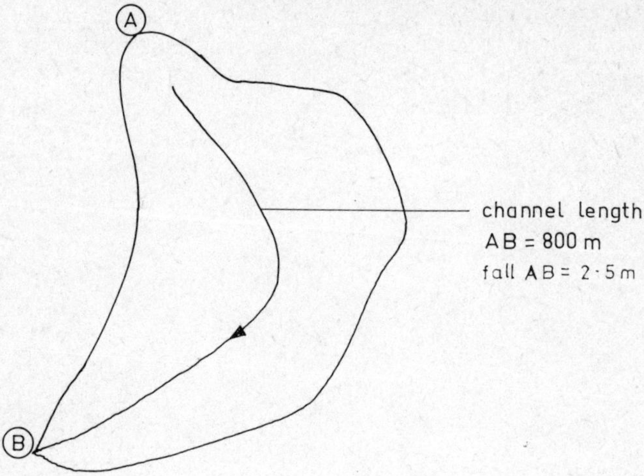

FIGURE 8.17 *Small rural basin of 87 ha*

rational formula and the curve number methods are most suitable. The procedure to be followed is illustrated, using the rational formula, for a small rural basin of 87 ha, located in Southern England, which is to be developed as a light industrial estate (see figure 8.17).

The required information is:

(a) Time of concentration (T_c):

- rural basin: $T_c = \dfrac{L^{1 \cdot 15}}{3080\, H^{0 \cdot 38}} = 0{\cdot}5$ hours (using eq. 8.4)
- light industrial basin: gutters, sewers, etc, reduce the time of concentration by an estimated 0·2 hours so $T_c = 0{\cdot}3$ hours

(b) Runoff coefficients (C-values):

- rural basin: pasture, low infiltrability, slope $<5\%$ for which conditions (according to table 8.6) $C = 0{\cdot}4$, assumed to rise to a max. of $C = 0{\cdot}7$ at a return period > 10 years.
- Light industrial basin: $C = 0{\cdot}8$ (Annex 8.1)

(c) Rainfall intensity (i) during various T_c periods at frequencies of occurrence of between 1 and 100 years (given in table 8.13).

The development of the area as an industrial estate leads to higher discharges at all return periods due to the increased runoff co-efficients and the shortened time of concentration. The extent and damage of the flooding can be assessed provided the gradient and cross sectional areas of the disposal streams or canals are known as well as the topography and land use (see section 6.5).

Required reservoir volume

The required storage volume depends on the maximum permitted (safe) downstream discharge and upon the total discharge volume produced by the design storm. For a fixed return period of for example $T = 20$ years, the discharge volume varies directly

TABLE 8.13 *Calculation of peak discharges on the basis of the rational formula*

Return period years	*Rainfall intensity (mm.hr^{-1})* $T_c = 0{\cdot}5$ *hr*	$T_c = 0{\cdot}3$ *hr*	*C-value Rural*	*Industrial*	*Peak discharge (m^3.sec^{-1}) Rural* $T_c = 0{\cdot}5$ *hr*	*Industrial* $T_c = 0{\cdot}3$ *hr*
1	21	26	0·4	0·8	2·0	5·0
2	25	33	0·4	0·8	2·4	6·4
5	37	52	0·4	0·8	3·6	10·1
10	42	60	0·4	0·8	4·1	11·6
20	50	70	0·5	0·8	6·0	13·5
40	59	84	0·6	0·8	8·6	16·2
100	76	110	0·7	0·8	12·9	21·3

with the storms duration. The storage provided should protect areas downstream against storms of specified return periods, for agricultural areas usually taken as 15 to 25 years and for residential/industrial areas as 35 to 100 years. The safe downstream discharge may either be taken as the downstream canal/stream capacity allowing for 0·3–0·4 m freeboard, or be taken equal to the peak discharge occurring at a frequency of one of twice a year before development of the area.

The principles of the estimation of the required storage are illustrated in figure 8.18. For the case of the off-stream reservoir, the maximum controlled downstream discharge is kept virtually constant (see figure 8.18a). Upstream discharges above this level are simply diverted into the reservoir. For the case of the on-stream reservoir in figure 8.18b, the downstream discharge increases with the rising water level in the reservoir. The required storage in the two cases is equal to the hatched areas of the hydrographs. These simple principles enable the identification, as a first step, of the approximate storage volume required and also the critical storm duration. This may suffice for small reservoirs (farm ponds, etc) but more elaborate analyses are required for larger reservoirs, examples of which are presented below.

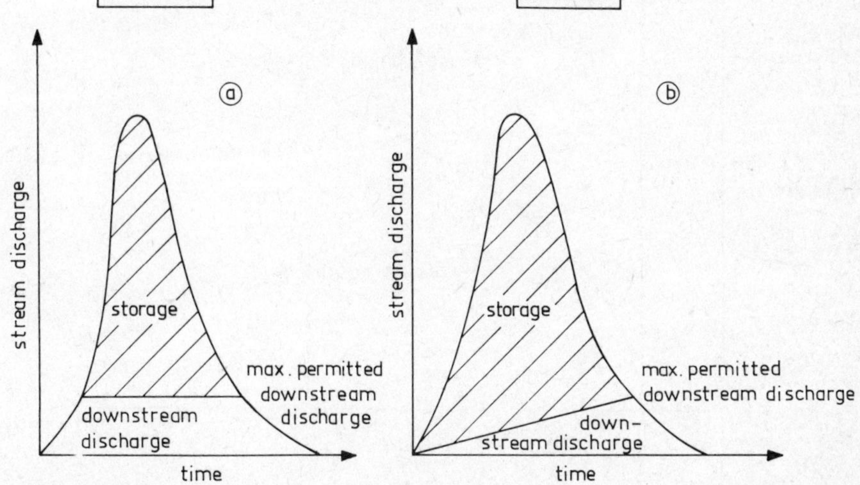

FIGURE 8.18 *Principles of storage estimation for off-stream and on-stream retention reservoirs*

Storage calculations for an off-stream reservoir

The critical storage required for an off-stream reservoir may be calculated fairly simply by evaluating the discharge from storms of variable duration as demonstrated below for the case previously discussed of the light industrial basin of 87 ha. Using the curve number method discharge volumes are calculated as in table 8.14 for the selected return period of 20 years. The basin is assumed to be in moisture condition III at the time of the critical event and is further assumed to be composed of 56·6 ha (65% of the basin) of industrial estate area with CN = 96 and 30·4 ha (35%) of estate area with CN = 82.

TABLE 8.14 *Estimation of discharge volume on the basis of the curve number method*

Storm duration (hrs)	*Storm depth (mm)*	*Rainfall intensity* ($mm.hr^{-1}$)	*Discharge volumes* (m^3)		
			56·6 ha *CN* = 96	*30·4 ha* *CN* = 82	*87 ha* *Total*
0·1	15	150	4004	76	4080
0·2	19	95	5875	290	6165
0·5	25	50	8858	840	9698
1·0	30	30	11439	1450	12889
1·5	34·5	23	13816	2090	15906
2·0	38	19	15684	2651	18335

The maximum permitted downstream discharge is 2·0 $m^3.sec^{-1}$, equal to the 1 × 1 year peak discharge before the development of the area.

Further analysis proceeds as in figure 8.19. The discharge volumes for the various

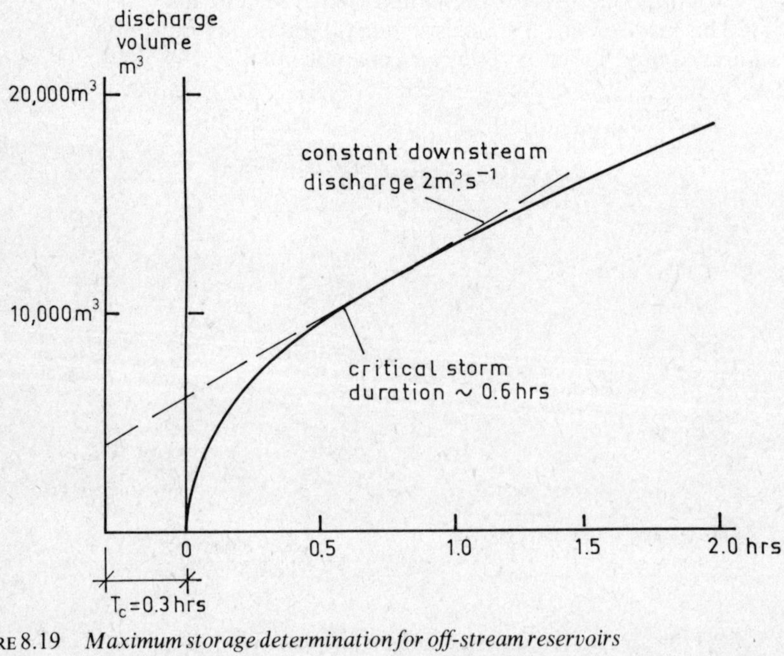

FIGURE 8.19 *Maximum storage determination for off-stream reservoirs*

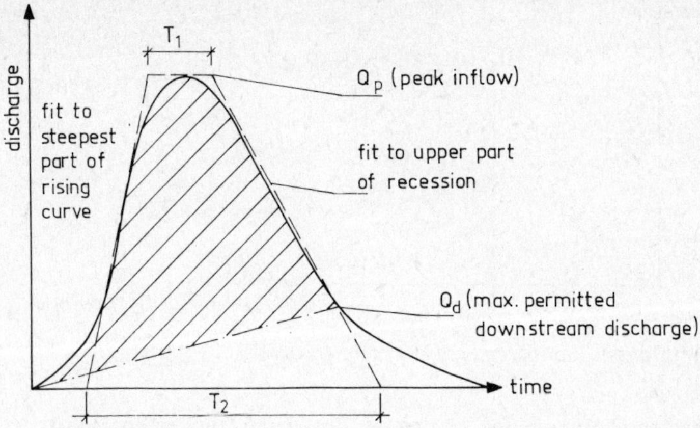

FIGURE 8.20 *An approximate method to estimate reservoir capacity of on-stream reservoirs*

storm durations as calculated in table 8.14 are plotted, forming a curved line. Tangent to this curve, a line is drawn with a slope equal to the permitted downstream discharge of 2·0 m^3.sec^{-1}. The origin of this line is set back by the time of concentration (T_c = 0·3 hr) as the permitted downstream discharge will be reached within minutes of the commencement of the storm whereas the total volume of discharge from the same storm will only be accumulated over a period of at least T_c + storm duration. The required storage can now be read from the set-back vertical axis, in this case amounting to ~4 000 m^3.

Storage calculations for an on-stream reservoir

Once again the required storage depends on the storm duration and may be evaluated by analysing the consequences of storms of different durations (range from T_c to 5 T_c) at the selected return period (eg 20 years).

The basic simplifying assumptions for one approximate, though often sufficiently accurate, procedure for estimating the required reservoir capacity are sketched out in figure 8.20. The inflow hydrograph is approximated by a trapezium and the rising limb of the outflow hydrograph by a straight line (outflow is assumed to increase linearly in time) up to the permitted peak discharge (eg taken equal to the downstream canal/stream capacity with 0·3–0·4 m freeboard). The required storage (hatched area of the inflow hydrograph in figure 8.20) may be approximated as:

$$S\,(\text{storage}) = Q_p\left(\frac{T_1 + T_2}{2}\right) - \frac{Q_d\,T_2}{2}$$

the various parameters being indicated in figure 8.20. This should be repeated using hydrographs of storms of various durations, and the storage in each case should be plotted against the respective storm duration as in figure 8.21 to enable the critical storm duration and the corresponding maximum storage required to be identified.

Spillways

The downstream discharge for an off-stream reservoir is dependant on the head of

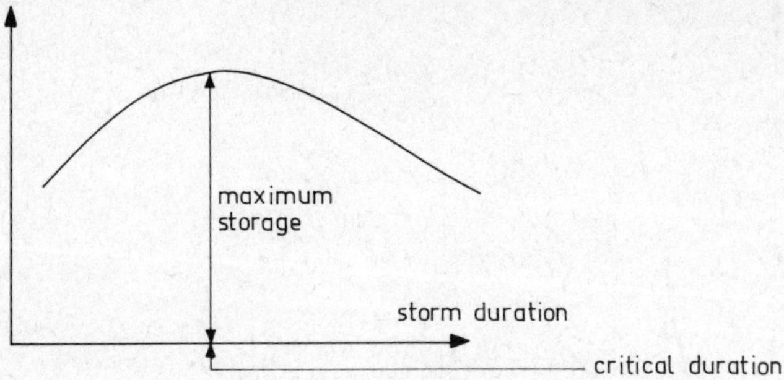

FIGURE 8.21 *Optimisation of required reservoir storage*

water above the control pipe or orifice. This head may be limited to a maximum value by installing a long side-weir some distance up-stream of the control which serves to divert all discharges above the permitted maximum into the reservoir. This side-weir should be of sufficient length to cope with all peak discharges up to, for example, a 1×20 year peak without causing the head just above the control to increase appreciably. Its sill should actually be set so as to begin diverting flow before the maximum downstream discharge is attained thus ensuring that the reservoir will function once or twice per annum. On rare occasions a particularly severe storm or series of storms might occur which could jeopardise the reservoir and the downstream control structure. Provision should therefore be made to lead these very high discharges through the reservoir and back to the channel below the control structure. The simplest way to achieve this is to construct a section of the reservoir embankment at a lower level. This section which should be located as near as possible to the downstream channel to which water will flow should be grassed to prevent it being eroded (*green spillway*).

8.8.2 Canal storage

For the flat basins in which canal storage can be considered, the following simple water balance applies:

$$I - O = \Delta S \text{ (change in storage S)} = \frac{A_w}{A} . \Delta d \qquad \text{......(eq. 8.6)}$$

where: A = basin area
A_w = canal surface area (or more generally 'openwater surface area')
Δd = permitted (temporary) rise of canal level in the design situation (rise above the HW-line as defined in section 9.1.2 or, in the case of a polder, rise above the polder level)
I = inflow into the canal system (outflow from field drain)
O = outflow from the canal system (outlet through sluices, pumps, etc).

Limits are imposed on the area A_w (costly large canals) and Δd (damaging to the land use). A compromise should be sought between the open water area, permitted rise in open water level and the outlet capacity for the basin. One extreme is to provide a lot of open water area in the basin and permit considerable rise of water level, in which case the outlet capacity can be low; the other is to have a minimal A_w

(canals only for water transport) and permit no rise in water level, in which case a large outlet capacity is required.

Storage calculations

The relevant considerations and calculation procedures are outlined below:

(a) *Inflow:* the field drain discharge into the main system may be estimated as outlined in section 8.4.1 (groundwater drainage) and section 8.4.2 (shallow drainage). For groundwater drainage, this may be done on the basis of the non-steady state drainage formula (eq. 7.18). Alternatively, simple rules of thumb may be used (the $\frac{1}{3}$-$\frac{1}{3}$-$\frac{1}{3}$ rule for groundwater drainage; the $\frac{1}{2}$-$\frac{1}{2}$ rule for shallow drainage). The influence of discharge transformation can generally be neglected when the water balance (eq. 8.6) is applied to daily periods. In the example elaborated in table 8.15, the design storm and drain discharge are the same as those in table 8.2 determined by the rule of thumb (groundwater drainage).

(b) *Pump outlet:* In table 8.15 three different cases are considered for each of which the required maximum storage is calculated:

	Pump capacity	*Pump start*	*Maximum storage*
Case 1	7 mm.day^{-1}	day 1	9 mm
2	7 mm.day^{-1}	day 2	16 mm
3	10 mm.day^{-1}	day 2	10 mm

In this example the maximum storage requirement is estimated for a known mode of pump operation. Alternatively, the procedure can be used to establish the required pumping capacity for a predetermined maximum storage.

TABLE 8.15 *Calculation of canal storage requirements for the case of a pumped outlet*

	Day 1	*Day* 2	*Day* 3	*Day* 4	*Day* 5	*Day* 6
– Design rainfall (mm)	20	10	8			
– Drain discharge (mm): day 1 rain	7	7	6			
day 2 rain		4	3	3		
day 3 rain			3	3	2	
daily inflow	7	11	12	6	2	
cumulative inflow	7	18	30	36	38	
– Outflow by pumping (capacity = P)						
(1) P = 7 mm.day^{-1}; start on day 1						
– cumulative outflow	7	14	21	28	35	
– to be stored	0	4	9	8	3	
(2) P = 7 mm.day^{-1}; start on day 2						
– cumulative outflow	0	7	14	21	28	35
– to be stored	7	11	16	15	10	3
(3) P = 10 mm.day^{-1}; start on day 2						
– cumulative outflow	0	10	20	30	40	
– to be stored	7	8	10	6	0	

Storage requirements may also be estimated by graphical means as shown in figure 8.22. The inflow line corresponds to the cumulative drain inflow as calculated in table 8.15. The pumping line is straight and has a slope equivalent to the pumping capacity (in this case P = 7 mm.day^{-1}) whilst its point of origin coincides with the point in time that pumping commences (in this case the beginning of day 1). At any moment in time the vertical separation between the pumping line and the inflow line corresponds to the storage requirements at that time. By adjusting the origin (start of pumping) and the slope (pump capacity) of the pumping line, a suitable combination of mode of pump operation, pump capacity and required storage may be selected, furthermore the duration of the period with extra high canal levels may be taken into account.

If the maximum storage is known, its value may be plotted as the ordinate at the start of pumping, and the required pumping capacity may then be determined from the slope of the tangent line to the inflow curve (see figure 8.22).

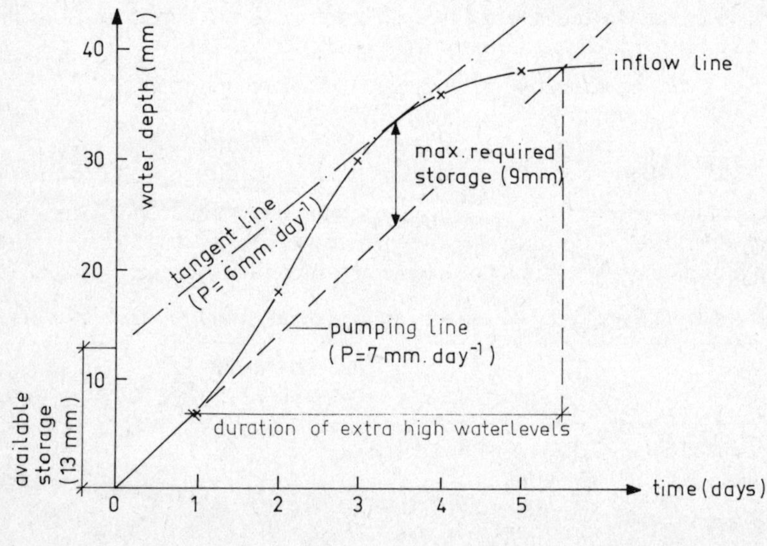

FIGURE 8.22 *Graphical determination of storage requirements for the case of a pumped outlet*

(c) *Tidal sluice outlet:* in this case water is only able to discharge for part of the day (during the tide free period only, roughly lasting half a day, see figure 9.11).

The discharge through the sluice is directly related to the inner water level just upstream of the sluice, which is itself related to the current canal storage (S) in the basin. Storage and discharge normally begin to increase in parallel with the onset of rain, rising ultimately to peak values, after which they begin to decline. This is shown in figure 8.23. The sluice line is shown as a broken line, being horizontal during tide locked periods and sloping during tide free periods. The slope of the sluice line (= rate of discharge through the sluice) is related to the current storage (slope steepens when storage increases the latter corresponding to the vertical separation between the sluice and the inflow line). This illustrates the principles of the storage

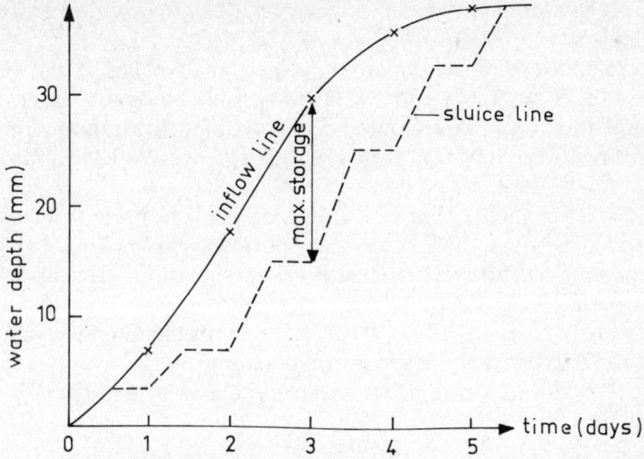

FIGURE 8.23 *Graphical determination of storage requirements for the case of a tidal sluice outlet*

requirement analysis although actual determinations are best done by tabular calculations as in table 8.15 with the slight difference that in this case half-day intervals should be used and also the current storage (S) should be determined (using the balance eq. 8.6).

References and further reading

ASCE 1969: 'Design and construction of sanitary and storm sewers', Manuals and Reports on Engineering Practices no 37, American Society of Civil Engineers

BUREAU OF RECLAMATION 1974: *Design of Small Dams*, US Department of Interior Washington DC

CHOW, VEN T (Ed.) 1964: *Handbook of Applied Hydrology*, McGraw-Hill, New York

DUNNE, T and LB LEOPOLD 1978: *Water in Environmental Engineering*, Freeman, San Francisco

FAO 1972: 'Farm water management seminar', Irrigation and Drainage Paper no 12, Food and Agriculture Organisation of the United Nations, Rome

VAN DER GOOR, GAW 1974: 'Drainage of rice fields', In: *Drainage Principles and Applications*, volume IV, publication 16, International Institute for Land Reclamation and Improvement, Wageningen

GRAY, DM (Ed.) 1970: *Handbook of the Principles of Hydrology*, Canadian Committee for the Int. Hydrol. Decade, WRB Canada, Ottowa

HALL, MJ and DL HOCKIN 1980: 'Guide to the design of storage ponds for flood control in partly urbanised catchment areas', Technical Note 100, Construction Industry Research and Information Centre, London

DE JAGER, AW 1965: *Hoge afvoeren van enige Nederlands stroomgebieden*, Centrum vor Landbouwpublicaties en Documentatie, PUDOC, Wageningen (in Dutch)

KRAIJENHOFF VAN DER LEUR, DA 1973: 'Rainfall-runoff relations and computational models'. In: *Drainage Principles and Applications*, volume II, publication 16, International Institute for Land Reclamation and Improvement, Wageningen

POIREE, M et C OLLIER 1978: *Assainissement Agricole*, Eyrolles, Paris (in French)

RAKOCZI, L and K SZESZTAY 1968: 'Surface water hydrology', Lecture Notes of the International Post-Graduate Course on Hydrological Methods for Developing Water Resources Management, Research Institute for Water Resources Development, VITUKI, Budapest

ROCHE, M 1963: *Hydrologie de Surface*, Gauthiers-Villars, Paris (in French)

RODIER, J A and C AUVRAY 1965: 'Flood computations in West Africa', Proceedings of the Symposium of Budapest, International Association of Scientific Hydrology, publication no 66 (1): 12–38

SOIL CONSERVATION SERVICE 1972: *Hydrology*, National Engineering Handbook, Section 4, US Department of Agriculture, Washington DC

WANIELISTA, M 1978: *Stormwater Management, Quantity and Quality*, Ann Arbor Science, Michigan

WILSON, E M 1974: *Engineering Hydrology*, MacMillan, London

WMO 1974: 'Guide to hydrological practices', Technical Publication no 168, 3rd edn. World Meteorological Organisation, Geneva

DE ZEEUW, J W 1973: 'Hydrograph analysis for areas with mainly groundwater runoff'. In: Drainage Principles and Applications, vol. II, publication 16, International Institute for Land Reclamation and Improvement, Wageningen

Annex 8.1 Guidelines for the determination of the discharge coefficient C in the rational formula

1 *General* (Gray 1970)

$C = 1 - \Sigma C'$

Topography		*Soil conditions*		*Cover*	
– Flat (0·02–0·06%)	$C' = 0·3$	Tight clay	$C' = 0·1$	Cultivated	$C' = 0·1$
– Rolling (0·3–0·4%)	$C' = 0·2$	Clay loam	$C' = 0·2$	Woodland	$C' = 0·2$
– Hilly (3–5%)	$C' = 0·1$	Sandy	$C' = 0·4$		

Example: rolling, cultivated land with clay loam soils, $C = 1-(0·2+0·2+0·1) = 0·5$

2 *African conditions* (OSTROM, see Rodier and Auvray 1965)

Land use or cover	*Average slope of the land*	*Soil conditions* permeable	moderately permeable	impermeable
Bare soil or only sparsely covered by vegetation; eroded land	< 5%	0·50	0·60	0·72
	5–10%	0·60	0·72	0·86
	>10%	0·72	0·83	0·95
Crop land, open scrub and woodland	< 5%	0·42	0·50	0·60
	5–10%	0·50	0·60	0·72
	>10%	0·60	0·72	0·86
Grassland, dense scrub land, wooded savanna and dense bush land	< 5%	0·25	0·30	0·36
	5–10%	0·30	0·36	0·43
	>10%	0·36	0·43	0·52
Mature forest, dense bush with subgrowth	< 5%	0·15	0·18	0·22
	5–10%	0·18	0·22	0·26
	>10%	0·22	0·26	0·31
Primary tropical forest	< 5%	0·13	0·15	0·18
	5–10%	0·15	0·18	0·22
	>10%	0·18	0·22	0·26

Note: The given C values apply to basins of 100 ha; somewhat higher values should be taken for smaller basins and somewhat lower values for larger basins (for example when $C = 0·60$ for a 100 ha basin, the value would be about 0·65 for 25 ha basin and 0·55 for 500 ha basin)

3 *Urban areas* (ASCE 1969)

Residential area: $C = 0·30–0·50$ (single family homes)
$C = 0·50–0·75$ (apartments)
Industrial area: $C = 0·50–0·80$ (light) to $C = 0·60–0·90$ (heavy)
Pavements: $C = 0·70–0·95$

Chapter 9 Hydraulic design of drainage canals and associated structures

The design of a drainage canal system may be divided into:

(a) selection of type and layout of the system (type and alignment of the canals, type and location of the structures, etc)
(b) determination of the hydraulic dimensions of the different components of the system (*hydraulic design*)

The various considerations underlying the choice of type and layout of a drainage canal system have already been discussed in chapter 5. Now the main aspects of the hydraulic design are given. In general, this part of the design is fairly straightforward, based on well established hydraulic engineering principles and practices. For a more elaborate treatment reference is made to the relevant handbooks.

9.1 Drainage canals

The hydraulic design of drainage canals may be based in almost all cases on *steady uniform flow* (discharge is constant in time and also constant along the length of the canal section considered). Under these conditions, the following formulae apply:

$$v = C\,R^{\frac{1}{2}}\,I^{\frac{1}{2}} \qquad \text{(Chézy 1769)......(eq. 9.1)}$$

$$C = K_m\,R^{\frac{1}{6}} \qquad \text{(Manning 1889)......(eq. 9.2)}$$

$$v = K_m\,R^{\frac{2}{3}}\,I^{\frac{1}{2}} \qquad \text{......(eq. 9.3.A)}$$

$$Q = v\,A = K_m\,R^{\frac{2}{3}}\,I^{\frac{1}{2}}\,A \qquad \text{......(eq. 9.3.B)}$$

where:
v = flow velocity ($m.sec^{-1}$)
Q = discharge rate ($m^3.sec^{-1}$)
R = hydraulic radius (m)
I = hydraulic gradient ($m.m^{-1}$)
A = wet cross-section (m^2)
C = Chézy roughness coefficient ($m^{\frac{1}{2}}.sec^{-1}$)
K_m = Manning roughness coefficient ($m^{\frac{1}{3}}.sec^{-1}$)

Eq. 9.3 will be referred to as the Chézy-Manning formula (although it is more commonly known as the Manning formula). The meaning of the different parameters is illustrated in figure 9.1.

Given the discharge (Q) to be carried by the canal and having established the hydraulic gradient (I) and some characteristics of the canal (K_m, x and b/d ratio), the main dimensions of the required canal can be calculated with the Chézy-Manning formula as in the following example (see also flow chart figure 9.2).

Given:

$Q = 3{\cdot}5\ m^3.sec^{-1}$ $\qquad K_m = 30$

$I = 0{\cdot}02\%$ $\qquad x = 1\frac{1}{2}$

$b/d = 3$

Calculations:

$$A = d.\tfrac{1}{2}(2b+2dx) = d\,(b+dx)$$

$$P = b+2\sqrt{d^2+(dx)^2} = b+2d\sqrt{1+x^2}$$

$$R = \frac{A}{P} = \frac{d\,(b+dx)}{b+2d\sqrt{1+x^2}} = \frac{b+dx}{b/d+2\sqrt{1+x^2}}$$

for $x = 1\tfrac{1}{2}$ and $b/d = 3$: $A = d\,(b+dx) = d(3d+1\tfrac{1}{2}d) = 4{\cdot}5\,d^2$

$$R = \frac{b+dx}{b/d+2\sqrt{1+x^2}} = \frac{3d+1\tfrac{1}{2}d}{3+2\sqrt{3{\cdot}25}} = 0{\cdot}68\,d$$

$$Q = K_m\,R^{\frac{2}{3}}\,I^{\frac{1}{2}}\,A$$

$3{\cdot}5 = 30\,(0{\cdot}68\,d)^{\frac{2}{3}}\,(0{\cdot}0002)^{\frac{1}{2}}\,(4{\cdot}5\,d^2) \rightarrow d = 1{\cdot}40\,\text{m}; b = 4{\cdot}2\,\text{m}$ (rounded)
Check: $v = Q/A = 3{\cdot}5/8{\cdot}82 = 0{\cdot}40\,\text{m.sec}^{-1}$

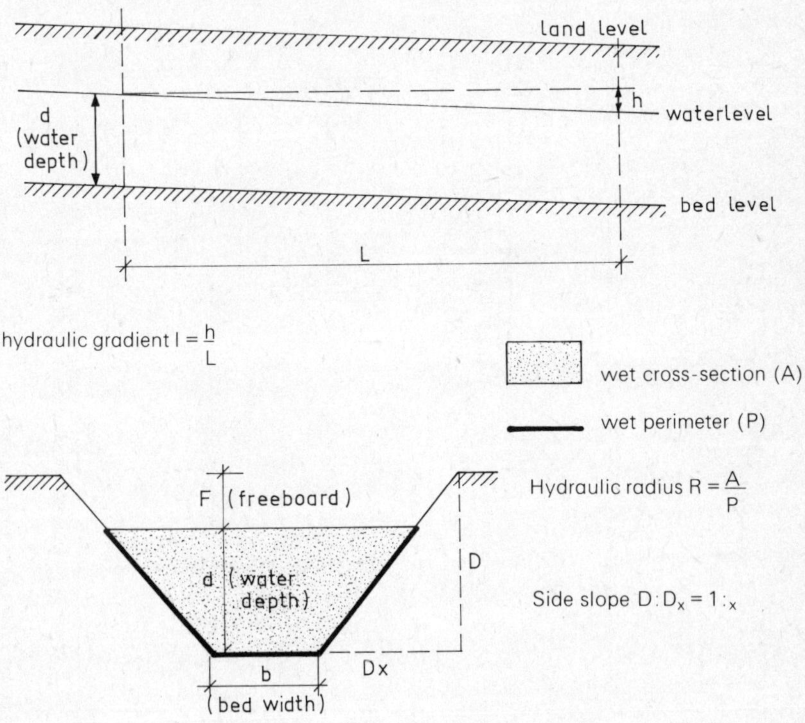

FIGURE 9.1 *Main parameters in the Chézy-Manning formula for canal design*

Use may be made of a small electronic calculator. The Chézy-Manning formula, however, may also be conveniently solved using nomographs, an example of which is presented in figure 9.3. The determination of the different input parameters of the Chézy-Manning formula are discussed in the following sections.

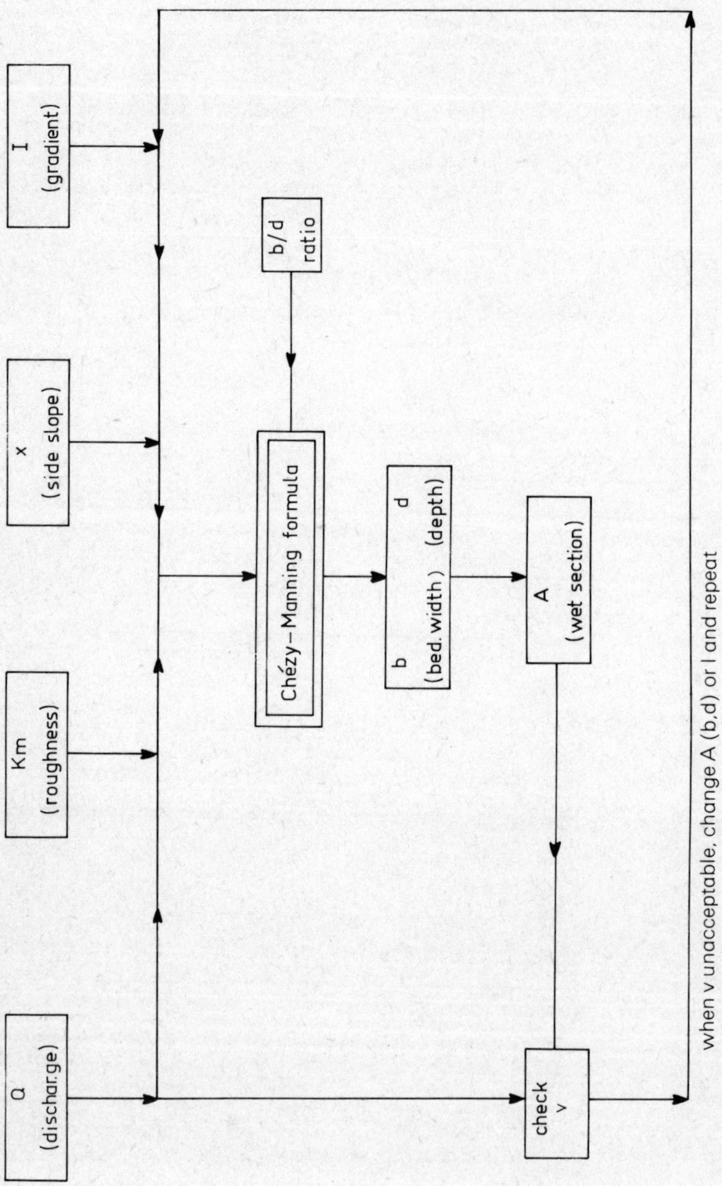

FIGURE 9.2 *Flow chart for canal design*

Given: $Q = 3{\cdot}5\,m^3.sec^{-1}$ $K_m = 30$

$I = 0{\cdot}02\% = 0{\cdot}2\,m.km^{-1}$ side slope $1{:}1\frac{1}{2}$

Find: bed width (b) and water depth (d) of the required canal

Procedure: connect $Q = 3{\cdot}5\,m^3.sec^{-1}$ on the Q-scale with $I = 0{\cdot}2\,m.km^{-1}$ on the I-scale (for $K_m = 30$). Extend this line until it connects with the b-d graph and enter this graph horizontally. On this horizontal $R^{\frac{2}{3}}$. A is constant and of the desired value, so all b, d combinations on this line fulfil the Chézy-Manning equation. The combination of $d = 1{\cdot}40\,m$ and $b = 4{\cdot}15\,m$ is selected in this case as it gives a b/d ratio closest to 3, which is the desired ratio for a canal of this depth (see section 9.1.3). The wet cross-section of the canal may be determined by following the direction of the dashed lines from the b-d point to the A-scale (A is constant along these dashed lines). This leads in this case to $A = 8{\cdot}8\,m^2$. The flow velocity may then be readily calculated as $v = Q/A = 3{\cdot}5/8{\cdot}8 = 0{\cdot}40\,m.sec^{-1}$. When the velocity exceeds the permissible limits as given in Tables 9.1 or 9.2, the hydraulic gradient should be reduced so as to arrive at a velocity within the permitted range.

Note: this nomograph is for canals with side slope $1{:}1\frac{1}{2}$. Similar graphs may be prepared for other side slopes (three types with side slopes 1:1, $1{:}1\frac{1}{2}$ and 1:2 will cover most requirements). Such graphs are commercially available as an album to the reference ILRI 1964.

9.1.1 Discharge rate

The discharge rate (Q) is normally taken equal to the design discharge (Q_{design}) when the Chézy-Manning formula is used to design new canals. The determination of design discharges is extensively discussed in chapter 8. The methods indicated establish the range, but the final selection and judgement should be left to the designer. Somewhat lower or higher values than those calculated may be taken, depending on how the costs, risks and damage are evaluated in each particular case. For example, design discharges for structures are often taken some 25 to 50% higher than those for canals, the argument being that cross-sections in structures are generally more constrained than in the canals and underdesign may lead to unacceptably high flow velocities and backing-up of the flow. Also, it is generally more costly to increase the discharge capacity of a structure than that of a canal once the system is constructed and the assumed design discharge proves to be too low.

9.1.2 Hydraulic gradient and water levels

The hydraulic gradient should be neither too small nor too large. When the gradient is small, a large cross-section (A) is required to carry the flow (high excavation costs). Values should generally not be less than 5–$10\,cm.km^{-1}$ ($0{\cdot}05$–$0{\cdot}10\%$) but in flat land lower values may have to be used as the total available head between field and outlet is small and will partly be expended in unavoidable head losses in the structures in the canal system (culverts, bridges, etc).

High values for I, on the other hand, lead to high flow velocities in the canals with the attendant danger of scouring of beds and erosion of banks. Permissible flow velocities are discussed in section 9.1.3. Drop structures may be used to dissipate surplus head where canals run down steep slopes (eg in the case of the collector ditch in figure 4.6).

FIGURE 9.3 *Nomographs for the design of trapezoidal canals based on the Chézy-Manning formula*

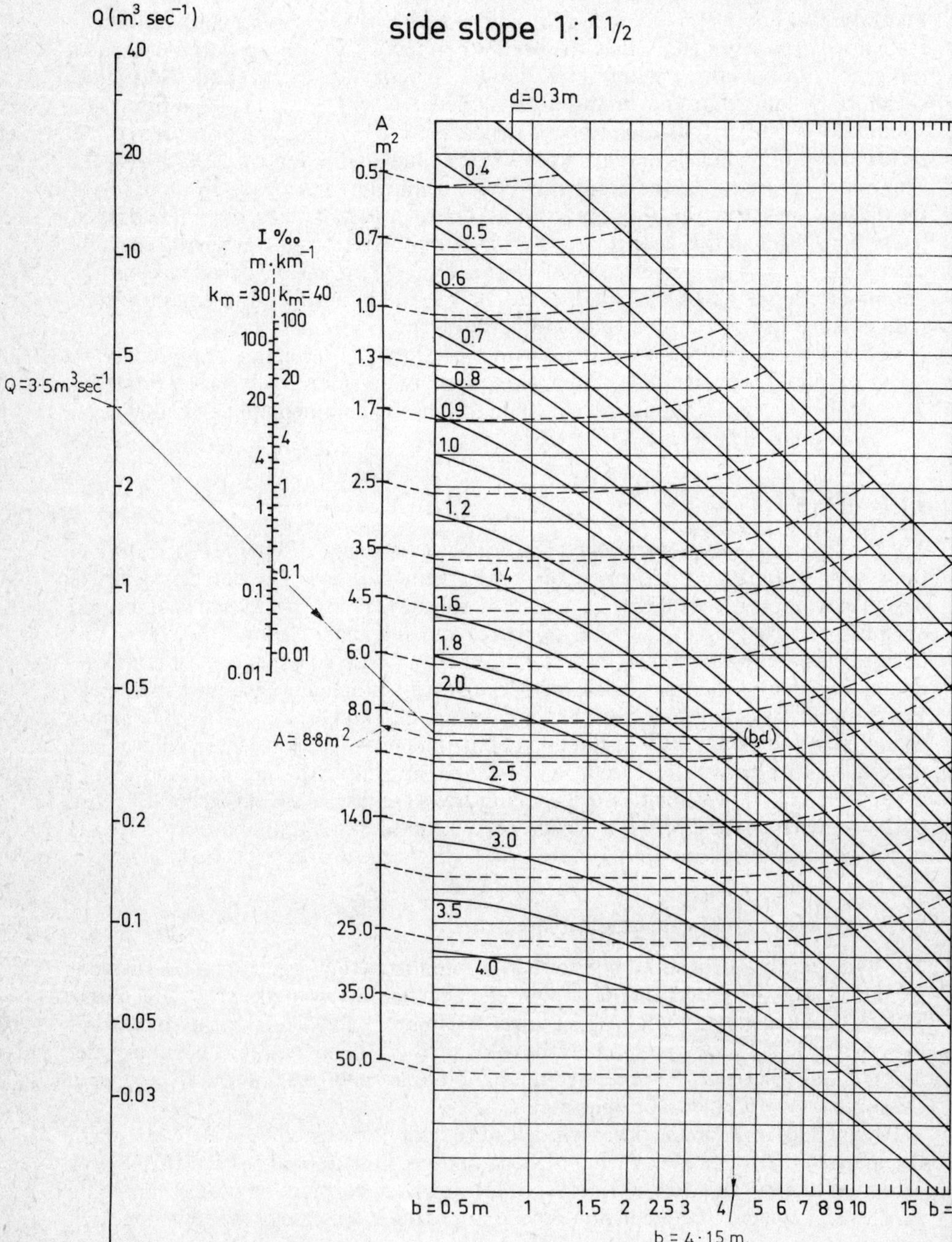

In practice, hydraulic grade lines often closely follow the slope of the land in the direction of discharge. Also, in the case of steady uniform flow, the land slope, the hydraulic grade line and the bed slope will all be parallel. The *velocity head* ($v^2/2g$ with $g = 9{\cdot}8$ m.sec^{-2}) in drainage canals is normally negligible so that the hydraulic grade line and the water surface line may be assumed to coincide. In these circumstances, the water depth in the canal section under consideration will be constant and the water surface will be at a constant depth below the land surface (this situation is depicted in figure 9.1).

Water levels

The water level in drainage canals must meet certain freeboard requirements (see section 9.1.6) and in the case of lower order canals serving as collectors, the water level must in principle remain 5–10 cm below the field drainage base to permit unimpeded outflow from the field drains. For deep groundwater drains, a temporary impedance in outflow is generally not harmful and the hydraulic grade line at the design discharge can be allowed to run above the outlet levels of such drains.* Collector and main canals are in fact commonly designed on the premise that there should be unimpeded outflow from the field drains for the much more frequently occuring event of $Q = 0{\cdot}5\ Q_{design}$. Under normal conditions a discharge of this magnitude may on average be expected to occur several times per year whereas Q_{design} would occur only once per year or even less frequently.

The hydraulic grade line at $Q = Q_{design}$ is designated as the *HW-line* (high water level) while the grade line at $Q = 0{\cdot}5\ Q_{design}$ is designated as the *NW-line* (normal water level). The NW-line may be determined with the Chézy-Manning formula assuming a flow equal to $0{\cdot}5\ Q_{design}$ passing through the canals designed to carry Q_{design}. For many drainage canals, however, water depth roughly varies with $\sqrt{Q}$.** So, a canal which fills to a depth 'd' while carrying flow Q, will fill to 0·7 d when the flow is reduced to 0·5 Q. This rule may be applied to approximately locate the NW-line in a canal, knowing the HW-line.

Longitudinal sections, showing the slope and elevations of the land along the canal alignment, are useful in establishing grade lines. Various water level requirements to be met at various places can be marked on the section (field drainage bases, freeboard requirements, levels of existing structures, etc). Such sections are almost indispensable in fixing grade lines of canals running through sloping, irregular land. An example is given in Annex 9.1.

9.1.3 Permissible flow velocities

For turbulent flow as normally encountered in drainage canals, the tractive force exerted by the flowing water on the bed material is proportional to the square of the average flow velocity (v^2). When the tractive force exceeds those forces retaining the bed material, the constituent particles will begin to move with the water causing the bed to erode.

* High water levels in the collectors raise the field drainage base. In the case of pipe drains, the water will rise above the pipe up into the trench. The pipes will coninue to discharge, although at a reduced rate, as long as there is head, ie as long as the watertable in the land is above the water level in the collector.

**This rule holds for small to medium sized, trapezoidal canals with side slopes close to 1 : 1½, see example of calculation by hand of canal dimensions in section 9.1.

The movement of non-cohesive (single grain) bed material is resisted by friction which increases with the weight of the particle. Safe, non-erosive flow velocity limits may be established upon the basis of the particle size distribution of the bed material. For cohesive material, encountered in most drainage canals, there is no straightforward relationship between erosivity and size of the bed material and limits are based mostly upon experience. Permissible flow velocities for drainage canals as used in the Netherlands are given in table 9.2. The well known limits set by Fortier and Scobey (1926) are presented in table 9.1.

TABLE 9.1 *Maximum permissible mean flow velocities in unprotected canals (in m.sec^{-1})*

Bed material	*Clear water*	*With suspended load*
Fine sand	0·45	0·75
Sandy loam, silty loam, silt	0·50 – 0·60	0·75 – 1·00
Loam	0·70 – 0·80	1·00 – 1·10
Clay	1·10 – 1·20	1·50
Gravel (fine→coarse)	0·75→1·20	1·50→1·80

Note: adapted from Fortier and Scobey, 1926; values apply to mature, straight canals with small grades

The erosion hazard is always most severe in newly constructed canals. Higher velocities can generally be permitted once the canals have matured and the bed has become vegetated. Higher velocities can also be permitted when the canals are large and deep since in these the actual velocities along the bed are considerably lower than the mean. Erosion hazards are increased by curves in the canal (permissible flow velocities to be reduced by 25–40% when protection is not to be provided).

9.1.4 Cross-section

Steep side slopes are desirable in that they save on excavation as well as on land loss. However, they can only be used in cohesive, well aggregated soils or where bank protection measures are adopted. Unprotected side slopes in unstable soils should generally not be steeper than 1:2 to avoid serious caving-in. Some general guidelines for design are presented in table 9.2.

TABLE 9.2 *Limitations on flow velocity and on side slope in drainage canals* (adapted from ILRI 1964)

Soil type	*Maximum permissible mean flow velocity (in m.sec^{-1})*	*Maximum permissible side slope*
Fine sand	0·15–0·30	1:2 to 3
Coarse sand	0·20–0·50	1:1½ to 3
Loam	0·30–0·60	1:1½ to 2
Heavy clay	0·60–0·80	1:1 to 2

Note: highest velocities and steepest side slopes apply to well vegetated canals.

Side slopes generally become more stable once they have become vegetated but design should mostly be based on the initial period after excavation. Factors other than soil type should also be considered. Seepage inflow into the canal for example, promotes caving in, and gentle side slopes (or even toe protection) are to be recommended in situations where this occurs. It is also sound practice to adopt gentler side slopes with increasing canal depth and for canals with widely fluctuating water levels. Special conditions, eg a road running along the canal or the provision of drinking access for cattle, may also dictate gentler side slopes.

Canals with a semi-circular wet cross-section are hydraulically the most efficient (minimum excavation per unit wet cross-section). Other considerations (construction, maintenance, etc), however, also exert an influence so that, all in all, canals with a trapezoidal cross-section have been found to be most suitable for drainage canals. As a compromise between hydraulic efficiency and other considerations, the following ratios between bed width (b) and water depth (d) should approximately be adhered to:

	Recommended b/d ratios for trapezoidal canals
Small canals ($d < 0{\cdot}75$ m)	$b/d = 1$ (clay)–2 (sand)
Medium canals ($d = 0{\cdot}75$–$1{\cdot}5$ m)	$b/d = 2$–3
Large canals ($d > 1{\cdot}5$ m)	$b/d > 3$–4

For reasons of construction and maintenance, canals should not be made too small. Depending on the method of construction, minimum sections should have bed widths of 0·50–1·00 m, side slopes of $1{:}\frac{3}{4}$, 1:1 or $1{:}1\frac{1}{2}$ and water depths of 0·30–0·50 m. Minimum sections typically apply to ditches bordering fields, carrying flows up to $0{\cdot}5\ m^3.sec^{-1}$. These ditches are often constructed without grade (no fall in bed line).

9.1.5 Roughness coefficient*

The roughness coefficient K_m should be understood as the constant of proportionality between the average flow velocity (v) and the combination $R^{\frac{2}{3}}\ I^{\frac{1}{2}}$ (see eq. 9.3). As eq. 9.2 shows, K_m not only depends on the bed roughness (as its name implies), but also on the shape of the canal (expressed by the parameter R). The bed roughness, however, is the most important influencing factor so that K_m depends primarily on the type of bed material and bed vegetation. Other factors are the canal alignment (number and smoothness of curves) and the size of the canal (K_m is somewhat lower for small canals). Values vary from as high as $K_m = 70$–80 for a good sized concrete lined canal to as low as $K_m = 10$ for a small heavily overgrown ditch.

The value of K_m used for design is of considerable importance as canal capacities and excavation costs depend directly on the value taken. In spite of this, its estimation can seldom be done objectively as there is usually little assurance as to how well the canals will be maintained. Most design is based on the assumption of fair maintenance, to which the general guidelines of table 9.3 apply. Medium sized, earthen drainage canals are commonly designed with $K_m = 30$–40, the values which are incorporated in the nomograph of figure 9.3.

* In some countries, notably in the USA and the UK, the roughness coefficient K_m is not normally used; instead roughness is expressed by the coefficient n, where $n = 1/K_m$

TABLE 9.3 *Guidelines for the selection of K_m values for the design of drainage canals* (adapted from ILRI 1964)*

Canal description	*K_m-value*	*(n-value)*
(a) Small canals (waterdepth <0·75 m)		
– sandy soil	20	(0·050)
– clayey soil	15	(0·065)
(b) Medium canals (waterdepth 0·75–1·50 m)		
– sandy soil	30	(0·035)
– clayey soil	20	(0·050)
(c) Large canals (waterdepth > 1·50 m)	40–50	(0·020–0·025)
(d) Concrete lined canals	60–80	(0·0125–0·017)

Note: these guidelines apply to canals with their wet cross-section well covered by a weedy or grassy vegetation, kept reasonably short by fair maintenance (the kind of conditions on which the designer generally bases his calculations). Lower values should be taken when poor maintenance is anticipated and when the canal system is composed of short lengths with variable cross-sections and sharp curves/junctions. Higher values may be used for well maintained, continuously full running canals (limits growth of plants).

9.1.6 Freeboard

Freeboard is meant to provide a margin of safety in case the actual discharge exceeds the designed discharge capacity of the canal (Q_{design}). As such, the provision of freeboard conforms to the common engineering design practice of introducing a safety factor in the design. It follows that less freeboard needs to be provided for the higher design discharges associated with higher return periods.

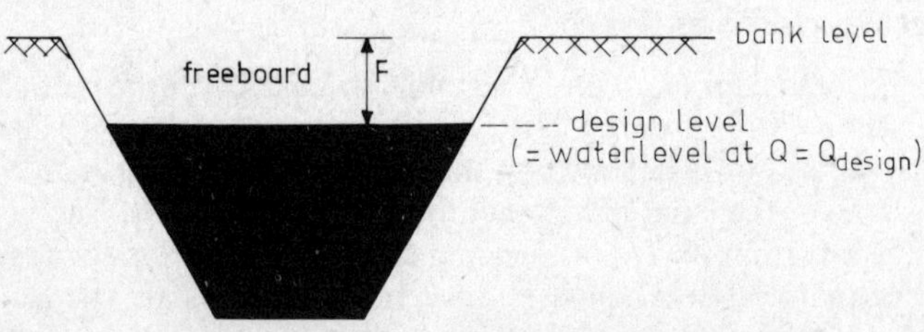

FIGURE 9.4 *Freeboard in drainage canals*

The freeboard required is often established on the basis of experience, typical guidelines being:

- small canals ($Q < 1–2\ m^3.sec^{-1}$).... freeboard F = 30 cm
- large canals ($Q > 5\ m^3.sec^{-1}$)....... freeboard F = 50 cm

A more rational approach towards defining freeboard requirements is to specify the *bankfull capacity* of the canal. In situations where the Q_{design} is taken for example

* This publication also gives pictures of canals with different K_m – values

as the 1×5 year event, the bankfull capacity could be based on the 1×25 year frequency. Alternatively, freeboard may be chosen to provide adequate over-capacity for the canal to carry $2 \times Q_{design}$ without over-topping of banks. For a Q_{design} based on a 1×5 year frequency, $Q = 2 \times Q_{design}$ may well represent the 1×25–50 year event. The water depth at $Q = 2 \times Q_{design}$ would be roughly $1{\cdot}4 \times$ the water depth of the canal when carrying Q_{design} (so the freeboard for a canal with a waterdepth of 125 cm would be some 50 cm).

The concept of freeboard is most applicable to shallow drainage where the actual discharge can be a multiple of the Q_{design}. Groundwater discharge is very much steadier and flooding risks are low. Water levels in groundwater drainage canals are in the first instance determined by the field drainage base level, and are in fact often deeper than required simply on the basis of freeboard considerations (for guidelines see section 9.1.2).

9.2 STRUCTURES IN DRAINAGE CANALS

Common structures in drainage canals are culverts and bridges. Drop structures are used where gradients are too steep. Weirs may be installed to retain water in the canals during dry periods. At the outlet point, sluices, gates or pumps may be encountered.

9.2.1 Culverts and bridges

The flow through culverts and bridges may in most drainage design work be calculated with the simple formula:

$$Q = \mu A \sqrt{2gh} \qquad \text{......(eq. 9.4)}$$

where: Q = discharge rate ($m^3.sec^{-1}$)
A = wet cross-section (m^2)
g = gravitational acceleration ($= 9{\cdot}8\ m.sec^{-2}$)
h = headloss (m)
μ = coefficient

Given Q and having set a value for h, the wet cross-section (A) can be calculated. The permitted head loss depends very much upon the total available head in the canal section in which the structure is located. When this head is limited, culverts and bridges are usually designed for only 5 cm (small structures) to 10 cm (large structures) head loss.

The value of μ depends mainly on the prevailing entry/exit conditions for the water into/out of the structure. High values apply when the contraction of the streamlines upstream and their divergence downstream are gradual and smooth, while values are lower when these transitions are accompanied by considerable turbulence in the water. High friction encountered in the flow of the water through structures also depresses the μ-value. Under normal conditions (smooth in- and outflow; smooth concrete lining), the following values may be used:

- short culverts (length <10 m)...... $\mu = 0{\cdot}8$
- long culverts (length 20–30 m)...... $\mu = 0{\cdot}7$

Common values used for bridges are $\mu = 0{\cdot}8$–$0{\cdot}9$. For bridges with a wet cross-section almost equal to that of the canal, hardly any headloss needs to be accounted for.

A culvert usually consists of pipes when the discharge is small ($Q < 0{\cdot}5\,m^3.sec^{-1}$). For larger discharges, more pipes may be installed but when the discharge exceeds 1–2 $m^3.sec^{-1}$, box type culverts are generally preferred. The invert of the culvert is usually laid about flush with the canal bed whilst the top of the culvert should normally have a minimum 10 cm clearance, especially when the water carries debris (see figure 9.5). Unreinforced pipe culverts should be buried at least 50 cm deep when they have to support normal wheel loads due to on-farm or rural traffic.

In the design, the outflow velocities should always be checked and if found to be too high, either anti-scouring measures should be provided for or the cross-section should be enlarged (the latter is usually the cheapest solution). Outflow velocities into unprotected earthen canals should be no more than 1½–2 times the maximum velocities given in tables 9.1 and 9.2.

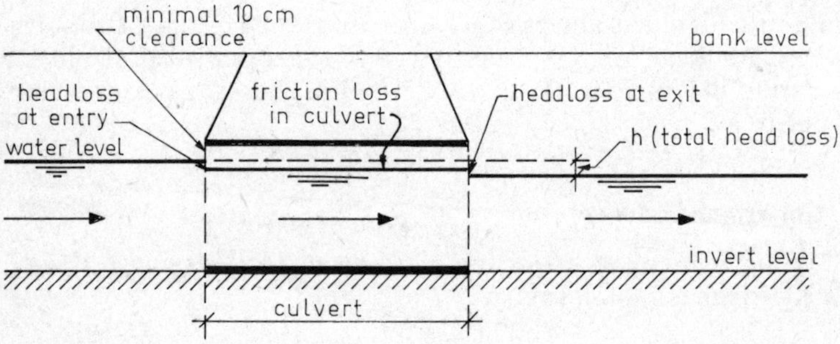

FIGURE 9.5 *Longitudinal section of a typical drainage culvert*

9.2.2 Weirs

Weirs are installed in drainage canals for different purposes. They may serve as drop structures to dissipate excess head in the canal system. More commonly weirs are used to prevent water levels in certain canal sections from falling too low during periods of low discharge, thus preventing 'over-drainage' of the adjacent land or even promoting lateral inflow of water from the canals into the land (see uses described in section 5.2.1).

The free flow discharge over a rectangular weir as depicted in figure 9.6 may be calculated as:

$$Q = \mu \tfrac{2}{3}\sqrt{\tfrac{2}{3}g}\, b h_1^{\frac{3}{2}} = 1{\cdot}7\mu\, b h_1^{\frac{3}{2}} \qquad \text{......(eq. 9.5)}$$

where: Q = discharge rate ($m^3.sec^{-1}$)
μ = coefficient
g = gravitational acceleration ($9{\cdot}8\,m.sec^{-2}$)
b = crest width (m)
h_1 = upstream water level above crest level (m)

Eq. 9.5 actually applies to broad crested weirs only (in which the streamlines over the crest are essentially horizontal) but may also be used for sharp crested weirs (in which the streamlines over the crest are quite curved) by introducing suitable values of the coefficient μ. For the semi-sharp crested weirs normally encountered in drainage canals, μ-values are of the order of 1·0–1·1. The higher values apply when the width of the weir is almost the same as that of the approaching canal and when the water training edges of the structure are smooth and well-rounded. When these edges are rather sharp and the weir is of the so-called *contracted* type (ie the weir crest $\ll$ canal width as is the case in figure 9.6), the lower values of μ should be used.

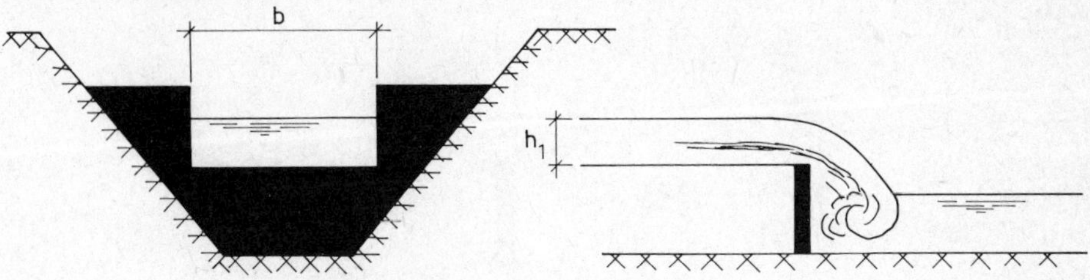

FIGURE 9.6 *Free flow discharge across a weir*

Submerged discharge occurs when the flow over the weir is hindered by a high downstream water level (the weir is 'drowned', see figure 9.7). For this type of discharge the following formula applies:

$$Q = \mu b h_2 \sqrt{2g(h_1 - h_2)} \approx \mu b h_2 \sqrt{2gz} \qquad \text{......(eq. 9.6)}$$

Actually, the *specific energy head* $H = h + \frac{v^2}{2g}$, should be used rather than simply the water depth (h). However, provided the flow velocity remains low (as is usually the case, seldom exceeding 0·5–1·0 m.sec^{-1}), there are no significant errors made when using water depths in eqs. 9.5 and 9.6 instead of specific energy heads.

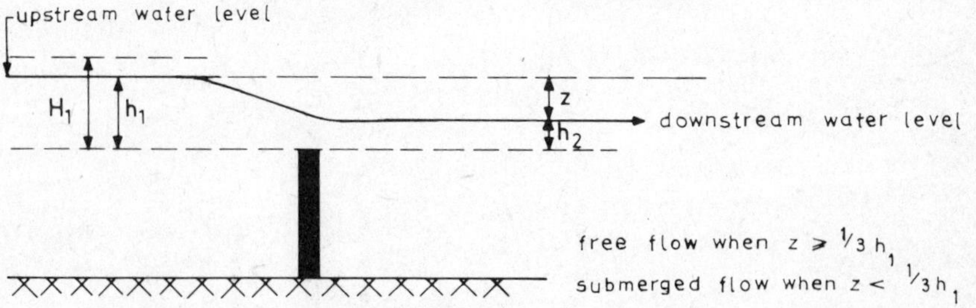

FIGURE 9.7 *Conditions for free and submerged flow over a weir*

Movable weirs are most suitable when the main purpose is to retain water during the dry season. Crest levels can be lowered during the wet season to permit unimpeded drainage flow. Conversely, during the dry season the weir crest may be raised. The most simple form of a movable weir is the stoplog weir. To save on costs (construction and operation), movable weirs may be installed at the most critical locations, supplemented with fixed crest weirs at various other locations.

9.2.3 Backwater curves

Structures in a canal system will often create bottlenecks in the discharge. The canal levels will adjust to this by allocating additional head to the flow through these obstructions so as to arrive at equal flow rates throughout the system. This extra head will create a backwater curve as in figure 9.8, which merges asymptotically with the normal hydraulic grade line (ie the grade line which would become established in the absence of structures).

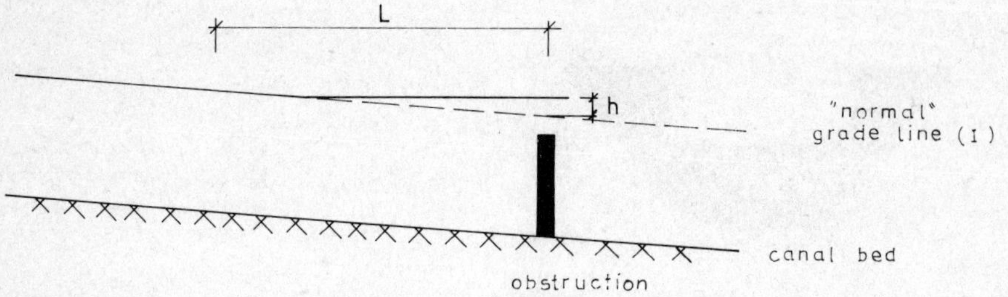

FIGURE 9.8 *Backwater curve upstream of a structure*

The reach of the backwater curve (= distance L in figure 9.8) may be approximated as:

$$L = \frac{2\,h}{I} \qquad \text{......(eq. 9.7)}$$

where: L = reach of the backwater curve (m)
I = hydraulic gradient/bed slope ($m.m^{-1}$)
h = headloss over the structure (m)

Example calculation: h = 0·10 m

$$\left.I = 0{\cdot}02\% = 0{\cdot}0002\,m.m^{-1}\right\}L = \frac{2 \times 0{\cdot}10}{0{\cdot}0002} = 1\,000\,m$$

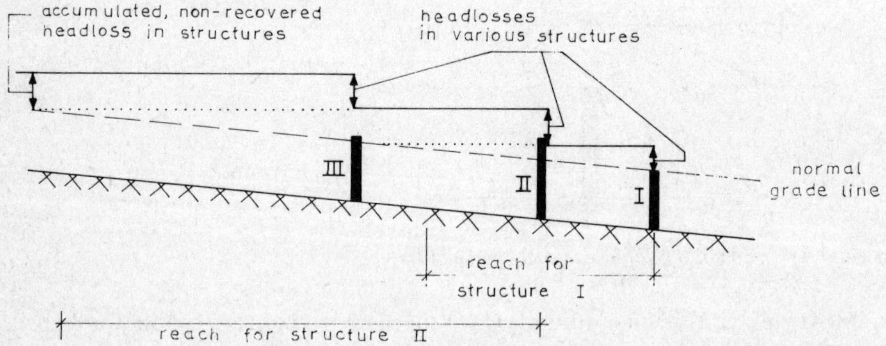

FIGURE 9.9 *Overlapping backwater curves*

When the nearest upstream structure is outside the reach of the backwater curve, the headloss h will be fully recovered in the canal section between the two structures.

When the backwater curves overlap as in figure 9.9 the headloss is not fully recovered in the canal sections and so part of the total available head in the system must be specifically allocated to the flow through the structures. A rule of thumb states that head does not need to be allocated to structures located at the head of canals and to small structures designed for $h \leqslant 5$ cm when there is no adjacent structure in the immediate upstream vicinity (say within 500 m). For larger structures, roughly half of the available head in a canal section between two such structures should be allocated to the flow through the structure located at the downstream end of the section, the other half to the canal flow in the section.

9.3 Tidal outlets

The normal tidal movement of the sea level, resulting from the mutual attraction between the moon and the earth and the rotation of the earth, generally conforms to a sinusoidal type of curve as shown in figure 9.10 with two high and two low levels within each lunar day. The lunar day is longer than a calendar day (24 hours and 50 minutes compared to 24 hours), resulting in the about one hour daily shift in the time of occurrence of the highest and lowest levels. The sun also has an influence. At full and at new moon the forces exerted by the moon and the sun on the earth reinforce each other, generating the largest tidal range (see figure 9.10). This situation is termed *spring tide.* At half moon the forces oppose each other, resulting in the smallest tidal range in a lunar cycle, termed *neap tide.* Figure 9.10 presents an example of the variation of the highest and lowest tide levels and of the variation in tidal range during one lunar cycle.

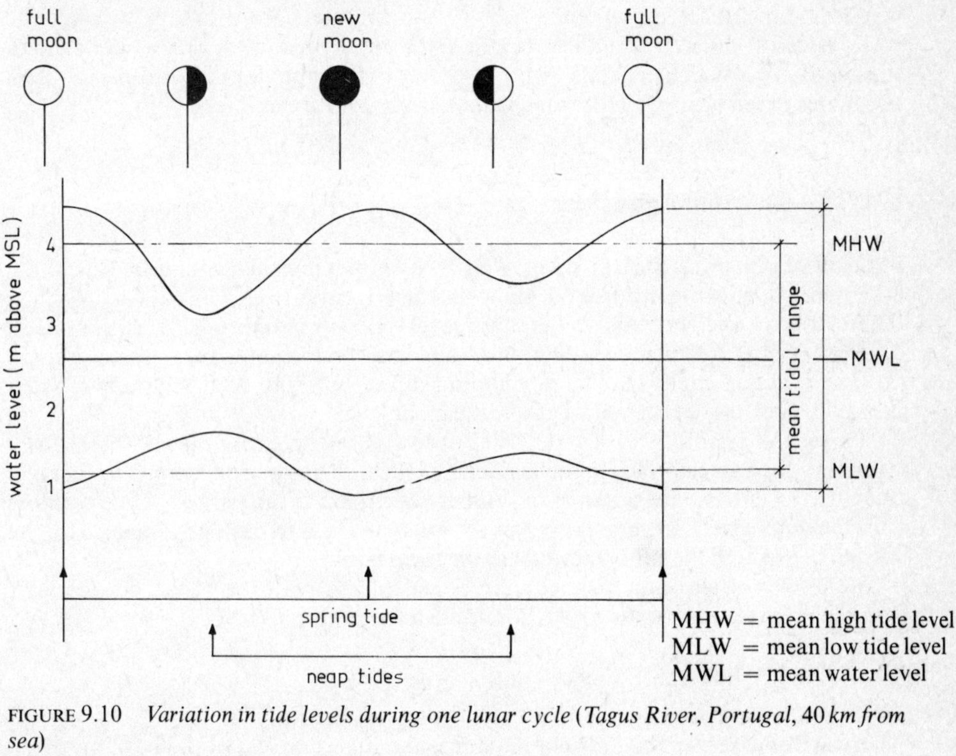

FIGURE 9.10 *Variation in tide levels during one lunar cycle (Tagus River, Portugal, 40 km from sea)*

9.3.1 Local variations in tide

Tides vary a great deal between different sea locations due to the influence of the differing shapes and sizes of the oceans and their location relative to the equator, factors which damp as as well as amplify the tide. Mean tidal ranges are mostly between 1 and 4 m with high tide levels rising nearly equally above MSL (mean sea level) as low tide levels fall below it. During spring tides sea levels commonly rise 0·5–1·0 m above the normal high tide levels. Tides may also be superimposed upon seasonal variations in sea level (eg corresponding to changes in wind direction) or incidental variations (eg due to wind surges). These can add several metres in extreme cases where hurricanes pass along the coast. Tidal effects are also always much more pronounced in funnel shaped bays and estuaries.

Tidal effects may travel considerable distances up rivers, becoming in the process damped and ultimately phased out (the more so for low tide levels than for high levels, see figure 9.10). This tidal penetration is accompanied by a parallel intrusion of salty sea water. The extent and magnitude of the saline intrusion depends to a large extent on the river discharge, the penetration always being much more extensive at low discharges than at high discharges. The morphology of the river, however, also has an influence. In some rivers tide and salt intrusion are still noticeable some 100–200 km inland.

Critical water levels at the coast can generally be predicted reliably on the basis of either a short measuring programme set up in conjunction with available tide tables, or using analytical techniques which have been developed. The prediction, however, becomes difficult for outlets into tidal rivers. Methods are available to calculate water levels at different distances up the river under various discharge conditions, the reliability of which depends on the accuracy of the input for the formula adopted. Estimates based on locally recorded values are always preferable.

9.3.2 Discharge through a sluice

Figure 9.11 illustrates the principles of a gravity outlet in a tidal situation. Discharge takes place during the tide free period which lasts from A to D. The gate opens when the inner level is slightly above the outer level (point A in figure 9.11), a 10 cm head difference being sufficient for outlet into the sea when the gates are well maintained (in fresh water a 5 cm head difference should suffice). Closure occurs at point E when the outer level has risen a few cm above the inner level.

During the period from A to B, the outflow is restricted by the still high outer levels and the same applies to the last part of the tide free period from C to D when the outer level has once again risen. In between there is the period of free outflow with the outer levels being so far below the inner levels as to have no influence on the discharge through the sluice (period lasting from B to C).

The discharge through a sluice may be calculated with the broad crested weir formula presented earlier as eq. 9.6 (for notation see figure 9.7):

$$Q = \mu b h_2 \sqrt{2g(h_1 - h_2)}$$

where in this case b = the width of the sluice and the value of μ would be of the order of 0·9–1·1. Free flow occurs when $h_2 \leqslant 2/3\,h_1$. Inserting $h_2 = 2/3\,h_1$ in the above formula yields:

$$Q = 2/3\,\mu b h_1 \sqrt{2/3\,g h_1}$$

which is similar to eq. 9.5

For discharge into the sea, the water depth h_2 should be increased by 3·5% to account for the higher density of sea water.

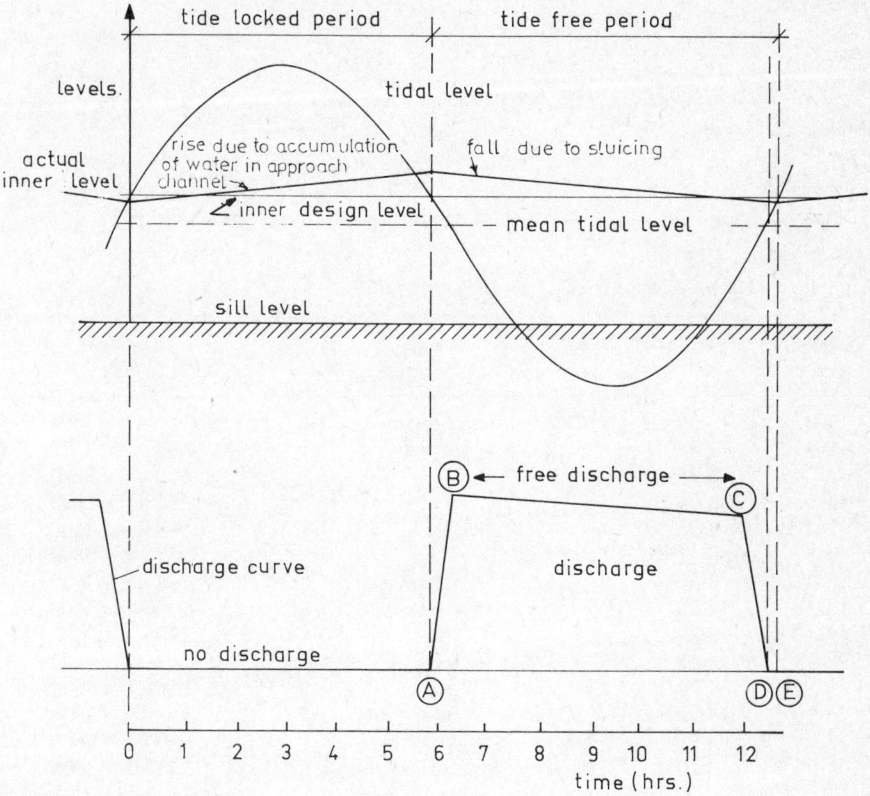

FIGURE 9.11 *Gravity outlet under tidal conditions*

9.3.3 Example of calculations

The variations of the inner water level (h_1) during a sluicing cycle may be determined by water balance calculations, treating the lower part of the main drainage system behind the sluice as a reservoir (storage of backed-up discharge in the canals, see section 8.8.2). Given the 'reservoir area' and the inflow and outflow rates, the variation in h_1 may be determined as the variation of the reservoir level. The inflow may be taken equal to the design discharge of the basin draining through the sluice, and is assumed to remain constant throughout the sluicing cycle. The outflow which is zero during the tide locked period, varies once the sluice gates have opened, depending especially on the value of h_1. This implies repetitive calculations in which due account is taken of the relationships between h_1, reservoir storage and outflow. However, outflow calculations are often sufficiently accurate when the inner water level is assumed to fall linearly as in figure 9.11 (and also in the calculations in table 9.4).

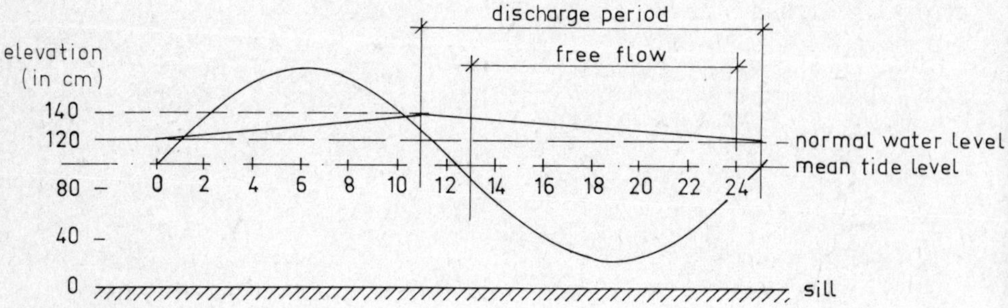

FIGURE 9.12

TABLE 9.4 *Calculation of discharge through a tidal sluice*

Period (hours)	h_1* (m)	h_2* (m)	*Type of discharge*	μ	*Outflow* (m^3)
11–12	1·39	1·18	submerged	1·1	2·63 × b × 3600
12–13	1·37	1·00	submerged	1·1	2·96 × b × 3600
13–14	1·36	0·82	free	1·1	3·00 × b × 3600
14–15	1·34		free	1·1	2·92 × b × 3600
15–16	1·33		free	1·1	2·88 × b × 3600
16–17	1·31		free	1·1	2·82 × b × 3600
17–18	1·30		free	1·1	2·78 × b × 3600
18–19	1·28		free	1·1	2·72 × b × 3600
19–20	1·27		free	1·1	2·70 × b × 3600
20–21	1·25		free	1·1	2·63 × b × 3600
21–22	1·24		free	1·1	2·58 × b × 3600
22–23	1·22		free	1·1	2·52 × b × 3600
23–24	1·21	0·64	free	1·1	2·50 × b × 3600
24–24^{50}	1·20	0·87	submerged	1·1	2·43 × b × 3000
24^{50} → negligible					135 420 × b

* Values read from figure 9.12 and, in the case of h_2 increased by 3·5%

The example applies to a case in which the design discharge is 10 $m^3.sec^{-1}$ (area = 8 000 ha, drainage coefficient = 1·25 $l.sec^{-1}.ha^{-1}$). The reservoir area is 200 ha ($2\frac{1}{2}$% of the basin area). During the 11 hours tide locked period, the inner water level rises 20 cm [$(11 \times 10 \times 3600)/(200 \times 10^4)$] above its normal level. Sufficient sluice capacity should be installed in order to lower the level of the reservoir down to the normal level during a single discharge period. The calculations proceed as in table 9.4 using average levels during hourly periods (this is generally satisfactory for the limits of accuracy required). The discharge through the sluice is calculated for each hourly period using either eq. 9.6 (submerged flow) or eq. 9.5 (free flow). The total outflow during the $13\cdot\frac{50}{60}$ hours discharge period amounts to $135420 \times b\,m^3$. This should be equal to the total discharge load per sluice cycle of $24\cdot\frac{50}{60} \times 10 \times 3600\,m^3$, so that it follows that a sluice with a width b = 6·60 m is required. In this case the level of the sill was fixed in advance although it could of course be a design variable. Different combinations of sill level and sluice width are possible, each satisfying the required discharge capacity, enabling a suitable choice to be made.

9.4 PUMPS

Pumped drainage is required in low lying areas from which water is unable to drain by gravity. These typically are flat with poor natural outlets, often with the drainage base level lying below the outlet level (as in the case of polders, see section 5.2).

The duty of a pump is the term used for the work which the pump expends, as determined by the quantity of water to be discharged and the difference between inlet and delivery level (the lift, see figure 9.13 and 9.16). A common feature of the duty of a land drainage pump is the combination of a fairly high discharge with a low lift, eg $>0{\cdot}5\,m^3.sec^{-1}$ and <5 m respectively. The lift may vary due to changes in the outlet level as for example in the case of an outlet into a river with a variable regime or an outlet into the sea.

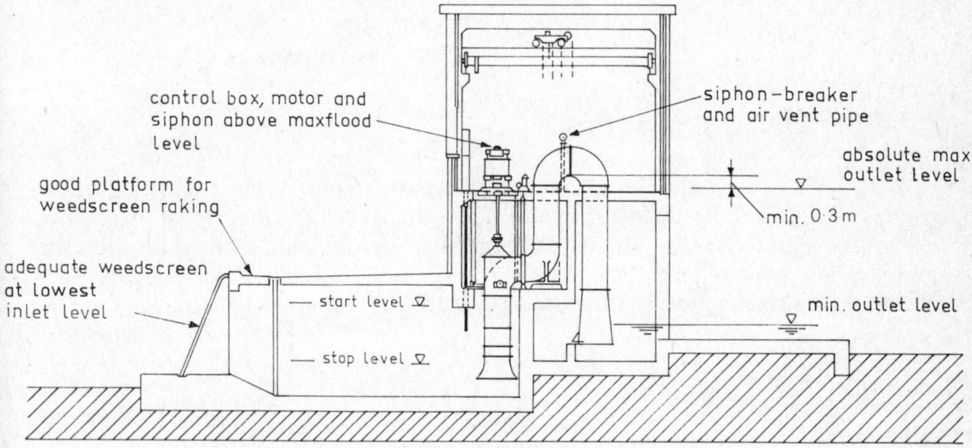

FIGURE 9.13 *Main characteristics and layout of a land drainage pumping station* (courtesy A P E Allen Ltd, England)

The time during which the pump operates may vary from only a few days annually, to more extended periods of continuous operation (eg 1 to 2 months during a cool wet season). The overall efficiency of the pump is important in the latter case while it is relatively unimportant in the former case where a pump operates for short periods only in response to rather infrequent heavy rainfall. Reliability is then of much greater concern, particularly when the site is remote and unattended.

The pump, drive-unit and control works may be sited in various ways to suit local conditions. A fairly standard arrangement is shown in figure 9.13. The pump itself is normally placed in or close to a sump in which the drainage water from the area collects by gravity flow. Pumps normally operate in cycles, starting and stopping automatically in response to preset water levels in the sump (or in the approach canal).

9.4.1 Types of pumps

Several types of pumps are used for land drainage. The *Archimedean screw pump* consists of an inclined spindle fitted with a surrounding, spirally wound blade which rotates within a fixed semi-circular casing. Water trapped between the blades is lifted from the lower to the higher level (see figure 9.14). Rotation speeds normally vary between 20 and 120 rpm.

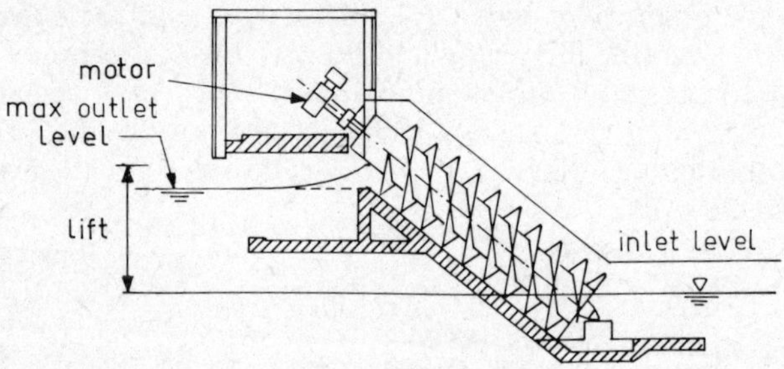

FIGURE 9.14 *Archimedean screw pump*

The much more widely used *rotodynamic pumps* consist of an impeller which rotates within a totally enclosing casing. There are three principal types: *radial flow*, *mixed flow* and *axial flow*, the terms describing the direction of flow through the pump. These have different shaped impellers and casings as shown in figure 9.15, each type being suited to a particular range of duties.

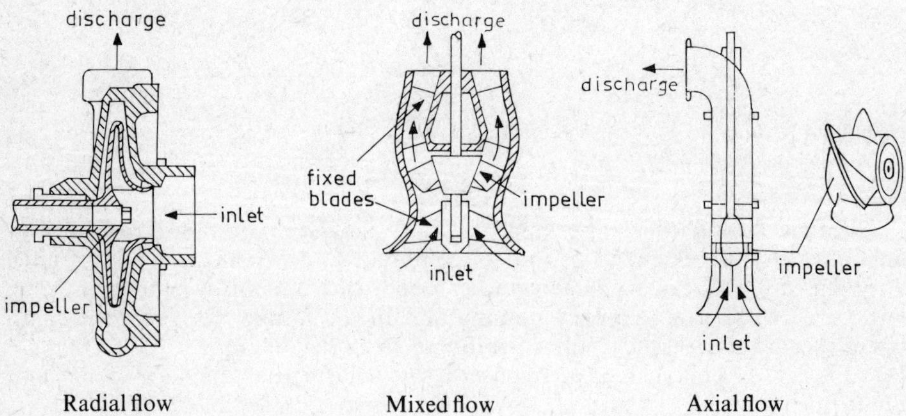

FIGURE 9.15 *Shapes of rotodynamic pumps* (PROSSER 1977)

9.4.2 Pumping head and characteristics

Pumps raise water from a lower to a higher elevation through a vertical distance termed the *lift* (*static head*, see figure 9.16). During pumping additional head is expended as frictional losses in the suction line leading the water into the pump and in the pressure line leading water away from the pump, and as a loss of kinetic energy when the water discharges into the delivery bay. The pump has to be capable of imparting a total head (*dynamic head*) to the water which is equivalent to the sum of the lift plus these losses:

$$H = H_s + h_{fs} + H_d + h_{fd} + \frac{v_d^2}{2g} \qquad \text{......(eq. 9.8)}$$

where:
H = total dynamic head (m)
H_s = suction head (m)
H_d = delivery head (m)
h_{fs}, h_{fd} = frictional losses in the suction and pressure lines
v_d = flow velocity at the outlet of the pressure line

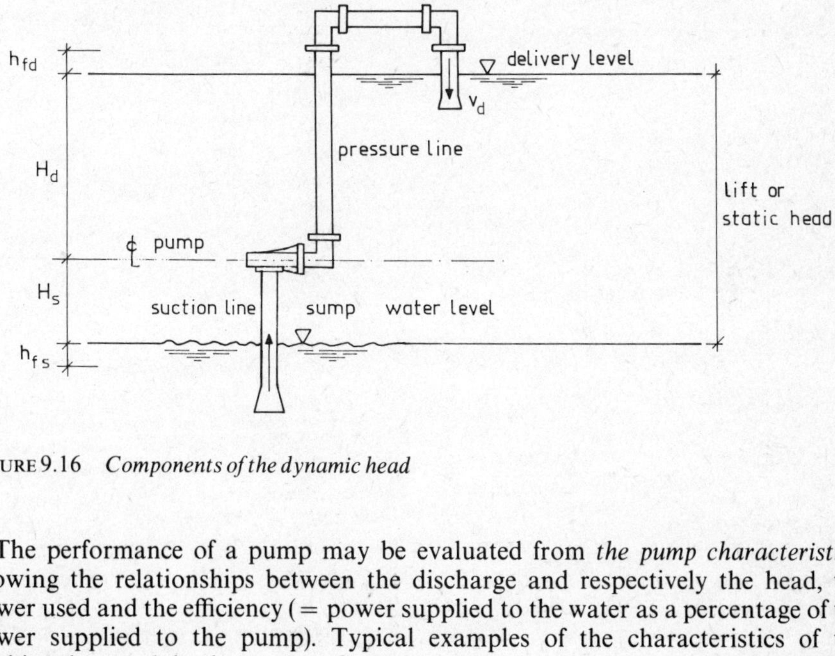

FIGURE 9.16 *Components of the dynamic head*

The performance of a pump may be evaluated from *the pump characteristics*, showing the relationships between the discharge and respectively the head, the power used and the efficiency (= power supplied to the water as a percentage of the power supplied to the pump). Typical examples of the characteristics of the Archimedean and the three types of rotodynamic pumps are presented in figure 9.17.

Archimedean pump

The head and efficiency characteristics of the Archimedean screw pump are quite different from the corresponding characteristics of rotodynamic pumps due to the fact that the depth of submergence of the screw impeller changes with the level of water at the inlet. The discharge increases from zero, when the screw is wholly above the inlet water level, up to a maximum when the central spindle is fully submerged (the latter case being depicted in figure 9.14). The discharge then remains constant with further submergence of the screw. The maximum outlet level should coincide with the level of the invert of the casing at its highest point since higher levels would require the installation of expensive non-return valves to prevent backflow when the pump stops. With the fixed casings generally used, all water is lifted to this high level, even though the outer level, for example a river, may be much lower. Consequently, variations in lift are almost wholly dependant upon the variations of the inlet level so that the head characteristic is fairly flat (see figure 9.17).

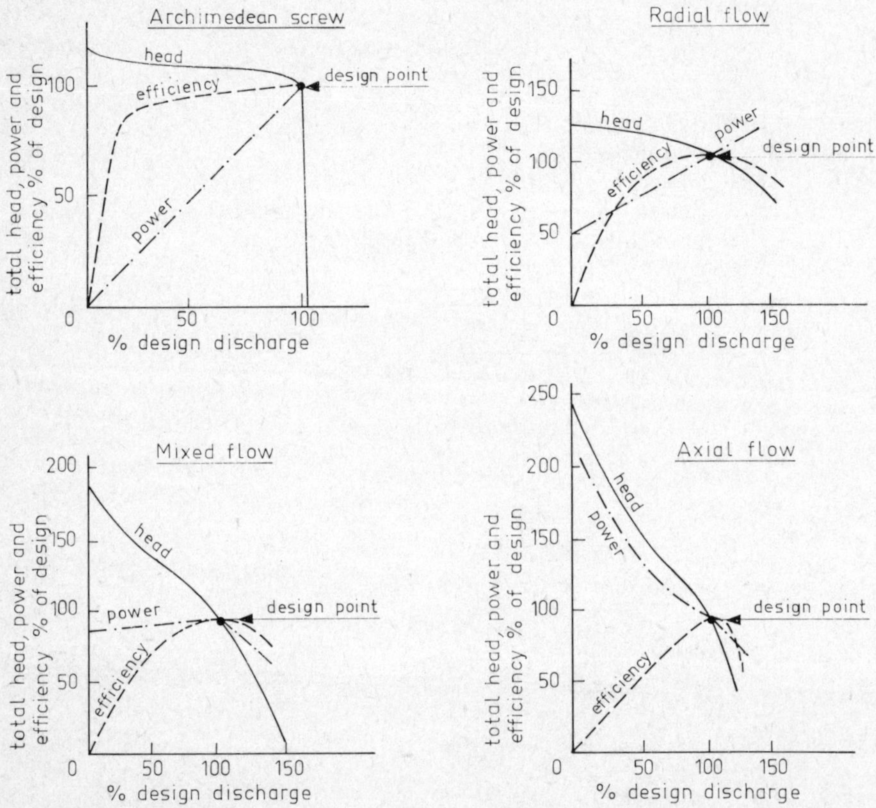

FIGURE 9.17 *Typical pump characteristics of the major types of pumps*

The power consumed increases steadily with increasing discharge to a maximum which coincides with the design duty. The drive unit, designed for this maximum, is thus unlikely ever to be overloaded – a desirable feature of the pump. The pump is also able to deliver a wide range of discharges with rather similar overall efficiency (range 60–75%) thus giving the pump a certain operational flexibility.

Rotodynamic pumps

The characteristics of these types of pumps indicate that the head developed by the pump decreases as the discharge increases, the maximum head coinciding with zero discharge (ie pumping against a closed delivery valve). The efficiency increases with increasing discharge up to a maximum (usually of the order of 75–85%) which ideally should coincide with the design duty.

The power characteristics of the three types of rotodynamic pump are quite different. For a radial flow pump the power consumed increases steadily with the discharge (the pump is so called self-regulating), whereas in contrast the peak power demand of an axial flow pump occurs at zero discharge. This condition exists when the pump is started and a precaution adopted for this type of pump, to prevent overloading of the motor, is the avoidance of the use of any valve in the pipeline which might inadvertantly be left in a permanently closed position (ie gate valves). Some provision must, however, be included to prevent reverse flow siphoning back through the pump once it has stopped. This may take the form of either a good flap valve placed at the end of the pressure line or, for large installations, a siphon-breaker placed at the highest point in the pressure line (see figure 9.13).

9.4.3 Pump selection

Pump selection should be based on a careful analysis of the prevailing Q (discharge) and H (head) combinations under which the pump is to operate, in relation to the pump characteristics. A pump may in fact be regulated by means of valves to develop any combination of discharge and head from its unique characteristic, although there is only one combination which coincides with the peak efficiency. This is obviously an important combination, since the corresponding discharge and head represent the optimal pump duty.

In the selection of pumps future discharge requirements should also be considered as well as such practical points as facilities for maintenance. Long term costs, not just the initial capital outlay, should be considered bearing in mind that an initially cheap pump may in the long term prove to be more expensive due to rapid wear and high costs of maintenance.

The Q_{design} values as determined in chapter 8 apply to the design of canals and related structures (including outlet sluices). As pointed out above, pump selection should not be based on a single Q value, but rather the expected range and the frequency of occurrence should also be considered. Where the expected discharge is fairly uniform as, for example, is usually the case with groundwater drainage, the Q_{design} used for the canal system, would also apply to pump design. Accordingly, a pump should be selected which has its peak efficiency at this discharge (and the corresponding, required lift). When the discharge is composed of groundwater drainage and shallow drainage water, peak discharges will be high but of short duration. The pump should of course be able to cope with these high discharges (unless storage is provided, see section 8.8), although it need not operate at its highest efficiency. Instead, the peak efficiency of operation should normally be based on a lower, more frequently occurring discharge.

Number of units

The number of pump units to be installed in a station should also be based on a discharge analysis. More units allow more flexibility of operation although one full capacity pump is usually cheaper to install and to operate than two half capacity pumps. If the discharge generally ranges between Q and 2Q (= maximum), two pumps, each with a capacity of Q, would be suitable. If standby capacity is required, three units with a capacity of Q each could be used, thus permitting one unit to be non-operational, eg for overhaul, while preserving the required maximum capacity.

Type of pump

The Archimedean type of pump is particularly suitable for situations where the outlet level is approximately constant and the converse follows that it is unsuitable in situations where this level varies considerably. It can raise discharges ranging from 0·05 $m^3.sec^{-1}$ up to 6 $m^3.sec^{-1}$ through lifts of 2–10 m but is most commonly used in small to medium sized pumping stations with lifts in the range 2–4 m. It is sturdy and is able to raise weed and debris (a weed screen is not needed) which makes it suitable for remote unattended stations.

A rotodynamic pump should, other things being equal, be selected primarily on the basis of its suitability for the design duty. Radial flow pumps should preferably be used in situations with low Q and high H, while axial flow pumps are to be preferred in situations with high Q and low H. For the typical duty of a land drainage pump, consisting of high discharge ($Q > 0{\cdot}5\,m^3.sec^{-1}$) and a low head ($H < 5$ m), the axial flow pump is well suited. This type of pump is also attractive on the basis of its initial costs for the pump itself is compact and the small propeller shaped impeller operates at high speed, lending itself to the use of relatively cheap electric motors. This high speed, however, increases the risk of cavitation (see section 9.4.4). These and other considerations have led to selection charts such as figure 9.18. Particular conditions may prevail justifying deviation from the general guidelines laid down in this chart.

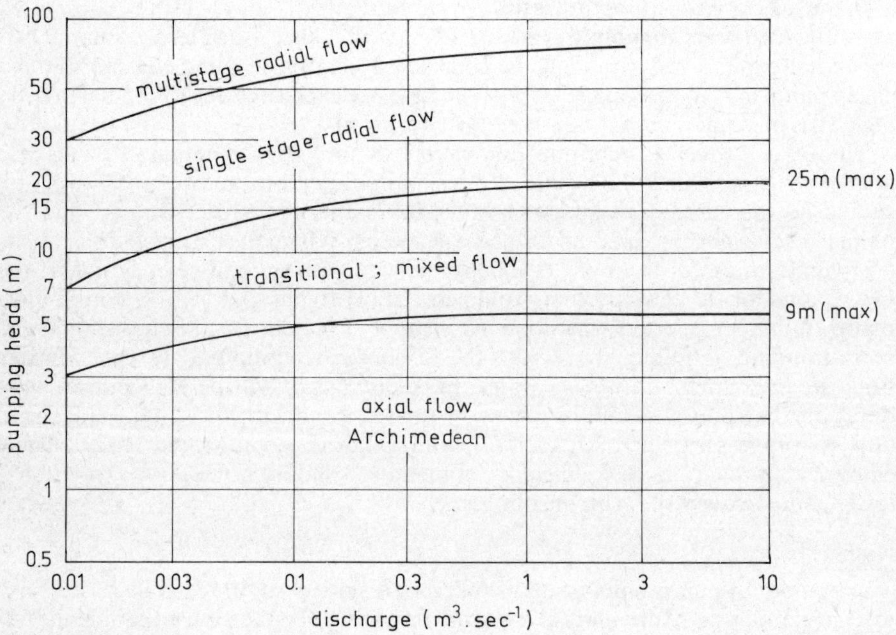

FIGURE 9.18 *Selection chart for pumps*

9.4.4 Sump and intake design

The design of the sump depends on the type of installation and the number of pumps, although the example presented in figure 9.19 illustrates most of the principal features of a well-designed sump. Generally, the last 100–200 m of the approach canal should be straight and have no inflow from the side, to make the flow towards the centrally installed pump uniform and undisturbed. The height of the bellmouth inlet above the floor and the minimum depth of the water above this inlet are also critical factors, all affecting the pumping efficiency.

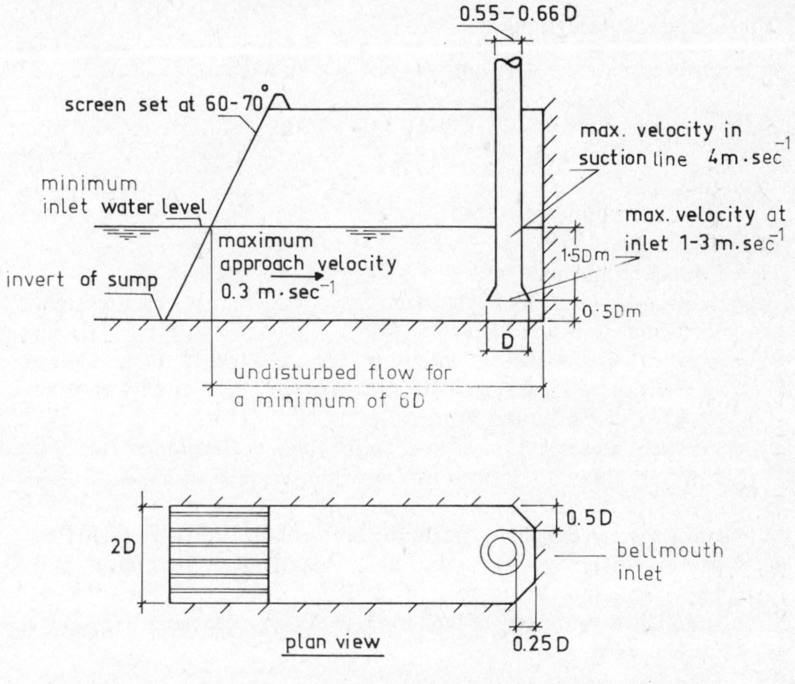

FIGURE 9.19 *Principal dimensions of a well designed intake sump* (PROSSER 1977)

Rotodynamic pumps may be situated either above or below the water level in the sump. In the former situation, the water pressure in that section of the suction pipeline above the inlet water level is sub-atmospheric. This partial vacuum situation may lead to separation of the water column, so that the pump fails to draw water, or to 'boiling' of the water causing inefficient operation and cavitation (ie vapour bubbles collapsing with violent force near the tips of the impellers, causing damage to the blades, noise and vibration). This places a limitation on the maximum height between the pump and the lowest inlet water level in the sump, a limitation which varies with the type of pump. The speed of operation of the pump also has an influence and axial flow pumps which operate at a higher speed than radial flow pumps, are more susceptible to cavitation and therefore have a lower 'suction capability' than the latter (respectively <1·5 m compared to >7·5 m). Axial flow pumps are often installed permanently submerged which has the added advantage that the pump is primed at all times (pump must be full of water to start).

The sump and the positioning of the inlet pipe to a rotodynamic pump should be such as to prevent the formation of air entraining vortices (which reduce the pumps efficiency) and to exclude weed and debris (which might clog the impellers), see figure 9.19. Debris and weed should be intercepted by a weed screen constructed of flat bars (0·6 cm wide, normal to the flow direction, and ~6 cm deep), with clear spaces of between 3 cm and 6 cm. The total clear area for flow should be sufficient to maintain the velocity through the screen below 0·75 m.sec^{-1}. The screen should be set at an angle to facilitate the raking of weed up to a platform from which it can be collected.

9.4.5 Power and cost calculations

The power requirements (E) of a pump may be calculated as:

$$E = \frac{9{\cdot}8\,Q \times \rho \times H \times e_1 \times e_2}{\eta_t \times \eta_p} \qquad \text{......(eq. 9.10)}$$

where E = power consumption (kW)
Q = discharge (m^3.sec^{-1})
H = total dynamic head (m)
ρ = density of water (kg.l^{-1}); normally $\rho = 1{\cdot}0$ but values are higher when the water is salty or dirty
e_1 = site correction factor; value is 1·0 at MSL and 15°C (reference) increasing by 1% for each 100 m rise in elevation and by another 1% for each 5°C rise in temperature
e_2 = safe load factor, taking into account the fact that the engine should not run fully loaded continuously; normally taken as $e_2 = 1{\cdot}3$–$1{\cdot}4$ (engine 70–75% loaded)
η_t = transmission efficiency, normally being of the order of 0·90 to 0·95
η_p = pump efficiency; for axial and mixed flow pumps $\eta_p = 0{\cdot}65$ for $H \sim 1{\cdot}0$ m, increasing to $\eta_p = 0{\cdot}80$ for $H = 2{\cdot}5$–$3{\cdot}0$ m and more; for radial flow pumps $\eta_p = 0{\cdot}60$ for $H = 1{\cdot}0$ m increasing to $\eta_p = 0{\cdot}80$–$0{\cdot}85$ for $H > 4{\cdot}0$ m.

Annual cost calculations are based on:

(a) Capital costs: depend on interest and depreciation rate (ranging from 10 000–50 000 running hours, or 4–25 years; on average take 10 years depreciation period)
(b) Energy costs: for electric power $E \times N \times P_1$
for fuel $E \times N \times P_2 \times U$

E = power consumption as calculated by eq. 9.10 ($e_2 = 1{\cdot}0$)
N = running hours per annum
P_1 = price per kWH
P_2 = price per litre fuel
U = fuel consumption per kWH (for gasoline 0·50–0·80 l/kWH, diesel 0·30–0·65 l/kWH, natural gas 0·65 m^3/kWH)

(c) Annual maintenance and repairs: as a rough approximation taken as 20–40% of the annual energy costs (20% for gasoline engines, 40% for diesel engines); alternatively estimated as 5% of the capital costs in (a))

(d) Other costs like personnel for operation, insurance, etc

For rough, quick estimates: $E = 13 \times Q \times H$ (kW)
Energy costs $= E \times N \times P$ (as in (b) above)
Total annual pump costs $= 2 \times$ energy costs

Example cost calculations are presented in Annex 9.2.

References and further reading

BRATER, EF and HW KING 1976: *Handbook of Hydraulics*, McGraw-Hill, New York

CHOW, Ven T 1959: *Open Channel Hydraulics*, McGraw-Hill, New York

FORTIER, S and FC SCOBEY 1926: 'Permissible canal velocities', *Trans Am Soc of Civil Eng* 89: 940–984

HENDERSON, FM 1966: *Open Channel Flow*, MacMillan, New York

ILRI 1964: 'Code of practice for the design of open watercourses and ancillary structures', Bulletin 7, International Institute for Land Reclamation and Improvement, Wageningen

ILRI 1974: *Drainage Principles and Applications*, volume IV, publication 16, International Institute for Land Reclamation and Improvement, Wageningen

KARASSIK, IJ *et al* (Eds) 1976: Pump Handbook, McGraw-Hill, New York

KINORI, BZ 1970: *Manual of Surface Drainage Engineering*, Elsevier, Amsterdam

LINSLEY, RK and JB FRANZINI 1979: *Water Resources Engineering*, McGraw-Hill, New York

PROSSER, MJ 1977: 'The Hydraulic Design of Pump Sumps and Intakes', British Hydromechanics Research Association, Cranfield, Bedford

SCHWAB, GO *et al* 1966: *Soil and Water Conservation Engineering*, Wiley, New York

SIMON, AL 1981: *Practical Hydraulics*, Wiley, New York

SOIL CONSERVATION SERVICE 1971: *Drainage of Agricultural Land*, National Engineering Handbook, Section 16, US Department of Agriculture, Washington DC

SOIL CONSERVATION SERVICE 1977: 'Design of open channels', Technical Release no 25, US Department of Agriculture, Washington DC

THORN, PB (Ed.) 1959: *The Design of Land Drainage Works*, Butterworths, London

UHDEN, O (Ed.) 1978: *Taschenbuch Landwirtschaftlicher Wasserbau*, Franckh'sche Verlagshandlung, Stuttgart (in German)

WALKER, R 1972: *Pump Selection*, Ann Arbor Science, Ann Arbor, Michigan

WEBBER, NB 1974: *Fluid Mechanics for Civil Engineers*, Chapman and Hall, London

Annex 9.1 Examples of canal design

1 *Designation*

A suitable, simple designation system for drainage canals is shown in figure 9.20 (refers to the basin discussed earlier in section 5.1, figure 5.1). The canal system is divided into sections, bounded by junction points, structures or any other condition changing the input used for the canal design. The numbering of the sections starts with the section farthest removed from the outlet point, a principle repeated for each tributary joining.

Structures may be designated independantly, or be coded with reference to the canal section in which they are situated.

2 *Canal design practices*

Procedures commonly used are illustrated for the canal line composed of the sections 1, 3, 5, 7, 19 and 21 as in figure 9.20. Two different methods are followed, respectively based on longitudinal section and tabular calculations.

(*a*) *Longitudinal section*

Longitudinal sections help to establish the hydraulic grade lines for the entire canal as well as for the different sections. This visual presentation enables the designer to assess how well the grade line meets the different requirements imposed upon it, such as:

- depth of waterlevel in the canal relative to the land level (freeboard requirements, outlet for field drains).
- minimum and maximum grades and outlet conditions for the main drains.
- minimisation of earth movement.

Longitudinal sections are especially helpful in finding a balance between cut and fill for a canal traversing irregular land. For the case under consideration where the land along the canal alignment has very regular and gentle slopes, a longitudinal section is not strictly required.

The projected grade line in figure 9.21 closely follows the slope of the land. The inner water level at the outlet is 6·60 m$^+$ which is 10 cm above the maximum outer level for gravity outlet.

(*b*) *Tabular calculations*

Tabular calculations are quicker to make and are still quite adequate for the case under consideration. A suitable format for these calculations is shown in table 9.5. As with the longitudinal section, the calculations start at the upstream end and move down the canal, section by section. The calculations are self-explanatory. A 10 cm headloss has been allowed for the culvert between sections 7 and 19, while the headlosses due to the small culverts in the sections 3 and 5 have been assumed to be recovered.

Annex 9.2 Example calculations of costs of an electrically driven pump

A pump station is required to drain 650 ha of flat arable land in the west of England. The pump station is to be capable of evacuating the total rainfall occurring over a 5 day consecutive period and at a return period of 10 years which is 60 mm. The average winter excess rainfall of 190 mm occurs on average over a 145 day time period. The average peak monthly excess rainfall of 93 mm occurs in December. The water has to be raised to an embanked river which is approximately 4 m above the normal level in the channel leading to the pump station.

Discharge requirements

Design peak discharge:

$$Q = \frac{60}{1\,000 \times 5} \times 650 \times 10\,000 \times \frac{1}{3\,600 \times 24} = 0{\cdot}90\ \text{m}^3.\text{sec}^{-1}$$

FIGURE 9.20 *Designation system for drainage canals*

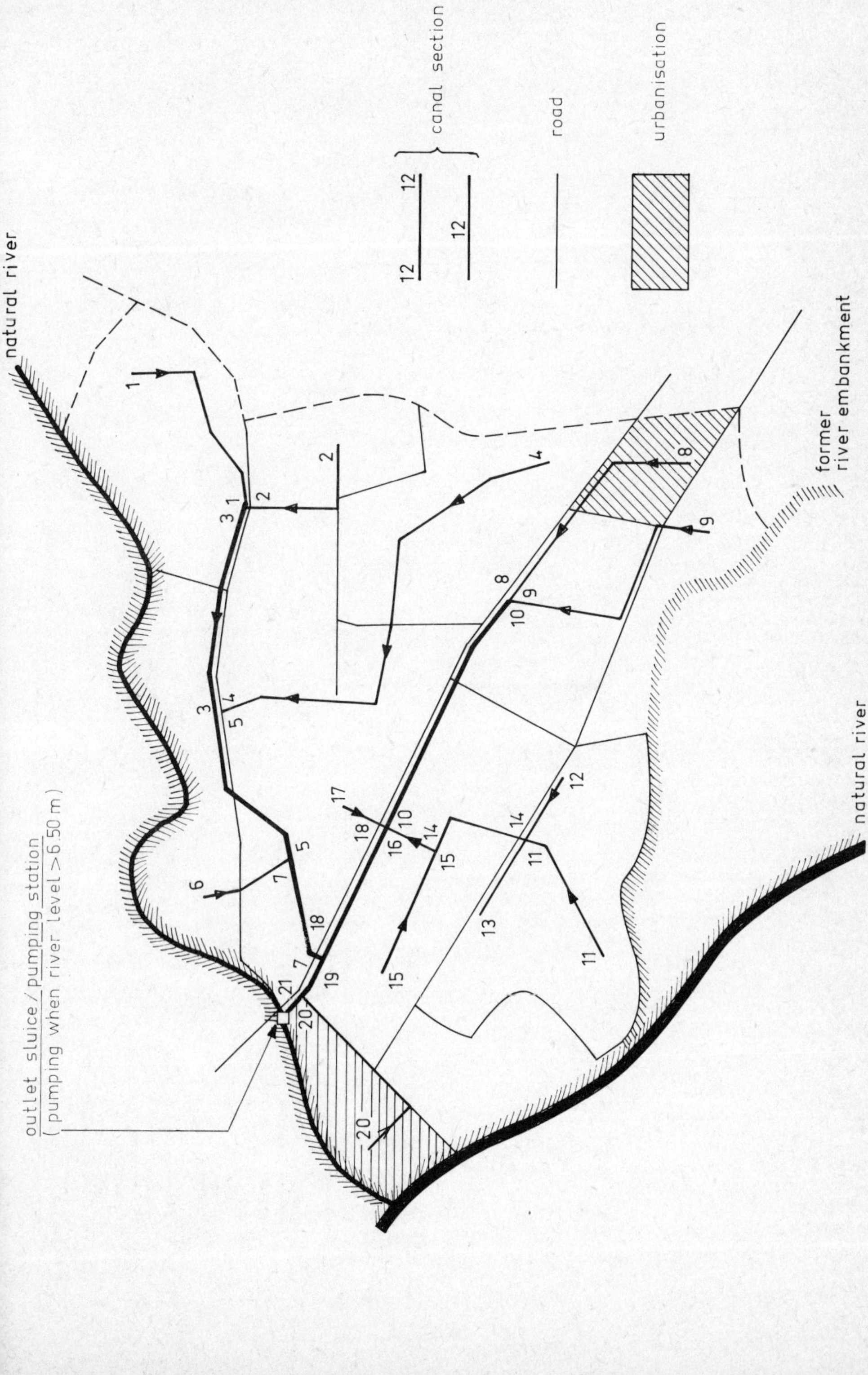

FIGURE 9.21 *Canal design using a longitudinal section*

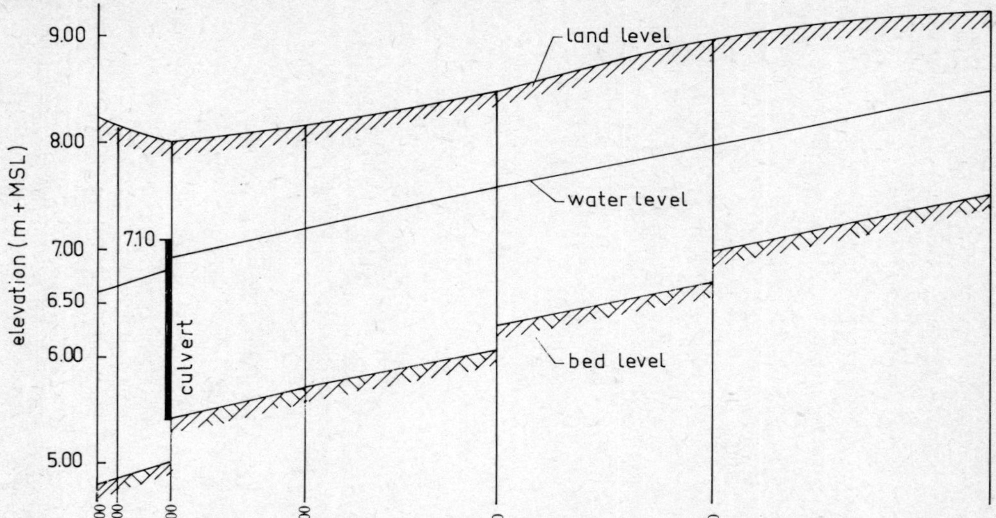

Station distance (m)	16 400	15 900	14 900		12 500		9000		5100		0
Canal section	21	19		7		5		3		1	
Section length (m)	500	1000		2400		3500		3900		5100	
Land level (m + MSL)	8·25	0·00			8·15 8·15		8·50 8·50		9·00 9·00		9·30
Sub-basins area (ha)	375		3420		420		2020		810	940	
drainage coefficient ($m^3.sec^{-1}.1000\ ha^{-1}$)	2.0				1·0		1·0		1·2	1·0	
discharge ($m^3.sec^{-1}$)	0·75		4·00		0·42		2·02		0·97	0·94	
Total area (ha)	7985	7610		4190		3770		1750		940	
Total discharge ($m^3.sec^{-1}$)	9·10	8·35		4·35		3·93		1·91		0·94	
Side slope roughness factor km	1:2/40			1:2/35		1:2/35		1:1½/30		1:1½/30	
Water level (m + MSL)	6·60		6·80 6·90		7·20 7·20		7·60 7·60		800		8·50
Hydraulic gradient ($m.\ km^{-1}$)	0·13			0·125		0·115		0·10		0·10	
Water depth (m)	1·80	1·80		1·50		1·50		1·30		1·00	
Bed width (m)	5·50	5·20		4·50		4·00		3·40		2·70	
Wet cross-section (m^2)	16·40	15·85		11·25		10·50		6·95		4·20	
Velocity ($m.sec^{-1}$)	0·55	0·53		0·39		0·37		0·27		0·22	
Bed level (m + MSL)	4·80		5·00 5·40		5·70 5·70		6·10 6·30		6·70 7·00		7·50

TABLE 9.5 *Canal design in tabular form*

Canal section		Contributing sub-basin			Hydraulic calculations								Levels (m + MSL)*		
designation	*length* (m)	*area* (ha)	*drainage coefficient* ($m^3.sec^{-1}.1000\ ha^{-1}$)	*discharge* ($m^3.sec^{-1}$)	*total discharge* ($m^3.sec^{-1}$)	*side slopes*	*roughness coefficient Km*	*grade* ($m.km^{-1}$)	*water depth* (m)	*bed width* (m)	*wet cross-section* (m^2)	*velocity* ($m.sec^{-1}$)	*land level*	*water level*	*bed level*
	5100	940	1·0	0·94	0·94	$1\frac{1}{2}$	30	0·10	1·00	2·70	4·20	0·22	9·30 9·00	8·50 8·00	7·50 7·00
	3900	810	1·2	0·97	1·91	$1\frac{1}{2}$	30	0·10	1·30	3·40	6·95	0·27	9·00 8·50	8·00 7·60	6·70 6·30
	3500	2020	1·0	2·02	3·93	2	35	0·115	1·50	4·00	10·50	0·37	8·50 8·15	7·60 7·20	6·10 5·70
	2400 (culvert)	420	1·0	0·42	4·35	2	35	0·125	1·50	4·50	11·25	0·39	8·15 8·00	7·20 6·90	5·70 5·40
	1000	3420		4·00	8·35	2	40	0·13	1·00	5·20	15·85	0·53	8·00 8·15	6·80 6·67	5·00 4·87
	500	375	2·0	0·75	9·10	2	40	0·13	1·00	5·50	16·90	0·55	8·15 8·25	6·67 6·60	4·87 4·80

* Upper lines give levels at the upstream limit of the canal section considered while lower lines give the levels at the downstream limit

Average flow in winter:

$$Q_A = \frac{190}{1\,000 \times 145} \times 650 \times 10\,000 \times \frac{1}{3600 \times 24} = 0{\cdot}10\,m^3.sec^{-1}$$

Average peak monthly flow in December:

$$Q_m = \frac{93}{1\,000 \times 31} \times 650 \times 10\,000 \times \frac{1}{3\,600 \times 24} = 0{\cdot}225\,m^3.sec^{-1}$$

The figures show that the pump station capacity (12 mm.day^{-1}) is much greater than the normal discharge (3 mm.day^{-1} in December, 1·3 mm.day^{-1} over the whole winter).

Analysis of the rainfall records indicates that the rainfall rate over durations shorter than five days but at the same frequency of occurrence, ie 10 years, exceeds the actual design rate, eg the 1 day rainfall event at $T_p = 10$ years is 30 mm. As the pump station is unable to evacuate more than 12 mm.day^{-1} some localised flooding may occur.

In this example the range variation of discharge is such that two similar pumps will be chosen although two pumps with capacities of 0·3 m^3.sec^{-1} and 0·6 m^3.sec^{-1} could also be considered. The selection chart, figure 9.18, indicates that the axial flow pump will be suitable for the required duties.

Power requirements and cost estimates

According to eq. 9.10, section 9.4.5:

$$E = \frac{9{\cdot}8 \times Q \times \rho \times H \times e_1 \times e_2}{\eta_t \times \eta_p}$$

$$E = \frac{9{\cdot}8 \times 0{\cdot}45 \times 1 \times 4{\cdot}4 \times 1 \times 1}{0{\cdot}9 \times 0{\cdot}8} = 27\,kW$$

$$\text{Expected annual running time} = \frac{190\,mm}{6\,mm.day^{-1}} \simeq 32 \text{ days or } 760 \text{ hours.}$$

Annual running costs based upon 1 000 hours operation at 27 kW:

$$\text{Electricity } 4{\cdot}1\text{ p/kWH (UK, 1981); cost} = \frac{27 \times 1\,000 \times 4{\cdot}1}{100} = £1\,107/\text{annum.}$$

The comparative running cost of diesel assuming this to be used at 0·4 litres/kWH and at a cost of 17·3 p/litre is: £1 868/annum.

In general diesel units are considered more economic for the larger pump stations (> 150 kW, or > 3–4 m^3.sec^{-1}) although much depends on the availability and costs of electrical power supply.

PART IV
DRAINAGE FOR SALINITY CONTROL

Chapter 10 Salty soils

All soils contain salts. These originate from rocks as they weather into soil. In this process the salts are carried downwards (*leached*) with the percolating water where they may either precipitate or continue to be transported in solution, ending up ultimately in the sea. In exceptional cases a high salt content in the soil may be directly related to the soil's parent material (this type of soil salinity being termed *primary* or *residual salinity*). The salinity of marine soils is a special case of residual salinity. Marine deposits may remain saline from past geological periods up to the present time in situations where there is very little leaching (arid climates, poor drainage).

By far the most common cause of high soil salinity however, is *salinisation*, ie the accumulation of salts in the upper layers of the soil from some outside source. Frequently, salinisation involves a reverse of the leaching process, ie the return into the upper soil layers of the leached salts, usually carried by water (therefore often termed *secondary salinity*). A great deal of contemporary salinisation is caused by man's activities, especially by incorrect irrigation practices (see chapter 11).

The occurrence of salty soils, although not restricted to hot, dry climates, is much more prevalent under these conditions than in the temperate, humid climates. In temperate climates there is usually sufficient excess water percolating downwards through the soil to maintain low salt levels in the upper layers. Soil and drainage conditions, however, also have a great influence for they largely determine the physical possibilities for leaching and for the removal of the excess salts from the land.

10.1 Forms of occurrence and distribution of salts in the soil

Salts occur in the soil in one of the following three forms:

- salt ions dissolved in the soil water (*the soil solution*)
- cations adsorbed on the negatively charged surfaces of the soil particles (*adsorption complex*)
- precipitated salts.

Dynamic exchange equilibria exist between the cations in the soil solution and those adsorbed on the complex, and also between dissolved and precipitated salts:

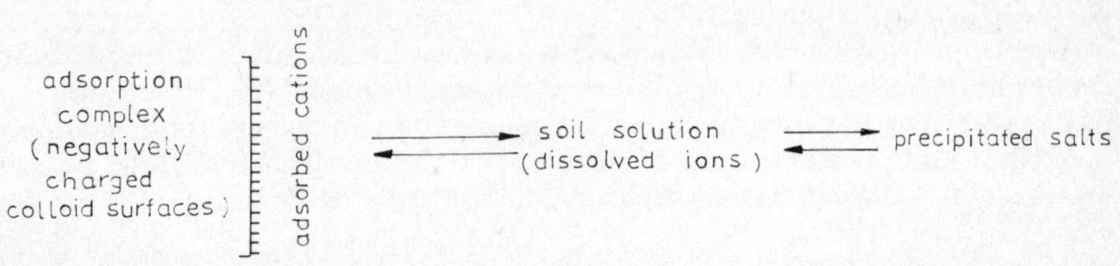

FIGURE 10.1 *Forms of occurrence of salts in the soil*

The salt composition in the soil is of course a reflection of the salt composition at its source of origin (parent-rock material, groundwater, seawater, etc) but various desalinisation-resalinisation cycles may have resulted in considerable changes in the original composition.

10.1.1 The soil solution

The main salt ions found in the soil solution of a salty soil are (see also figure 10.4 and table 10.3):

- cations: Na^+ Ca^{++} Mg^{++} (K^+)
- anions: Cl^- SO_4^{--} HCO_3^- (CO_3^{--})(NO_3^-).

A characteristic feature of many salty soils is the high concentration of Na^+. This contrasts with non-salty soils in which less than 5% of the soluble cations normally consist of Na^+, while Ca^{++} (or $Ca^{++} + Mg^{++}$) is by far the most dominant cation in the soil solution. In many salty soils the concentration of Na^+ exceeds that of Ca^{++} and Mg^{++}. The concentration of K^+ is usually only slightly higher in salty soils when compared to non-salty soils, remaining low in both cases. On the anion side, there is often a predominance of Cl^- and SO_4^{--} in salty soils. Some HCO_3^- is usually also present in salty soils at normal pH values (pH 6–8). At higher pH values this anion changes into CO_3^{--} although up to pH ~ 8·5 the quantity of CO_3^{--} formed remains insignificant.

The relatively high salt concentrations found in the soil solutions of salty soils frequently exceed the solubility limits of the poorly soluble alkaline earth carbonates ($CaCO_3$ and $MgCO_3$) and also that of gypsum ($CaSO_4 2H_2O$). In these circumstances the salts precipitate, withdrawing Ca^{++}, Mg^{++} and SO_4^{--} ions from solution.* Sodium (Na) salts and chlorides are generally highly soluble hence the preponderance of Na^+ and Cl^- in many salty soils. However, this high solubility ensures that their concentration also drops rapidly upon leaching of the soil. When the less soluble salts are leached, their concentrations in the soil solution fall more slowly as leached ions are replaced by others previously precipitated. Therefore, in a soil exposed to cyclic salinisation – desalinisation, the presence of Na^+ and Cl^- in the soil solutions is often of the same order or even less than that of Ca^{++}, Mg^{++} and SO_4^{--}

10.1.2 Adsorbed cations

Soils have the capacity to adsorb cations. This capacity is possessed by the negatively charged surfaces of the soil colloids which attract cations from the soil solution. It is quantified by the so-called *cation exchange capacity* of the soil. (CEC, expressed in meq/100 gm dry soil.)

A dynamic exchange equilibrium exists between the adsorbed cations and the cations in the soil solution. As the concentration of cation 'b' in the soil solution increases relative to cation 'a', the cation composition on the adsorption complex will adjust itself by exchanging adsorbed 'a' cations for 'b' cations from the soil solution until a new equilibrium becomes established (see figure 10.2).

* The solubility of $CaCO_3$ (lime) at pH > 7·5 is < 1 meq.L^{-1}; 1 meq = 50 mg). Its solubility sharply increases at pH < 7·5, especially in the presence of CO_2, eg from plants ($CaCO_3 + CO_2 \rightarrow Ca(HCO_3)_2$ which is highly soluble). The solubility of $MgCO_3$ is only slightly higher than that of $CaCO_3$. Gypsum ($CaSO_4.2H_2O$) has a solubility of 20–40 meq.$L^{-1} \simeq$ 1500–2500 mg.L^{-1}.

adsorption complex (adsorbed cations) *soil solution* (cations in solution)

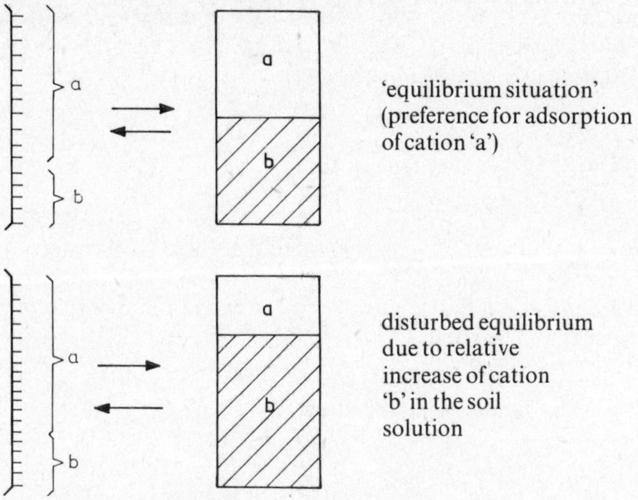

Adjustment by exchange of cation 'a' on the complex by cation 'b' from the solution:

$$]a + b \rightarrow]b + a$$

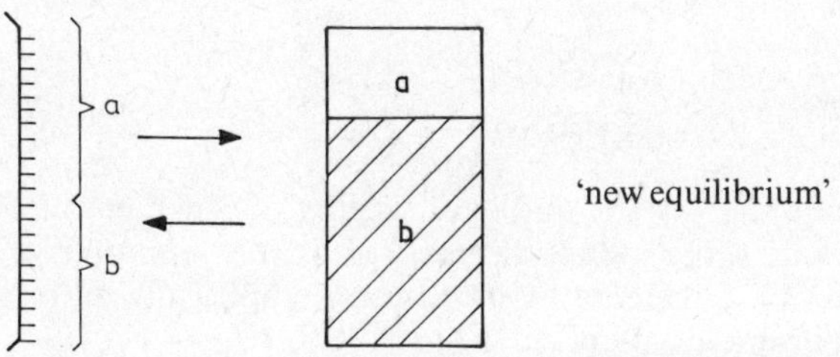

FIGURE 10.2 *Principals of the cation exchange process*

This cation exchange process is complicated by the fact that divalent cations are adsorbed more readily than monovalent cations, whilst in addition there is also a difference in adsorption preference between cations having the same valency. The following series gives the important cations encountered in the soil solution arranged in order of adsorption preference.

$$Ca^{++} > Mg^{++} > H^{+} > K^{+} > Na^{+}$$

Therefore, as long as Ca^{++} is the dominant cation in the soil solution, as it generally is in non-salty soils, little Na^{+} will be adsorbed, since it is very difficult for Na^{+} to displace Ca^{++} on the complex. This is fortunate because a high Ca^{++} occupancy of the complex is much more favourable than a high Na^{+} occupancy (see section 10.2.3). A significant Na^{+} occupancy is only to be expected when more than half of the cations in the soil solution consist of Na^{+}. In some strongly salty soils >90% of the cations in the soil solution consist of Na^{+} (see eg figure 10.4) in which case Na^{+} also becomes the dominant adsorbed cation.

10.1.3 Equilibrium relationships between adsorbed cations and cations in the soil solution

The equilibrium relationships between the adsorbed cations and those in the soil solution are described by the *Gapon equation* (Bolt and Bruggenwert 1976). For salty soils containing mainly Ca^{++}, Mg^{++} and Na^{+} cations, this equation reads:

$$\frac{Na^{+}_{ads}}{(Ca+Mg)^{++}_{ads}} = K\frac{Na^{+}_{sol}}{\sqrt{\frac{(Ca+Mg)^{++}_{sol}}{2}}} \qquad \text{......(eq. 10.1)}$$

with adsorbed (ads) cations are in meq/100 gm dry soil and soluble (sol) cations in meq. L^{-1}.

The constant K has a value of about 0·015. In salty soils it may be stated that $(Ca + Mg)^{++}_{ads} = CEC - Na^{+}_{ads}$. Further defining:

$$\frac{Na^{+}_{ads}}{CEC} \times 100\% = \text{ESP } (\textit{exchangeable sodium percentage}) \text{ and}$$

$$\frac{Na^{+}_{sol}}{\sqrt{\frac{(Ca+Mg)^{++}_{sol}}{2}}} = \text{SAR } (\textit{sodium adsorption ratio}) \text{ it follows that eq. 10.1}$$

may be rewritten as:

$$ESP = \frac{100\,(0.015\,SAR)}{1 + 0.015\,SAR} \qquad \text{......(eq. 10.2)}$$

Many factors interfere, limiting the validity and applicability of this theoretical relationship between the SAR-value of the soil solution and the corresponding ESP-value of the soil. Nevertheless a fair agreement exists with the empirically derived relationship between these two parameters depicted in figure 10.3.

The Gapon equation may be used to show that when the soil solution becomes more concentrated, Na^{+}_{ads} increases at the expense of $(Ca + Mg)^{++}_{ads}$, resulting in an increased ESP. In the case of dilution, the Na^{+}_{ads} will be replaced by $(Ca + Mg)^{++}_{ads}$ and the ESP will decrease.

10.1.4 Distribution of salts in the soil

The salinity level in the soil varies greatly in time and in space, the latter including variations in a vertical as well as in a horizontal direction. The principal cause of this variation is the migration of salts with the soil water movement and because field soil water regimes are highly dynamic it follows that salt distribution in soil is also highly dynamic. The upward and downward flow of water in the profile frequently changes the vertical distribution of salts. Following rain (or irrigation) the salt content may be found to increase with depth while the reverse situation is likely to be encountered at the end of a dry period (see figure 10.4a). At greater depths, soil moisture conditions change less, the salt content remaining more or less constant with only slight seasonal variations.

An example of horizontal variation in soil salinity is presented in figure 10.4b,

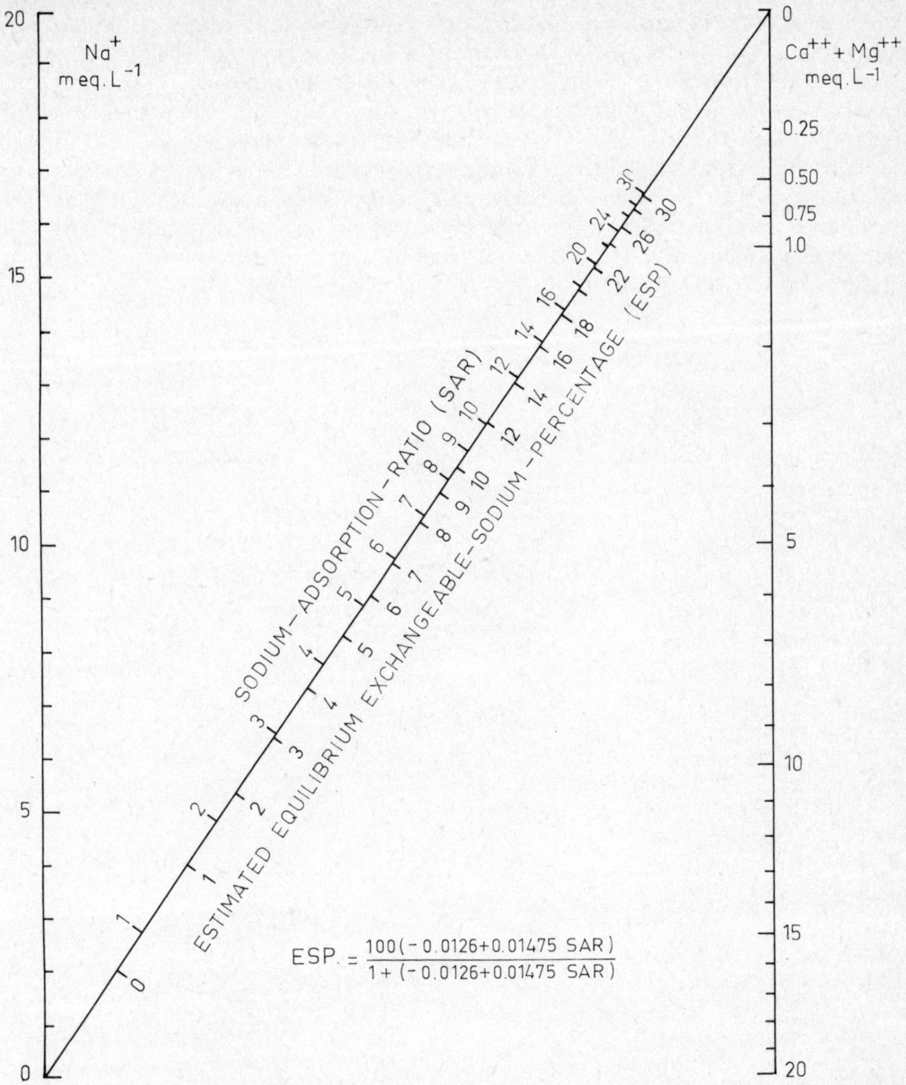

Determination of SAR and ESP

Plot the Na^+-value on the vertical left-hand scale and the $(Ca+Mg)^{++}$-value on the vertical right-hand scale; connect these two points and read the SAR-value at the point where the connecting line crosses the diagonal scale. The corresponding ESP-value is read on the opposite diagonal scale.

eg $Na = 12\ meq.L^{-1}$
$Ca = 3\ meq.L^{-1}$
$Mg = 2\ meq.L^{-1}$
read SAR = 7.5 and ESP = 9

Note: the correlation between ESP and SAR as in this graph actually holds only for the SAR_e values (i.e. the SAR-value of the saturation extract of the soil), see Annex 10.1

FIGURE 10.3 *Nomograph for the determination of the SAR-value of the soil solution/irrigation water and for the determination of the corresponding ESP-value of the soil* (USDA 1954)

which depicts the salt distribution in a cross-section through two cropped ridges in a furrow irrigated field, some days after the field has been watered. The applied water has infiltrated into the soil from the furrow and has pushed the salts downwards and sidewards. The highest salinity is found in the top of the ridge, the highest point of the soil surface, towards which the moisture from the surrounding soil is drawn by capillary forces and subsequently evaporates, leaving the salts behind. Variation in soil salinity is the rule rather the exception. Changes in soil hydrological regime, soil variations, root distribution, micro relief all contribute to this variation. Any soil salinity assessment should take this into account. The related sampling problems are discussed in section 10.3.3.

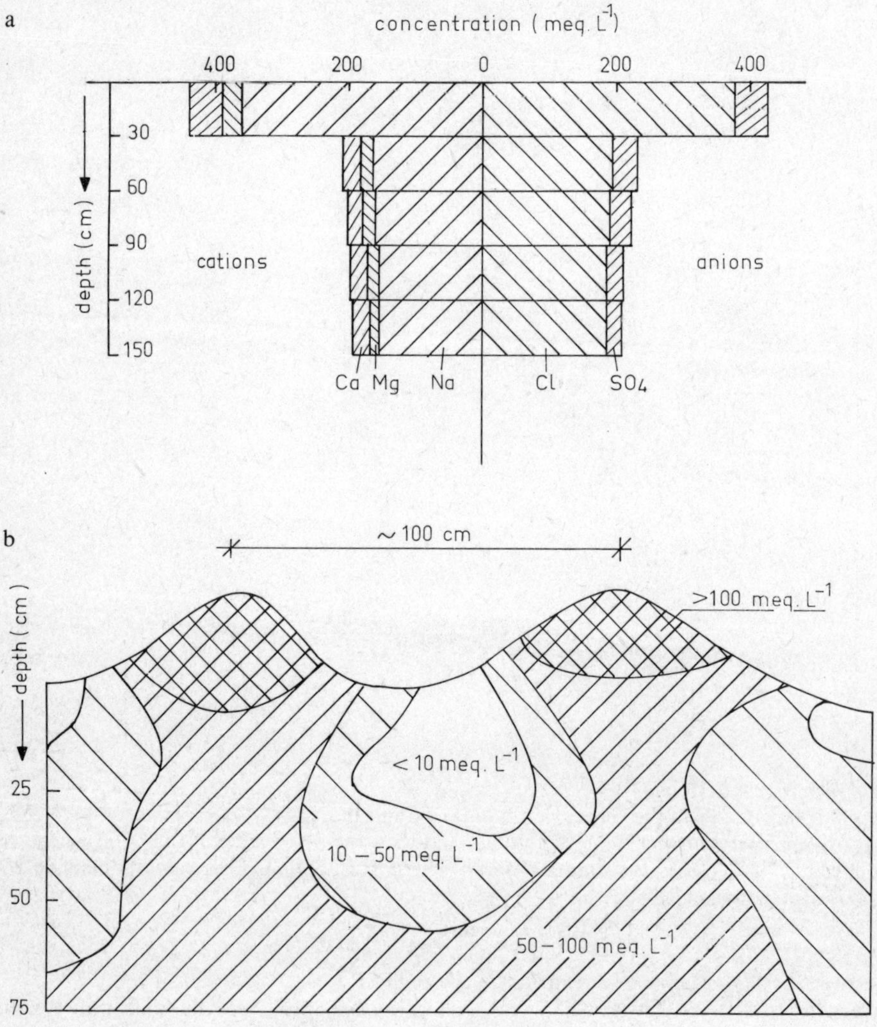

FIGURE 10.4 *Examples of the distribution of salts in the soil*

10.2 SALINITY PROBLEMS: DIAGNOSIS AND ASSESSMENT

The problems caused by soil salinity may be classified as follows, each type of problem being related to a particular aspect of soil salinity:

(a) *Osmotic problems*: caused by a high total salt concentration of the soil solution which raises the osmotic pressures that can be exerted by the soil solution. This makes it more difficult for the plant roots to uptake water from the soil.
(b) *Toxicity problems*: caused by a high concentration in the soil solution of some particular ion or by the imbalance between two or more ions, harming plant growth.
(c) *Dispersion problems*: caused by a relatively high percentage occupancy of the soil exchange complex by Na^+ which results in poor soil structure due to easy dispersion of the colloids in the soil.

Problems caused by the chemical aggressivity of some salts with respect to some construction materials are discussed briefly in section 10.2.4.

10.2.1 Osmotic problems

Salts dissolved in water exert binding forces on the water, these being termed osmotic forces. A plant extracting water from a salty soil must overcome these forces, in addition to the normal soil moisture retention forces (capillary forces etc, see section 1.3). Not only must the plant apply greater suction forces (or otherwise accept lower uptake rates) but also, a larger proportion of the soil moisture will be held by strong forces and thus be less readily available or even unavailable to the plants. As a result crops show early signs of moisture stress which is one of the principal causes of the poor growth of crops in salty soils.

Measurements

The osmotic forces increase in direct proportion to the total salt concentration of the soil solution for the range of salt mixtures normally found in salty soils. The total salt concentration may thus be used to indirectly assess the osmotic forces and problems to be expected. It should preferably be measured under the conditions under which the plant takes up most of its water from the soil, ie in the moisture range from field capacity (FC) to wilting point (WP). Since it is difficult to extract a soil solution sample from the soil in this moisture range, the salt concentration is determined at a standardised higher moisture content, *saturation point* (SP). This point was chosen, firstly because it is the nearest higher soil moisture content at which a soil solution sample can easily be extracted, and secondly because the soil moisture content at SP is easily reproducible.

Salt concentrations are normally expressed in $meq.L^{-1}$ or in $mg.L^{-1}$. Soil salinity concentrations are also commonly expressed in yet another, indirect, form using the *electrical conductivity* (EC-value) of the soil solution. For the salt mixtures normally found in salty soils, the EC-value and the salt concentration are linearly related and since the osmotic forces increase linearly with the salt concentration, a linear relationship also exists between the osmotic forces in the soil solution and its EC-value (USDA 1954, see figure 10.5). The unit of electrical conductivity is the $mho.cm^{-1}$ where mho = ohm^{-1} (reciprocal of the ohm), ohm being the unit of electrical resistance.* At the salt concentration levels prevailing in the soil, the $mmhos.cm^{-1}$ and the $\mu mhos.cm^{-1}$ are convenient units (1 mho = 10^3 mmho = 10^6 μmho). The EC measurement is temperature sensitive and is therefore standardised at 25°C.

The EC-value of the soil solution at saturation point is designated as the EC_e-value. It has proved to be a suitable diagnostic parameter for the osmotic problems caused by soil salinity. The main aspects of the determination of the EC_e-value of the soil are detailed in figure 10.6.

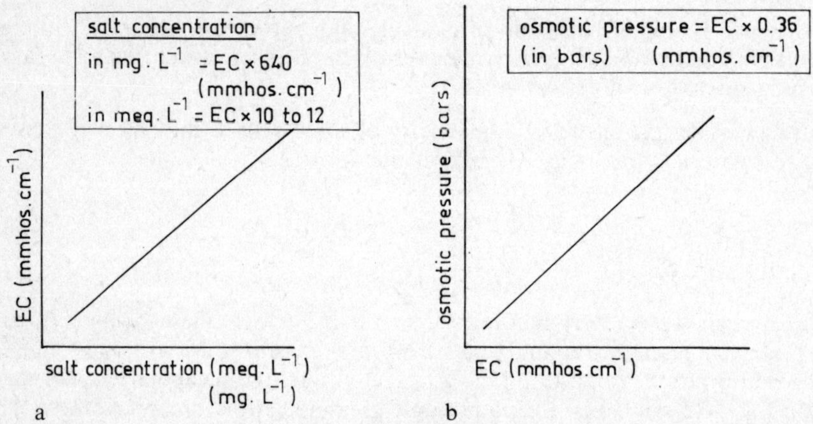

a b

FIGURE 10.5 *Linear relationships between soluble salt concentration, electrical conductivity and osmotic pressure (the given numerical relationships may be used to make rough conversions; they are valid for the range of salt mixtures and concentrations normally found in salty soils and are most accurate when used on the soil solution near SP, the saturation extract).*

The agricultural interpretation of EC_e-values

Good correlations between the EC_e-values and yields have been established for many crops. The following general criteria apply (*Schofield scale*):

EC_e = 0– 2 $mmhos.cm^{-1}$: negligible effect on all crops
2– 4 „ „ : slight effect on sensitive crops
4– 8 „ „ : significant effect on many common crops
8–16 „ „ : only salt tolerant crops can be grown
>16 „ „ : only highly resistant plants can survive.

Tolerance tables compiled by the *US Salinity Laboratory* give for various crops at which EC_e-value yield reductions of resp. 0, 10, 25 and 50% are to be expected (table 10.1). This table shows that the crops vary in their tolerance to high salt concentration of the soil solution. Beans, for example, are highly sensitive with yield reductions to be expected when $EC_e < 1.0$ $mmhos.c_m{}^{-1}$ while barley is able to tolerate EC_e-values up to 8 $mmhos.cm^{-1}$ without yields being affected. Tolerance may also vary between different varieties of the same plant.

* The SI unit for electrical conductivity in the Siemens (S) per metre (m) where $1.0S.m^{-1} = 10^{-5}$ $mmbos.cm^{-1}$, this unit is now becoming widely accepted.

(1) Make a saturated paste by adding water to the soil while stirring, until the saturation point is reached, ie the point at which the soil surface glistens although there is still no free water.

(2) Extract a soil solution sample from the paste (= *saturation extract*) using suction equipment (approx. 10 cc).

(3) Measure the EC-value of the saturation extract using a conductivity meter.

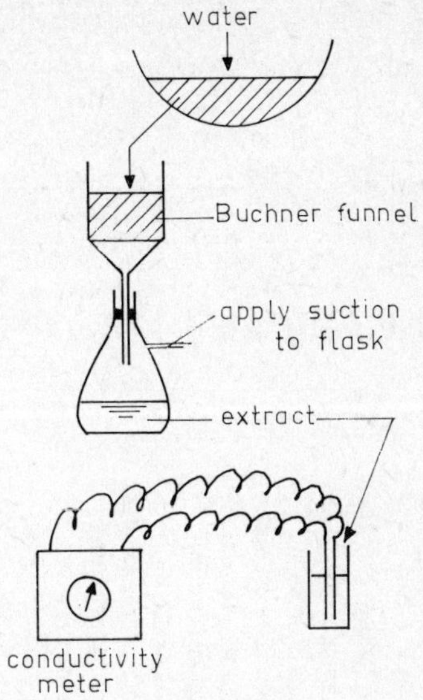

FIGURE 10.6 *Outline of the EC_e measurement*

Saturation point (SP) refers to a puddled soil and should not be confused with the moisture content of a saturated undisturbed soil. Sandy soils generally hold less water at SP than in saturated undisturbed conditions, while the opposite generally applies to clay soils. For medium/medium fine textured soils it roughly holds that (θ in %w):

$$\left.\begin{array}{l}\theta \text{ at SP} \sim 2 \times \theta \text{ at FC} \\ \theta \text{ at SP} \sim 4 \times \theta \text{ at WP}\end{array}\right\} \theta \text{ at SP} \sim 3 \times \begin{pmatrix}\text{average soil moisture} \\ \text{content in the available} \\ \text{moisture range, } AM\text{-}range\end{pmatrix}$$

As the salt concentration of the soil solution is inversely proportional to the soil moisture content, it follows for medium textured soils that the soil solution in the AM range is on average 3 × as concentrated as at SP, or EC faced by the plants $\simeq 3 \times EC_e$. For sandy soils a somewhat higher concentration ratio applies; for clays it would normally be slightly lower. The inverse proportionality between salt concentration and soil moisture content of course applies only when all salts remain dissolved. In the presence of large quantities of only slightly soluble salts (lime, gypsum), somewhat lower concentration ratios apply since precipitation occurs upon concentration.

Salinity conditions in the rootzone and thus the osmotic problems faced by the crop, are seldom uniform. Conditions vary spatially within the rootzone as well as in time. Plant response reflects the integration of these variable conditions experienced by the plant. As the root activity and water uptake generally decrease with depth, conditions in the upper rootzone may normally be expected to carry more weight than conditions in the lower rootzone. Which rootzone salinity parameter best relates to crop response is still the subject of further research (RHOADES 1974).

10.2.2 Toxicity problems*

Whereas osmotic problems are caused by the high total salt concentration of the soil solution, the cause of the toxicity problems appears to be either a high concentration

* For further information on tolerance of different crops to different kinds of toxicity see FAO 1976b.

TABLE 10.1 *Crop tolerance tables for soil salinity* (FAO 1976b)
(gives yield reduction to be expected at the EC_e-value indicated in the table, mmhos.cm^{-1})

	Yield reduction			
FIELD CROPS	0%	10%	25%	50%
Barley (Hordeum vulgare)	<8	10	13	18
Cotton (Gossypium hirsutum)	<7·5	9·5	13	17
Sugarbeet (Beta vulgaris)	<7	8·5	11	15
Wheat (Triticum aestivum)	<6	7·5	9·5	13
Safflower (Carthamus tinctorius)	<5·5	6	7·5	10
Soybean (Glycine max)	<5	5·5	6	7·5
Sorghum (Sorghum bicolor)	<4	5	7	11
Groundnut (Arachis hypogaea)	<3	3·5	4	5
Rice (paddy) (Oryza sativa)	<3	4	5	7
Sesbania (Sesbania macrocarpa)	<2·5	3·5	6	9·5
Sugar cane (Saccharum officinarum)		3	5	8·5
Corn (Zea mays)	<1·5	2·5	4	6
Flax (Linum usitatissimum)	<1·5	2·5	4	6
Broadbean (Vicia faba)	<1·5	2·5	4	7
Cowpea (Vigna sinensis)	<1·5	2	3	5
Beans (Phaseolus vulgaris)	<1·0	1·5	2·5	3·5
FRUIT CROPS				
Date palm (Phoenix dactylifera)	<4	7	11	18
Fig (Ficus carica), Olive (Olea europaea), Pomegranate (Punica granatum)	<2·5	4	5·5	8·5
Grapefruit (Citrus paradisi)	<2	2·5	3·5	5
Orange (Citrus sinensis)	<1·5	2·5	3	5
Lemon (Citrus limonea)	<1·5	2·5	3·5	5
Apple (Pyrus malus), Pear (Pyrus communis)	<1·5	2·5	3·5	5
Walnut (Juglans regia)	<1·5	2·5	3·5	5
Peach (Prunus persica)	<1·5	2	3	4
Apricot (Pyrus armeniaca)	<1·5	2	2·5	3·5
Grape (Vitis spp.)	<1·5	2·5	4	6·5

	Yield reduction			
VEGETABLE CROPS	0%	10%	25%	50%
Beets (Beta vulgaris)	<4	5	7	9·5
Broccoli (Brassica italica)	<3	4	5·5	8
Tomato (Lycopersicon esculentum)	<2·5	3·5	5	7·5
Cucumber (Cucumis sativus)	<2·5	3·5	4·5	6·5
Cantaloupe (Cucumis melo)	<2	3·5	5·5	9
Spinach (Spinacia oleracea)	<2	3·5	5·5	8·5
Cabbage (Brassica oleracea capitata)	<2	3	4·5	7
Potato (Solanum tuberosum)	<1·5	2·5	4	6
Sweet corn (Zea mays)	<1·5	2·5	4	6
Sweet potato (Ipomea batatas)	<1·5	2·5	4	6
Pepper (Capsicum frutescens)	<1·5	2	3·5	5
Lettuce (Lactuca sativa)	<1·5	2	3	5
Radish (Raphanus sativas)	<1	2	3	5
Onion (Allium cepa)	<1	2	3	4·5
Carrot (Daucus carota)	<1	1·5	3	4·5
Beans (Phaseolus vulgaris)	<1	1·5	2·5	3·5
FORAGE CROPS				
Tall wheat grass (Agropyron elongatum)	<7·5	10	13·5	19·5
Wheat grass (fairway) (Agropyron elongatum)	<7·5	9	11	15
Bermuda grass (Cynodon dactylon)	<7	8·5	11	14·5
Barley (hay) (Hordeum vulgare)	<6	7·5	9·5	13
Perennial rye grass (Lolium perenne)	<5·5	7	9	12
Trefoil, birdsfoot narrow leaf (L. corniculatus tenuifolius)	<5	6	7·5	10
Harding grass (Phalaris tuberosa)	<4·5	6	8	11
Tall fescue (Festuca elatior)	<4	6	8·5	13·5
Crested Wheat grass (Agropyron desertorum)	<3·5	6	10	16

Almond (Prunus amygdalus)	<1·5	2	3	4	Vetch (Vicia sativa)	<3	4	5·5	7·5
Plum (Prunus domestica)	<1·5	2	3	4·5	Sudan grass (Sorghum sudanense)	<3	5	8·5	14·5
Blackberry (Rubus spp.)	<1·5	2	2·5	4	Wildrye, beardless (Elymus triticoides)	<2·5	4·5	7	11
Boysenberry (Rubus spp.)	<1·5	2	2·5	4	Trefoil, big (Lotus uliginosis)	<2·5	3	3·5	5
Avocado (Persea americana)	<1·5	2	2·5	3·5	Alfalfa (Medicago sativa)	<2	3·5	5·5	9
Raspberry (Rubus idaecus)	<1·0	1·5	2	3	Lovegrass (Eragrostis spp.)	<2	3	5	8
Strawberry (Fragaria spp.)	<1·0	1·5	2	2·5	Corn (forage) (Zea mays)	<2	3	5	8·5
					Clover, berseem (Trifolium alexandrinum)	<1·5	3	6	10·5
					Orchard grass (Dactylis glomerata)	<1·5	3	5·5	9·5
					Meadow foxtail (Alopecurus pratensis)	<1·5	2·5	4	6·5
					Clover, alsike, ladino, red, strawberry (Trifolium spp.)	<1·5	2·5	3·5	5·5

Table 10.1 should be applied with caution, particularly keeping the following in mind:

(a) The table is based on results obtained in small plots, artificially salinised by irrigation water.
(b) The EC_e values indicated are average values over the active rootzone, although the salinity is normally somewhat higher in the upper than in the lower rootzone.
(c) The table applies to the growth period from seedling stage to early maturity. Generally, crops are slightly more sensitive during germination. Part of the salinity problems commonly experienced during germination are due to the fact that seeds are placed where soil salinity is highest (eg top of ridges) rather than due to a higher sensitivity to salinity during germination.
(d) The table is best applied to medium to medium fine textured soils. In light textured soils the soil solution between FC and WP is more concentrated than in the medium textured reference soils (see figure 10.6) and so slightly lower EC_e values should be adopted when crops are to be grown in light soils. In heavy clay soils the reverse argument applies and so slightly higher values may be adopted.
(e) The table applies when highly soluble salts dominate the soil solution. Somewhat higher EC_e-values may be adopted when only slightly soluble salts are present in large quantities (>20–30 meq.L^{-1}) or when the $CO_3^{--}+HCO_3^-$ concentration of the soil solution is high.
(f) Salt concentrations around the roots will generally be higher, the higher the evapotranspiration rates (E). Therefore it is prudent to take somewhat lower EC_e-values when E is high.

of a particular cation or anion, or an unfavourable salt composition in the soil solution resulting in an excess or imbalanced uptake by the plants.

Both soil and foliar analysis are used in the diagnosis of toxicity problems. These tests are often most conclusive if used in combination. Diagnostic parameters and criteria have been established for some forms of toxicity, notably toxicity resulting from excess uptake, but the salinity conditions underlying other forms of toxicity remain obscure. The occurrence of toxicity is often linked with a high salt concentration in the soil solution, thus occurring concurrently with osmotic effects. Part of the toxicity problems, therefore, are assessed jointly, although usually not explicitely, with the osmotic problems by the EC_e parameter, and the tolerance tables (table 10.1) which reflect a combination of toxicity and osmotic problems (although dominantly the latter).

Most serious toxicity problems are related to excess uptake of sodium, chloride and boron. Excess uptake leads to accumulation, especially in the leaves (concentration effect due to transpiration). Typical symptoms are leaf burn, scorch, dead outer leaf edges, especially occurring in the older leaves where the accumulation first reaches critical levels. With time the symptoms may progress between the veins towards the leaf centre. Fruit trees and other woody perennials are particularly sensitive to these problems while most annual crops are rather tolerant.

Na-toxicity

Sodium toxicity may be expected when the Na^+ concentration of the soil solution is high. The Ca^{++} concentration, however, is also important as sodium toxicity often reduces in the presence of soluble Ca in the soil. This is the reason why the ESP value is relevant for diagnosing Na-toxicity since this parameter reflects the Na^+/Ca^{++} ratio of the soil solution. Sensitive crops like deciduous fruits, nuts, citrus, avocado and beans are unable to tolerate ESP values in excess of 10–20%; most common crops are more tolerant but will suffer from the poor soil structure when ESP > 20% (see section 10.2.3.). Analysis of leaves from standing crops can help in the diagnosis of the problem. The Na content of leaves from affected and unaffected plants may be compared. Sensitive crops usually shows symptoms when the Na content of the leaf tissue is in excess of 0·25–0·50% (dry weight basis).

Cl-toxicity

Excess uptake of Cl affects sensitive crops like various kinds of berries, many fruit trees, some orange and grape spp. Damage may be expected when the Cl concentration in the saturation extract exceeds 10 $meq.L^{-1}$ or when the leaves contain > 0·3–0·5% Cl (dry weight basis).

B-toxicity

Plants require very little boron, but are sensitive to excess uptake (especially various citrus spp., grapefruit, many fruit trees, different nut spp. beans, many vegetables). Above normal boron concentration in the soil solution is seldom due to boron released *in situ* from the soil minerals, rather it is most likely to have originated from some external source, principally irrigation water. Few surface streams have significant concentrations of boron. It is, however, prevalent in well water and springs from geothermal areas or near earthquake faults. As a result boron toxicity is typically a local/regional problem. The problems can usually be assessed adequately by determining the boron content of the irrigation water (see section 11.5.3).

10.2.3 Dispersion problems

The adsorbed cations on the surfaces of the soil colloids (adsorption complex) form the so called *diffuse double layer.* Its thickness largely determines whether the soil colloids are in a state of coagulation or in a state of dispersion. When the layer is compressed the soil colloids are able to come close enough together to coagulate. When the layer expands, the attraction forces become too weak to hold the colloids together, as a result they disperse.

The thickness of the diffuse double layer depends mainly upon the following two factors:

(a) *The salt concentration of the soil solution*
A high salt concentration in the soil moisture compresses the layer while the layer expands when the salt concentration decreases. Thus soils are more liable to dispersion when the salt concentration of the soil solution is low than when it is high.

(b) *The composition of the adsorbed cations*
When the adsorbed cations are mainly divalent cations (especially Ca^{++}), the layer is compressed. The layer expands when there is a large proportion of monovalent cations (especially Na^+) on the adsorption complex. In the latter case, soils easily disperse.

It has been found that at normal salt concentrations the danger of dispersion in soils is closely correlated with the relative proportion of Na^+ compared to other cations adsorbed on the adsorption complex, this being expressed by the ESP-value (see section 10.1.3). Dispersion problems generally increase with the ESP-value.

The agricultural interpretation of ESP-values

A state of easy dispersion, as exists at high ESP-values, is liable to result in poor physical soil conditions such as:

(i) low hydraulic conductivity, conceivably due to blockage of pores by dispersed colloids. Downward movement by water of dispersed material may lead to the formation of an incipient clay pan, limiting root development and drainage;

(ii) unfavourable soil consistency: hard when dry and plastic-sticky when moist. Such soils are difficult to work;

(iii) low resistance to slaking, easily leading to the formation of surface crusts. These hamper the infiltration of water into the soil, and the mechanical strength of the crust is likely to hinder proper germination (irregular, patchy stands);

(iv) waterlogging, resulting from the general deterioration of the drainage characteristics of the soil associated with the above effects. However, care has to be taken in relating poor drainage in the soil to high ESP since, although high ESP frequently causes poor drainage, inherently poor drainage characteristics of the soil may also lead to high soil salinity, including high ESP-values.

In general problems are not experienced in soils with ESP-values $<15\%$. However, it has been pointed out that the salt concentration of the soil solution also has an influence on dispersion (see figure 10.7). This is why some soils having an

ESP > 15% do not disperse provided the salt concentration remains high. When this high salt concentration decreases, for example, due to leaching, dispersion problems arise. Of course, high salt concentrations in the soil are in any event harmful and the limit of ESP = 15% applies to acceptably low salt concentrations in the soil moisture ($EC_e \sim 1{\cdot}0$ mmhos.cm^{-1}, which corresponds with a salt concentration of approximately 10 meq.L^{-1}).

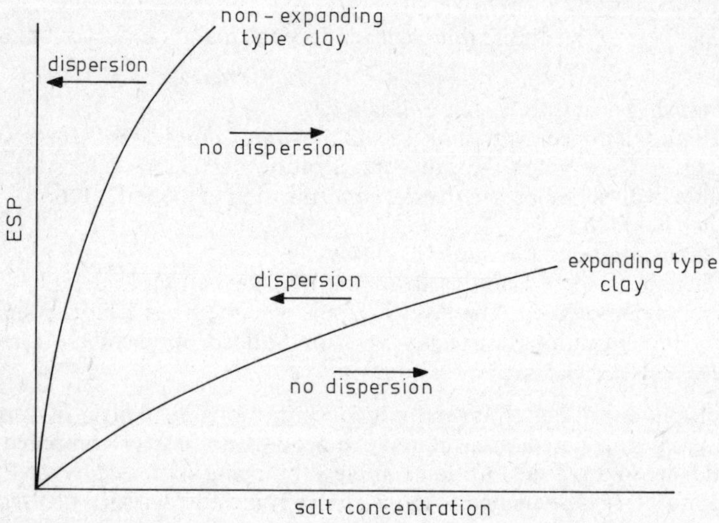

FIGURE 10.7 *Dispersion of soil as influenced by the salt concentration, ESP and type of clay mineral* (*adapted from* McNEAL 1974)

The limit of ESP = 15% is a general limit. Soils and crops vary in their sensitivity to dispersion problems. In general it can be stated that of two soils having the same ESP value, the soil with the highest percentage of colloids (especially when this involves clay minerals of the expanding lattice type) will experience the severest dispersion problems.* Rather than using the general limit of ESP = 15%, separate limits may be used for different textural classes (eg a limit of ESP = 10% for fine textured soils and ESP = 20% for coarse textured soils).

10.2.4 Corrosion problems

Certain salts may attack and corrode construction materials such as steel and concrete, requiring special protective measures when these are present in the soil. Concrete is especially affected by sulphates, Mg type sulphates (prevelant in sea water and as a result in coastal areas and in marine soils) being more aggressive than Ca type sulphates. Porous concrete is more sensitive to attack by salts than dense concrete. When the sulphate content of the soil is high, dense concrete made with sulphate-resistant cement should be used (generally when there is more than 1 000 mg.L^{-1} SO_4 in a 2:1 water extract of the soil).

* Although soils in Gezira, Sudan, do not seem to suffer from ESP values as high as 20–25% (Vertisols, often with > 50% montmorillonitic clay).

Steel is also corroded by sulphur in the water, especially by H_2S. Steel parts of tubewell pumps in particular are affected. The aggressivity of the water is enhanced by low pH-values, presence of CO_2 in the water and by low redox potentials of the water. Protective paintings may help, but in severe cases heavy duty steel or stainless steel should be used, rather than normal mild steel.

10.3 Classification and mapping of salty soils

10.3.1 Classification

In the taxonomic soil classification, soil salinity is recognised as a differentiating soil characteristic when the salinity is a more or less permanent inherent feature of the soil and has resulted in specific morphological or other soil conditions (groups of Halomorphic soil or Halosols). When the salts have not affected the soil permanently and can be readily removed by leaching, soil salinity is most appropriately classified at the phase level in a systematic soil classification system (eg soil type A, saline phase, to distinguish the soil in question from the standard soil type A which is non-saline).

Nowadays the system developed by the *US Salinity Laboratory* (Riverside, California) is most commonly used for the classification of salty soils. The system is suitable for phase level differentiation in a systematic soil mapping as well as for single value mapping. The diagnostic parameters in this system are the *EC_e-value* (indicative of the osmotic problems) and the *ESP-value* (indicative of the dispersion problems). The limits used conform to those discussed in sections 10.2.1 and 10.2.3.

TABLE 10.2 *The US Salinity Laboratory classification system for salty soils* (USDA 1954)

	$EC_e \leq 4$ mmhos.cm^{-1}	$EC_e > 4$ mmhos.cm^{-1}
ESP ≤ 15%	non-saline, non-sodic soils	saline soils[2]
ESP > 15%	sodic soils[1]	saline sodic soils

The pH-value of salty soils often fits the following pattern:

- saline soils pH < 8·5
- saline sodic soils pH ~ 8·5
- sodic soils pH = 8·5–10

} determined in the saturation extract.

This pattern is mostly due to the relationships discussed in Annex 10.1. Other factors may, however, interfere *viz* in the presence of lime the pH may well exceed 8·5 at low ESP, while in gypsiferous soils the pH seldom exceeds 8·2 regardless of the ESP.

Typical analytical results of the different types of salty soils are presented in table 10.3.

[1] Instead of *sodic* the term *alkali* is also used (older term).

[2] This rather specific and limited definition of the term *saline soils* precludes its general use for soils suffering from salinity. In this volume such soils are therefore referred to as *salty soils.*

TABLE 10.3 *Typical analytical data for different types of salty soils* (USDA 1959)

SP %w	CEC*	*Exchange Complex* Na$^+$	K$^+$	Ca^{++}	Mg^{++}	H$^+$	*Saturation extract* Ca^{++}	Mg^{++}	Na$^+$	K$^+$	Total	$CO_3^=$	HCO_3^-	SO_4^-	Cl$^-$	Total	EC$_e$ (mm hos. cm^{-1})	ESP %	SAR$_e$	pH	Gyp-sum (meq/100 gm soil)	Alka-line earth carbo-nates**
	(meq.100 gm^{-1} soil)						(meq.L^{-1})															
Non-saline, non-sodic soils																						
36	20	< 1	2	11	6	1	2·7	2·3	1·2	0·9	7·1	0	2·6	2·1	0·9	5·6	0·6	2	0·8	6·4	0	–
32	29	3	<1			0	3·3	1·9	12·2	0·7	18·2	0	6·1	4·3	4·9	15·4	1·7	10	8·0	7·8	0	+
40	17	< 1	<1			0	2·8	1·7	5·2	0·2	9·9	0	6·6	2·7	0·4	9·7	0·8	3	3·5	7·9	0	+
Saline soils																						
52	14	2	<1			0	31·5	37·2	102·0	0·2	170·9	0	4·5	90·0	78·0	172·6	13·9	13	17	7·9	7·1	+
47	17	2	<1			0	37·0	34·0	79·0	0·4	150·4	0	7·2	62·2	47·0	148·4	12·0	8	13	8·0	0	+
40	19	2	3			0	28·4	22·8	53·0	1·1	105·3	0	5·2	74·0	29·0	108·2	8·8	10	11	8·0	0	+
Sodic soils																						
59	33	6	1			0	1·1	1·4	15·6	0·4	18·5	0	6·5	8·5	2·9	17·9	1·7	18	14	8·3	0	+
61	34	8	1	11	10	4	1·4	1·0	21·5	0·3	24·2	0	3·3	3·8	16·7	23·8	2·5	24	20	7·3	0	–
39	22	10	7			0	1·1	0·3	29·2	4·1	34·7	8·4	18·7	4·6	7·5	39·2	3·2	46	35	9·6	0	+
Saline-sodic soils																						
62	36	9	1	10	12	4	6·7	9·9	79·5	0·5	96·6	0	2·4	20·1	72·0	94·5	9·2	26	28	7·3	0	–
60	40	11	1			0	32·4	38·3	145·0	0·5	216·2	0	3·3	105·0	105·0	213·3	16·7	26	24	7·8	42·2	+
36	26	17	2			0	0·6	0·9	58·5	1·6	61·6	5·0	19·9	21·5	16·3	62·7	5·6	63	68	9·3	0	+

* In principle: CEC = Na + K + Ca + Mg + H; in the table not all adsorbed cations are recorded.

**Present (+) or absent (–)

An older but still widely used form of classification distinguishes between:

Solonchak: this is essentially a saline soil but does not entirely exclude the saline sodic soils.
Solonetz: this classification is based on the presence of a pronounced columnar or prismatic structure in the B-horizon. Such a horizon, more or less pronounced, develops eventually in sodic soils but its development takes time (20–100 years). So a solonetz is almost always a sodic soil, but not all sodic soils by far are solonetzs.

10.3.2 Field appearance

Many salty soils have a normal field appearance. The salt content usually must be quite high before salinity becomes observable in the field. Soil salinity problems cannot therefore be properly assessed in the field but should be based on (laboratory) measurements. Field appearances of soils and vegetation are only able to give indications of soil salinity (usually of severe cases only). Typical symptoms in the soil are:

(a) *efflorescence phenomena*: powdery, crystalline salt deposits on exposed surfaces (from which water evaporates, leaving the salts behind) such as soil surfaces (especially high spots, eg top of ridges), side slopes of ditches, walls of soil pits, etc.
(b) *damp, oily looking soil surface* (due to hygroscopy of salts, especially $Ca\,Cl_2$);
(c) *mycelia* in the soil profile: salts precipitated in the fine pores, forming a pattern of thin white veins (usually carbonates);
(d) *crystals*, clustered or scattered, usually encountered at specific depths (especially with gypsum crystals);
(e) *crusts*: concentration of crystalline salt at certain depths (eg near or on the soil surface or any other evaporating plane in the profile) leading to the formation of a cemented layer;
(f) *dark film on the soil surface*, left by evaporating soil moisture containing dispersed organic matter (especially in the presence of $Na_2\,CO_3$).

The features (a), (b), (c), (d) and (e) mainly indicate a high salt content in the soil (high EC-value, saline soils) although the ESP may be high as well. Feature (f) indicates sodicity (high ESP and especially high pH). Poor soil physical conditions (hard consistency when dry, plastic/sticky when wet; surface crusting; poor aeration conditions; low hydraulic conductivity due to movement and blockage of pores by dispersed particles) are also commonly associated with high sodicity, although it must be remembered that poor physical conditions in the soil can be the result as well as the cause of the salinity problems.

Typical salinity symptoms in vegetation are: stunted growth, patchy stands ('slick spots'), dull/dark/bluish green colours and signs of toxicity as described in section 10.2.3. The salinity must be quite high for the signs to appear in crops. At low levels the crop growth is more likely to be rather uniformly depressed all over the field, without showing special signs. Such effects are easily overlooked; only yields will be reduced. In uncultivated land severe salinity is often apparent from (salt) indicator plants.

10.3.3 Mapping and sampling

The value of the use of the US Salinity Laboratory system of classification for mapping is its simplicity and the fact that the two classification parameters used, are diagnostic of the main salinity problems. The established limits used for EC_e and ESP, however, are very general. In detailed soil mappings the specific tolerances of the soil and the crops to be grown may be used to adapt the classification criteria accordingly. The general limits are, however, more appropriate for a preliminary assessment of the salinity conditions in an area.

Soil salinity is a highly dynamic soil characteristic, a fact that should be appreciated in salinity mapping. The time of the year, the crop, the irrigation schedule, soil management etc, all have an influence, and mapping results may differ as conditions vary. As salinity mapping is as a rule largely based on analysis of soil samples, considerable effort should be made to collect representative samples.

Random sampling is appropriate when all samples form one population. Locations are selected at random over the entire area. Normally, grid locations are sufficiently randomised to constitute a valid sampling. Stratified sampling is to be preferred to random sampling when it is thought that the salinity variation in the area is not entirely random but is influenced by identifiable factors (soil conditions, drainage conditions, topography, irrigation practices, soil management, etc).

It is seldom feasible or necessary to collect the large number of samples required for a proper statistical analysis of the results. It is usually sufficient to establish general patterns and the main underlying causal factors, which can then be used to inter or extrapolate the findings. A stratified sampling, skillfully carried out, can often provide this information with an acceptably limited number of samples.

Composite sampling (average sample composed by mixing samples from a number of sites) is as a rule of little use in salinity investigations. The salinity patterns in a soil profile and in an area are often fairly unique and considerable insight can be gained into the underlying salinisation processes from a careful study of these patterns. Composite sampling largely obscures these patterns.

Sampling can best be done at standard depths, unless there are special reasons to assume a particular vertical salt distribution. As a rule, salinity variations decreases with depth and therefore the sampling intensity should be highest in the upper profile zone. Suitable standard sampling depths are: 10 cm, 30 cm, 60 cm, 100 cm, and 150 cm. Salt crusts should be sampled separately. The required sampling intensity of the deeper layers (eg 250/300 cm) will rapidly become evident during the course of the survey. The same observation also applies to the sampling of groundwater for salinity evaluation.

Some correlations between salinity parameters which are useful to soil salinity survey work, are compiled in Annexes 10.1 and 10.2.

References and further reading

ALLISON, L E 1964: 'Salinity in relation to irrigation', Advances in Agronomy no 16: 139–179, American Society of Agronomy, Madison, Wisconsin

BERNSTEIN, L 1974: 'Crop growth in relation to salinity'. In: *Drainage for Agriculture*, J van Schilfgaarde (ed), Agronomy Monograph no 17, American Society of Agronomy, Madison, Wisconsin

BOLT, GH and MGM BRUGGENWERT (ed) 1976: *Soil Chemistry*, part A: Basic Principles, Elsevier, Amsterdam

FAO 1971: 'Salinity Seminar Baghdad', Irrigation and Drainage Paper no 7, Food and Agriculture Organisation of the United Nations, Rome

FAO 1976a: 'Prognosis of salinity and alkalinity', Soils Bulletin no 31, Food and Agriculture Organisation of the United Nations, Rome

FAO 1976b: 'Water quality for agriculture', Irrigation and Drainage Paper no 29, Food and Agriculture Organisation of the United Nations, Rome

FAO/UNESCO 1973: *Irrigation, Drainage and Salinity*: an international source book, Hutchinson, London

ILRI 1963: *Reclamation of Salt Affected Soils in Iraq*, publication 11, International Institute for Land Reclamation and Improvement, Wageningen

LOVEDAY, J (ed) 1974: 'Methods for analysis of irrigated soils', Technical Communication 54, Commonwealth Bureau of Soils, Harpenden, England

MCNEAL, BL 1974: 'Soil salts and their effect on water movement'. In: *Drainage for Agriculture*, J van Schilfgaarde (ed), Agronomy Monograph 17, American Society of Agronomy, Madison, Wisconsin

REEVE, RC and M FIREMAN 1967: 'Salt problems in relation to irrigation'. In: *Irrigation of Agricultural Lands*, Agronomy Monograph no 11, American Society of Agronomy, Madison, Wisconsin

RHOADES, JD 1974: 'Drainage for salinity control'. In: *Drainage for Agriculture*, J van Schilfgaarde (ed), Agronomy Monograph no 17, American Society of Agronomy, Madison, Wisconsin

USDA 1954: Diagnosis and improvement of saline and alkali soils, Agricultural Handbook no 60, US Department of Agriculture, Washington DC

YARON, B et al (eds) 1973: *Arid Zone irrigation*, Chapman and Hall, London

Annex 10.1 Determination of the ESP-value of the soil

Different methods can be followed to determine the ESP-value of a soil. Measured values may be expected to be more precise than the results of correlation methods. The accuracy of the SAR correlation method is, however, often good enough for routine work. The determination of the ESP by correlation with EC and pH is less accurate and should only be used to obtain rough, indicative values.

1 *Measurement*

The ESP measurement involves the following three steps:

(1) Measurement of Na^{+}_{ads} (in meq/100 gm soil), usually done by replacing Na^{+} and all other cations on the complex by NH_4^{+}, followed by measuring the displaced Na^{+} in the soil water extract. All this is standardised at SP. At a more diluted soil:water ratio, Na^{+}_{ads} will be less due to replacement by Ca^{++} and Mg^{++} (see section 10.1.3).

(2) Measurement of the CEC (in meq/100 gm soil), usually done by saturating the complex with one type of cation eg with Na^{+}, then replacing this cation, and measuring the displaced quantity in the soil water extract (standardised at pH = 7·0).

(3) Calculation: $ESP = (Na^{+}_{ads}/CEC) \times 100\%$.

2 *Correlation with SAR*

On the basis of the SAR_e (saturation extract), the ESP of the soil may be estimated with the aid of the empirical relationship shown in figure 10.3. SAR-values of other extracts (1:1, 1:2½, etc, soil:water ratio's) or even the SAR_i (irrigation water), should be converted into SAR_e values before being used in figure 10.3; the use of unconverted values, as is sometimes done, is only acceptable for rough estimates. Use of converted values may lead to quite erroneous results in the presence of considerable quantities of only slightly soluble Ca or Mg salts and CO_3^{--} or HCO_3^- as then precipitation and dissolution of ions will also affect the composition of the soil solution. This even applies when SAR_e is used directly, as these precipitation/dissolution effects are only partly accounted for in figure 10.3.

3 *Correlation with EC and pH* (ILRI 1963)

The following trends underlie this correlation.

(a) *At constant ESP the pH will tend to decrease as the EC increases*: as the salt concentration of the soil solution increases (ie EC increases), H^+ on the adsorption complex is exchanged for base type cations, thus lowering the pH of the soil solution

(b) *At constant EC, the pH increases with the ESP*: a high ESP corresponds to a high Na^+ concentration in the soil solution which leads to hydrolysis:
$2Na^+ + CO_3^{--} + 2H_2O \leftrightarrows 2Na^+ + 2OH^- + H_2CO_3$ or
$Na^+ + H_2O \leftrightarrows H^+ + Na^+ + OH^-$

(c) *The ESP increases with the EC and vice versa*: this tendency, although quite common, is not universal. It is inconceivable that the ESP will decrease with increasing EC, although the ESP may well remain fairly constant.

In case all three tendencies are operative and are fairly pronounced, the following type of plot emerges:

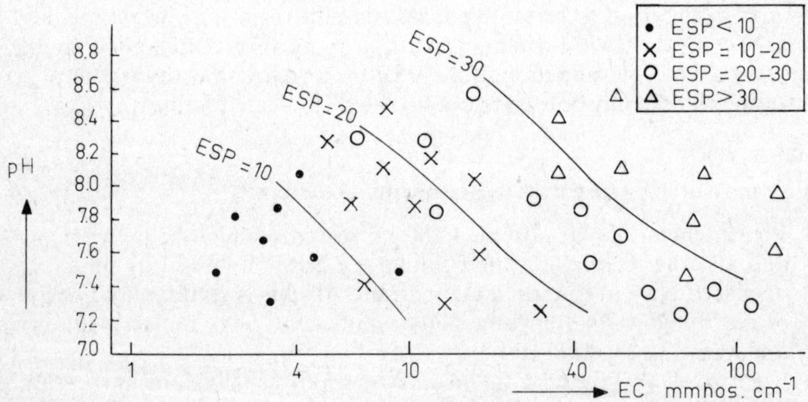

FIGURE 10.8 *Relationships between the pH, the ESP and the EC (salt concentration) frequently found in salty soils (example)*

This type of plot enables the ESP to be estimated on the basis of two simple measurements *viz* the pH and the EC (determined in a saturation extract, in a 1:1 or 1:2½ or 1:5 soil-water extract, or whichever extract the correlation graph is based on).

Annex 10.2 Relationships between EC_e and EC_1, EC_2, EC_5

Extraction of a soil moisture sample at SP requires vacuum suction equipment. At higher soil:water ratios there is enough free water to obtain such a sample by simple filtration. In large scale soil salinity survey work, therefore, EC values are often measured at lower soil:water ratios, eg

EC_1 value: measured in a 1:1 soil:water extract
(100 gm dry soil in 100 gm water)
EC_2 value: measured in a 1:2 soil:water extract
EC_5 value: measured in a 1:5 soil:water extract

These EC_x values are useful for a relative assessment of the soil salinity, eg the salinity distribution within an area. However, no set of standards exists as for EC_e-values to evaluate crop response. Establishment of a new set of standards is seldom done; in most cases conversions are made to EC_e-values, and the interpretation of the EC_x value is then done on the basis of the available EC_e standards (as given in table 10.1). Appropriate conversion factors can be derived either by correlation or by calculation.

(a) *Conversion by correlation*

This method is often used for salinity mapping in a large area. An 'EC_x/EC_e' conversion ratio is derived by determining the EC_e-value as well as the EC_x-value on a selected number of samples, usually at the start of the survey. On all other samples only the EC_x-value is determined and the established conversion ratio is used to estimate the corresponding EC_e-value:

$$EC_e = (EC_e/EC_x) \times EC_x$$

If the salt composition in an area is fairly uniform, this method can be quite satisfactory. In the example shown the EC_e/EC_2 ratio decreases with increasing soil salinity. This is quite common in the presence of poorly soluble salts. Compared to the EC_e value, the EC_2 value is measured in a more diluted extract so proportionally more of the poorly soluble salts are dissolved. When only highly soluble salts are present, a fairly constant ratio is to be expected.

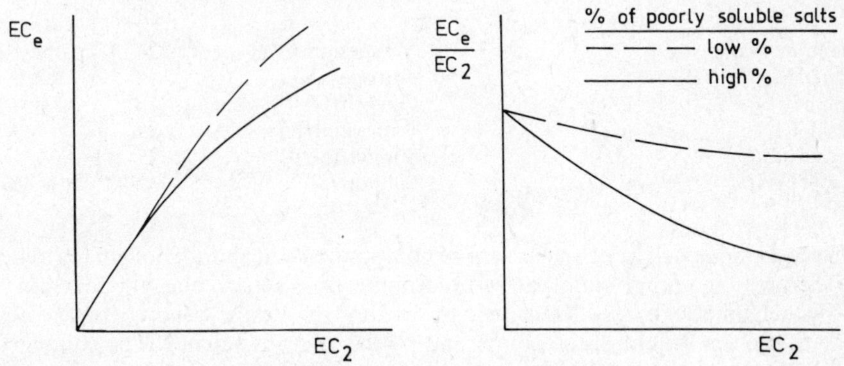

FIGURE 10.9 *Typical relationships between the EC-values of the soil solution for different extraction ratios*

(b) *Conversion by calculation*

Assuming an inverse linear relationship between the EC-value and the soil moisture content (θ), the following relationship holds

$$EC_e = \theta_x/\theta_{SP}.EC_x$$

Given the moisture content of a soil at saturation point (SP) conversion factors can be calculated.

Example

$EC_1 = 4\,\text{mmhos.cm}^{-1}; \theta_1 = 100\%\text{w}; \theta_{SP} = 40\%\text{w}$
so $EC_e = (100/40) \times 4 = 10\,\text{mmhos.cm}^{-1}$

Roughly the following conversion factors apply:

	Coarse	*Medium*	*Fine*
θ_{SP} (w%)	25–35	35–50	50–70
Conversion factors $EC_1 \rightarrow EC_e$	3–4	2–3	$1\frac{1}{2}$–2
Conversion factors $EC_5 \rightarrow EC_e$	14–20	10–14	7–10

Since the assumed inverse linear relationship between EC-values and θ clearly does not hold in the presence of poorly soluble salts, conversion by calculation should be applied carefully. For example, in a lime containing, heavy textured soil the following conversion factors were determined by correlation:

- at low salinity levels, $EC_e/EC_5 = 3{\cdot}5$
- at high salinity levels, $EC_e/EC_5 = 2{\cdot}6$

which compare to the factors 7–10 calculated above. Calculated factors can easily overestimate the soil salinity by a factor of 2 to 3. Chances of overestimation decrease:

(a) the closer the moisture content at the measurement is to SP
(b) the lower the salt content of the soil
(c) the lower the percentage of poorly soluble salts.

Annex 10.3 Selected atomic weights and their valencies

	Weight	*Valency*		*Weight*	*Valency*
Aluminium (Al)	27	2 or 3	Magnesium (Mg)	24	2
Boron (B)	11	3	Nitrogen (N)	14	3 or 5
Calcium (Ca)	40	2	Oxygen (O)	16	2
Carbon (C)	12	4	Potassium (K)	39	1
Chlorine (Cl)	35	1	Sodium (Na)	23	1
Hydrogen (H)	1	1	Sulphur (S)	32	2, 4 or 6
Iron (Fe)	56	2 or 3			

The equivalent weight of an element or compound is the number of gm (or mg) of that element or compound that will combine or displace one gm (or mg) of hydrogen. Numerically it is determined by dividing the atomic weight of the element or the molecular weight of the compound by the relevant valency. The equivalent weight of for example $CaCO_3$ is thus: Ca (40) + C (12) + 3 O (3×16) = 100 (molecular weight), divided by the valency (in this case the valency of Ca or CO_3, both 2), giving as equivalent weight 50 gm (or mg).

Chapter 11 Salinisation due to irrigation

Soil salinisation generally refers to the development of a non-salty soil into a salty soil, and especially to the development of a non-saline soil into a saline soil. It involves an increase of the soluble salt content of the soil, resulting in an increase of the salt concentration of the soil solution. The ESP, depending on the salt composition, may increase as well and sodic soils may develop (see section 11.3). Sodification almost never occurs in isolation but is usually triggered off by salinisation. Salinisation is the basic process underlying the development of almost all salty soils, whether saline or sodic.

Salinisation processes discussed in this chapter are those leading to high salinity in the upper soil layers and specifically in the rootzone (rootzone salinisation). Two main processes are discussed, both related to irrigation, thus emphasising the fact that most serious salinisation problems occur in irrigated areas.

11.1 Direct salinisation by irrigation due to inadequate leaching

All water for irrigation contains some salts. Most irrigation water originates as rainfall which percolates through the soil towards the groundwater and onwards towards the rivers, collecting salts on its way. The use of groundwater for irrigation poses especial problems because this water in particular may contain a considerable salt load. This is especially true of arid climates where, due to low rainfall and high evaporation, groundwater is not refreshed so frequently as in humid climates and salts tend to become more concentrated. Rivers often have a higher salt content during the low flow season than during the flood season, while salt conditions may also vary along the course of a river.

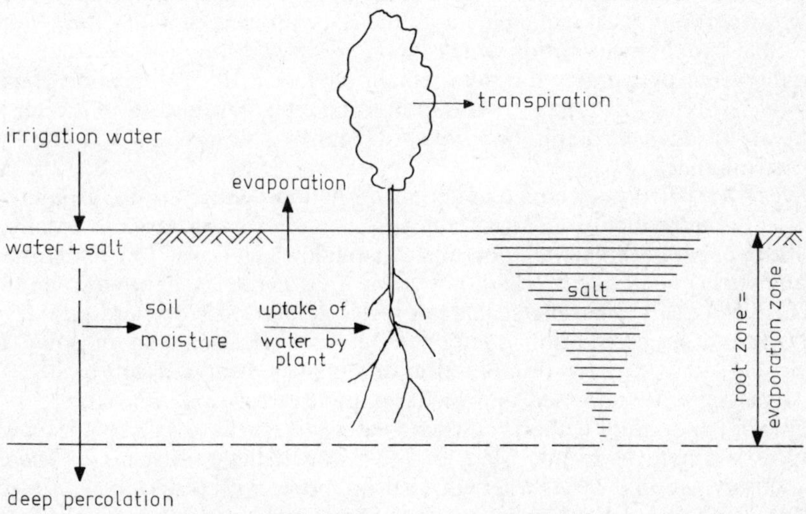

FIGURE 11.1 *Direct salinisation by irrigation*

Each irrigation brings a certain amount of salt into the rootzone. As the water is lost by evapotranspiration, the salts remain behind in the rootzone/evaporation zone (see figure 11.1) where they will accumulate unless an equivalent amount of salt is removed from this zone as is brought in by the irrigation water (*salt balance concept*). The salt uptake by crops is small, salt removal depending almost completely upon leaching by deep percolation, ie the washing out of salt by water percolating through the soil to depths below the rootzone.

This type of salinisation is most likely to occur under conditions where the salt influx to the rootzone is high and/or the salt efflux from the rootzone is low.

High salt influx conditions prevail when:

- the climate is hot and dry (high irrigation water requirements)
- the water is saline.

Low salt efflux conditions prevail when:

- the climate is hot and dry (low rainfall/high evaporation, so little excess rain goes into deep percolation)
- minimum irrigation practices are followed, ie just enough water is applied to meet the crop evapotranspiration, leaving no excess for deep percolation (under-irrigation)
- drainage conditions are poor, allowing insufficient percolation and drainage discharge

Salt balances for the rootzone under irrigation are extensively discussed in section 11.4.

11.2 Salinisation from groundwater (capillary salinisation)

Evaporation of (saline) groundwater from the soil is a common cause of soil salinisation. The groundwater may evaporate directly from the water-table when the latter occurs within the evaporation zone, or it may be drawn from deeper down as the evaporation itself will create a gradient for upward capillary flow from the watertable into the evaporation zone.

As the water evaporates the salts remain behind in the evaporation zone (see figure 11.2). So for this type of salinisation to take place, saline groundwater must occur within such a depth that upward capillary flow is able to reach the evaporation zone.

A great deal of irrigated land is underlain at shallow depth by saline groundwater. Prior to the introduction of irrigation, the groundwater recharge is usually low, consisting only of deep percolation due to rainfall which is also low under the arid climatic setting of a typical irrigation scheme. In general, the natural groundwater discharge can easily cope with such a low recharge, even with deep watertables (low head). The deep percolation component of the irrigation can in some cases (permeable soil, poor irrigation practices and management) amount to 20–30% of the irrigation water supplied and recharge under irrigation can easily rise to a multiple of the original recharge. Groundwater tables will rise up to a level which produces sufficient hydraulic head for the required higher discharge. There are numerous irrigation projects where this has happened with watertables rising from 20–30 m depth to 1–2 m depth below the soil surface in a period of 10–15 years after the start of the project.

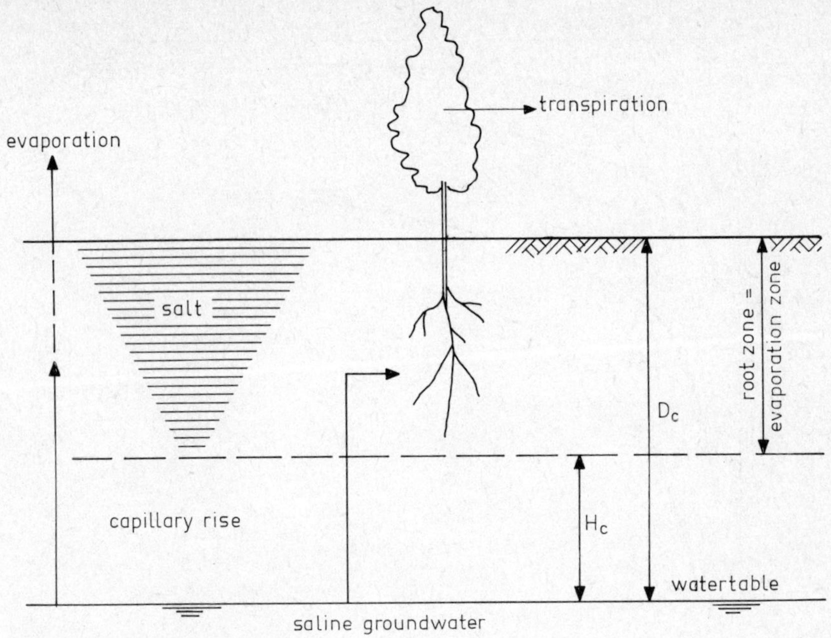

FIGURE 11.2 *Capillary salinisation*

11.2.1 Critical watertable depth

Capillary flow upwards from a watertable can reach to great heights but the rate of flow generally decreases with increasing height above the watertable. So the rate of upward salt movement, being proportional to the rate of upward flow, also decreases as the distance between the evaporation zone and the watertable increases (see figure 11.3). The distance at which the upward capillary flow becomes too small for any significant upward salt movement is called the *critical capillary height* (H_c). Its value depends on:

(a) *soil type*: soils with a large proportion of their pores of a small size, have large H_c values (fine sandy loam, silty loam). Coarse sand has a small H_c value whereas well structured medium to fine soils have intermediate values (see figure 11.3);

(b) *salt concentration of the groundwater*: upward salt movement is the product of 'flow rate' × 'salt concentration'. Therefore H_c values should increase with increasing groundwater salinity (see figure 11.4). In general, very little capillary salinisation will occur provided the salt concentration in the upper groundwater layer remains $< 1\,000$ mg.L^{-1} (EC $\leqslant 1{\cdot}5$ mmhos.cm^{-1}).

In situations where the groundwater inflow or outflow in the area is insignificant, the critical watertable depth (D_c) may be estimated as the depth to which the watertable falls towards the end of a long dry period. Some indicative values for H_c follow:

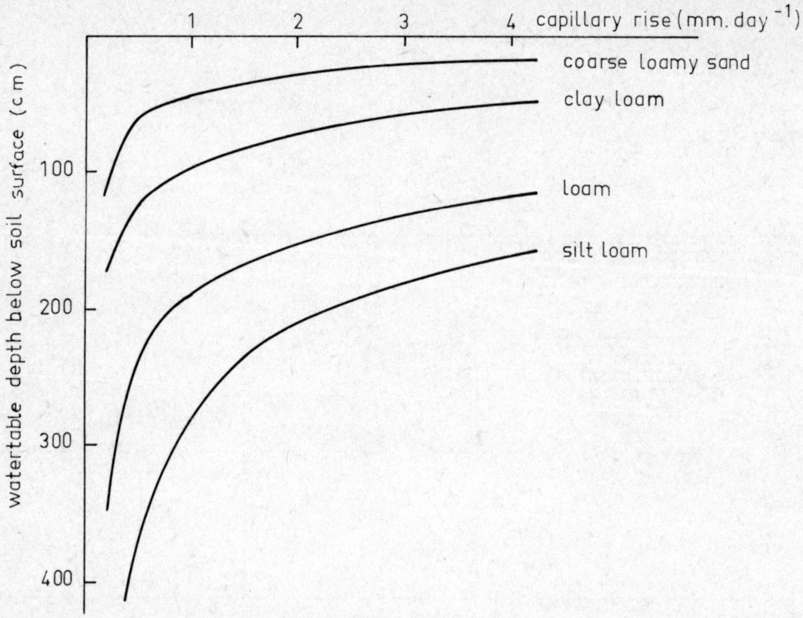

FIGURE 11.3 *Rates of capillary rise to the soil surface from stationary watertables at different depths when the soil moisture pressure at the soil surface is* – 16 *bar* (Rijtema 1969)

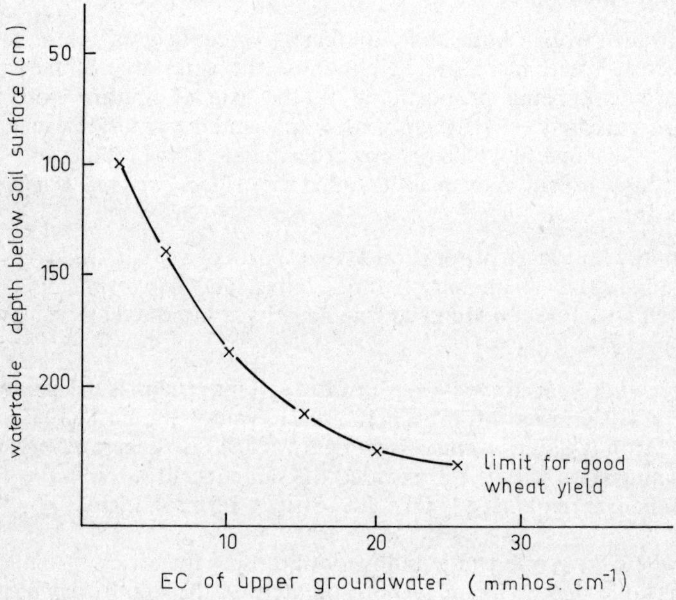

FIGURE 11.4 *The interaction between critical water table depth and salt concentration of groundwater (Lower Indus, Pakistan)*

- sand (coarse→fine) $H_c = 50 \rightarrow 75$ cm
- loamy sand, sandy loam $H_c = 100$–150 cm
- fine sandy loam, silt loam $H_c = 150$–200 cm
- loam, clay loam, clay $H_c = 100$–150 cm

These values apply to uniform profiles. Stratification will generally reduce the high values, but may well increase the low values.

11.2.2 Factors influencing capillary salinisation

Capillary salinisation from groundwater is influenced by a number of factors discussed in this section.

(a) *Influence of the evaporation rate* (*E*)

A high evaporation rate has a strong drying effect leading to a low moisture content in the evaporation zone and to correspondingly high suction forces. These enhance upward capillary flow although on the other hand this flow is restricted by the low moisture content (= low conductivity) in the evaporation zone. The net result is that after a short initial period of high soil moisture depletion, the dryness of the evaporation zone reduces the moisture losses from the layers below. As the evaporation zone becomes drier and thicker, losses will eventually become almost insignificantly small. Such a dry, protective layer is called a *mulch layer*.*

When the evaporation rate is low, poorly developed mulch layers are formed and evaporation losses from the soil can continue—although at a rather low rate—for a long period, drawing water from great depths. This explains why salinisation from groundwater is often less hazardous under conditions of high evaporation than under moderate evaporation. In the former case the initial salinisation is very pronounced but of short duration while in the latter case the salinisation rate is smaller but as it continues over a longer period, its effect is often greater.

(b) *Influence of vegetation*

In bare soil the evaporation zone is shallow (approx 15–20 cm). When there is vegetation, the water loss will occur over a much deeper zone, about equal in depth to the main rootzone from which water is taken up for transpiration.

From figure 11.5 it would appear that salinisation is a greater hazard in cropped land than in bare land, in view of the differences in D_c. This however is generally *not* the case. In arid climates crops are grown either under irrigation or in the rainy season, and in both cases there will be a leaching effect of the deep percolation component of the irrigation or of the rain water. It is mostly on fallow land during the dry season that capillary salinisation presents problems (enhanced by weed growth which negates the beneficial effects of the self-mulching).

(c) *Influence of ground water recharge*

As groundwater moves upward by capillary flow, the watertable will fall continuously to, or even beyond the depth D_c at which point capillary salinisation will virtually cease. For salinisation to be continuous, there has to be groundwater

* The process described is usually referred to as *self mulching*, indicating that the mulch establishes itself automatically as the soil dries out due to evaporation. This distinguishes it from the artificial establishment of a mulch layer by tillage or by spreading loose dry (organic) material on the soil surface.

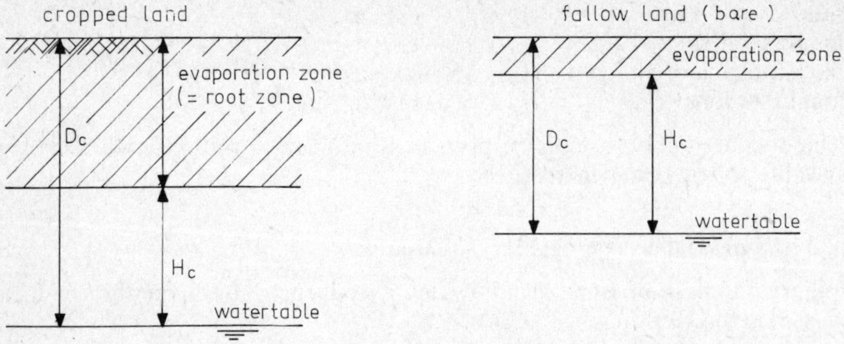

FIGURE 11.5 *Variation of the critical watertable depth with and without vegetation*

recharge compensating for the losses due to upward flow, and maintaining high watertables.

Groundwater recharge may occur as a result of deep percolation (either excess irrigation or rainwater) or as a result of seepage.

(i) *Recharge by deep percolation of irrigation/rainwater*

Deep percolation has a *recharge effect* and in addition it also has a *leaching effect*. It occurs intermittantly after irrigation or rain, ceasing at some stage between these events. The watertable thus fluctuates, conceivably being alternately below D_c (no salinisation) and above D_c (salinisation). Leaching also occurs in cycles. By their nature these two cycles are out of phase, with salinisation always following leaching. Figure 11.6 depicts a short term cycle related to a single irrigation or rain storm. However, the events and relationships depicted apply equally to long term cycles related to wet and dry seasons.

The balance between the salinisation and the leaching effects of deep percolation is of course of interest. In the long term (eg on an annual basis) there may well be an equilibrium, although serious salinisation may still occur for a temporary period within the interval considered. This is more likely to occur the longer the time interval between leaching events. Thus serious salinisation may occur when fields are irrigated say every two to three weeks but not when irrigated on a one-week schedule. Serious salinisation is also to be expected after the end of the rainy season when watertables are high and with little leaching in prospect until the following rainy season.

(ii) *Recharge by seepage*

The term seepage refers to groundwater flow into an area from outside the area (chapter 13). Typically, seepage water emerging in the area has travelled over considerable distances and depths through the soil or through the underlying rock, examples being shown in figure 11.7.

Areas subjected to seepage inflow are especially liable to salinisation by groundwater, because:

- they commonly have high watertables;
- upward capillary flow is compensated by seepage recharge so that watertables remain continuously high;

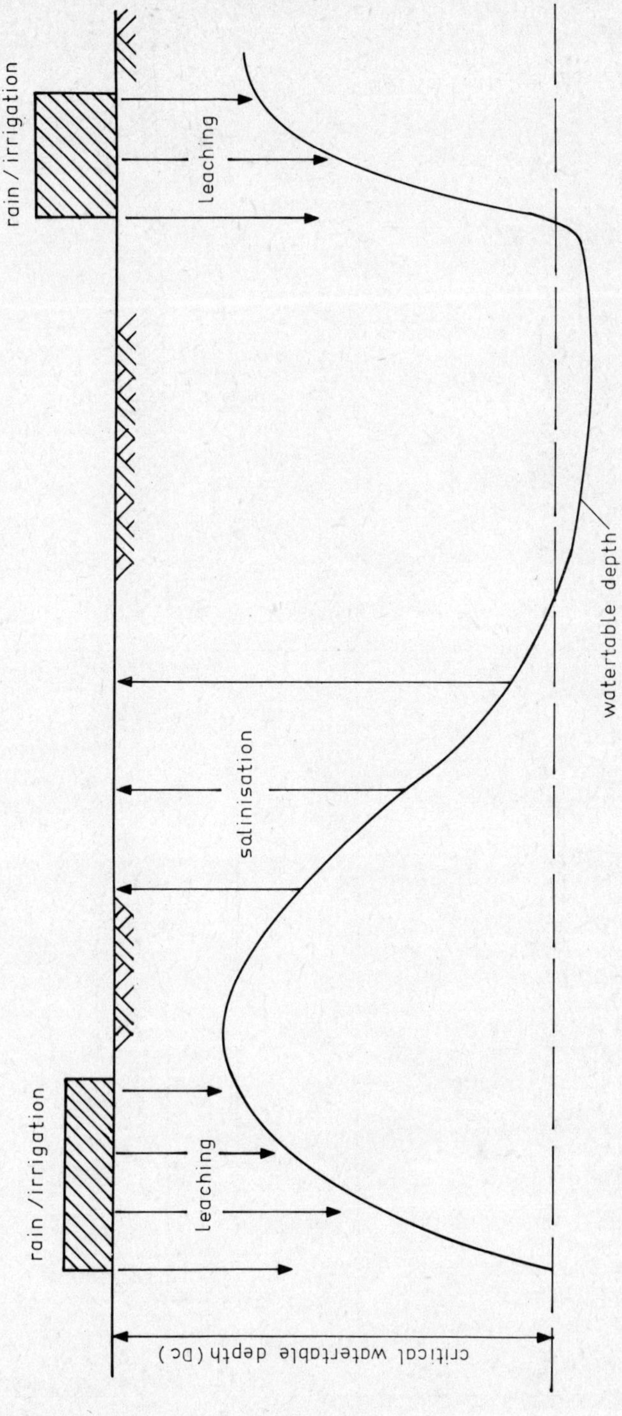

FIGURE 11.6 *Cyclic movement of salts due to watertable fluctuations*

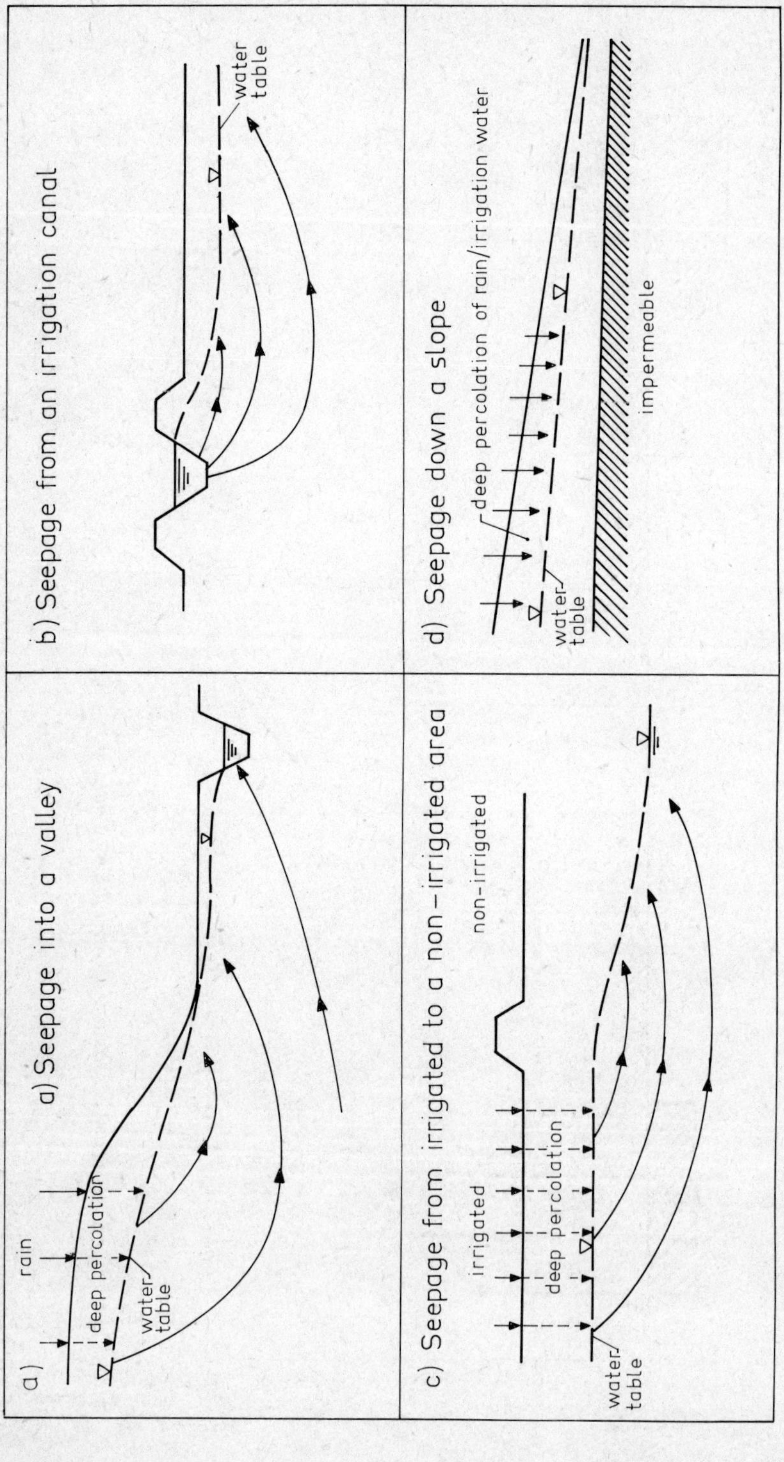

FIGURE 11.7 *Some typical seepage situations (adapted from* van der Molen 1973)

– the incoming seepage is often quite saline, having collected salts on its way through the soils or rocks.

The seepage rate and the salt content of the seepage water vary throughout the year, in line with variations at the source. This is especially evident in cases like that depicted in figures 11.7-b and 11.7-c. In the case of figure 11.7-a the seepage will continue, although at a diminishing rate, long after the supply to the source (rain) has stopped, until the reservoir of water held in the soil at the source has been depleted.

11.3 SODIFICATION

Sodification involves the replacement of other cations on the adsorption complex by sodium.* Significant replacement only occurs when Na^+ becomes the dominant soluble cation in the soil solution (high SAR-value, see sections 10.1.2 and 10.1.3). This may occur when the salinising source is Na^+ rich, when the soil solution becomes more concentrated or when for other reasons the salinisation processes favour the accumulation of Na^+. The presence of CO_2^{--} and HCO_2^- in the soil solution is especially important. These anions form salts with Ca^{++} which are only slightly soluble while the corresponding Na^+ salts are highly soluble. Their presence thus leads to a relative enrichment of the soil solution with Na^+ as the overall salt concentration increases because the Ca^{++} salts precipitate.

Sodic soils, having a low concentration of salts in the soil solution but a high Na^+ occupancy on the complex, also occur. The high ESP that these soils exhibit are unlikely to have developed under conditions of low salinity. Rather, these soils have almost always developed from saline-sodic soils by a leaching process in which the soluble salt concentration of the soil decreased more rapidly than the Na^+ occupancy on the complex.

11.4 SALT BALANCE OF IRRIGATED LAND

An equation expressing the salt balance for a rootzone under irrigation reads (see figure 11.8):

$$I.C_i + R.C_r + G.C_g = PC_p + \Delta S \qquad \text{......(eq. 11.1)}$$

where: I = irrigation water entering the rootzone
R = rainfall entering the rootzone
G = capillary flow from groundwater into the rootzone
P = deep percolation from the rootzone
C = salt concentration of the water
(subscripts i, r, g, p referring to irrigation, rainfall, etc)
ΔS = change in salt content of the soil solution in the rootzone

I, R, G and P may be expressed in various units although mm/period is most convenient (eg mm/month, mm/season, etc). The correct unit for C is (m)gm.L^{-1} but

* The older term alkalisation is also still commonly used to describe these same processes.

since C and EC are linearly related (see figure 10.5) (m)mhos.cm^{-1} may be used as well. Eq. 11.1 applies to the soluble salts only. Salts that precipitate or come into solution during the period under consideration are neglected. The salt uptake by common crops is small, either to be neglected or assumed to be offset by the fertilizer application.

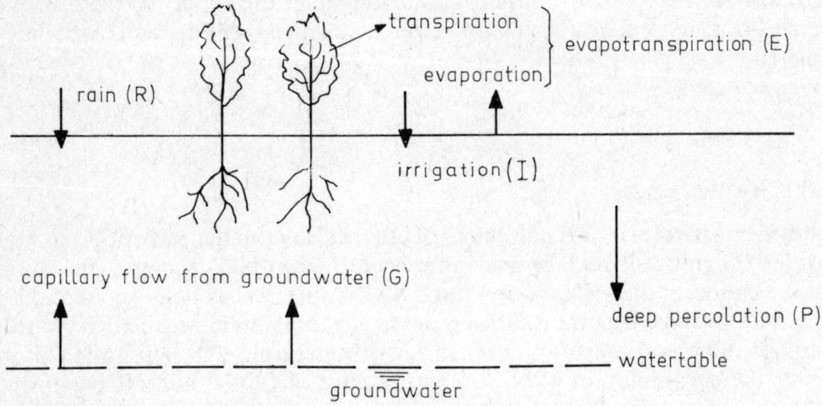

FIGURE 11.8 *Water balance of irrigated land*

Eq. 11.1 reduces to eq. 11.2 when:

(a) C_r is negligibly small as is the case for normal rainwater which has an EC-value of 0·02–0·05 mmhos.cm^{-1}; close to the sea, however, the salt content of the rainwater may rise to EC = 0·20–0·30 mmhos.cm^{-1}
(b) $C_g = C_p$, a reasonable assumption for averages over annual periods; less valid for short periods
(c) $\Delta S = 0$ ie the salt balance is in equilibrium (the salt content in the rootzone at the beginning of the period and at the end being equal).

$$\left.\begin{aligned} I.C_i &= (P-G).C_p = LR.C_p \\ I.EC_i &= (P-G).EC_p = LR.EC_p \end{aligned}\right\} \qquad \text{......(eq. 11.2)}$$

where *LR* = *leaching requirement* which is the excess of P over G. When a water quantity LR satisfying eq. 11.2 is drained from the rootzone, as much salt is leached from the zone as is brought in by the irrigation.

Eq. 11.2, expressed in EC-units, may be rewritten as:

$$LR = (EC_i/EC_p).I \qquad \text{......(eq. 11.3)}$$

Eq. 11.3 expresses LR as a fraction of I, ie the fraction of the infiltrated irrigation that must go into deep percolation in order to maintain a salt balance in the rootzone. The fractional factor EC_i/EC_p is called the *leaching fraction* (or *leaching percentage* when expressed as $EC_i/EC_p \times 100\%$).

In an equilibrium situation the following water balance also holds (figure 11.8):

$$I = (E-R)+(P-G) \qquad \text{......(eq. 11.4)}$$

The difference (E − R) is the rainfall deficit (= *net crop irrigation requirement*, designated as I_c) while the difference $(P-G)$ represents the leaching requirement, so:

$$I = I_c + LR \qquad \text{......(eq. 11.5)}$$

where I may be termed the *net soil irrigation requirement* (= nett crop irrigation requirement + leaching requirement). Combining eq. 11.3 and 11.5 gives:

$$LR = \frac{EC_i}{EC_p - EC_i} \cdot I_c \qquad \text{......(eq. 11.6)}$$

The required LR may be calculated, using this formula, on the basis of the known values of EC_i and I_c and a value of EC_p selected on the basis of an acceptable level of salinity in the rootzone (see next section).

11.4.1 Leaching requirement calculations

Values for EC_p may be determined experimentally by sampling the soil below the rootzone, the upper groundwater or the drain water. In planning, EC_p-values are usually set equal to the maximum salinity which can be tolerated in the rootzone, taking into consideration that:

(a) under leaching the soil salinity increases with depth (figure 11.9). In planning it is commonly assumed that EC_e-values at the lower rootzone boundary corresponding to 25 to 50% yield reduction ($EC_e25\%$ to $EC_e50\%$ as read from table 10.1) are still acceptable. The reason for this is that the weighted average EC_e-value for the entire rootzone (weighted according to root distribution) would still be only $EC_e0\%$ to $EC_e10\%$ while the salinity in the upper rootzone (on which crop response mainly depends) would be close to the $EC_e0\%$ value;

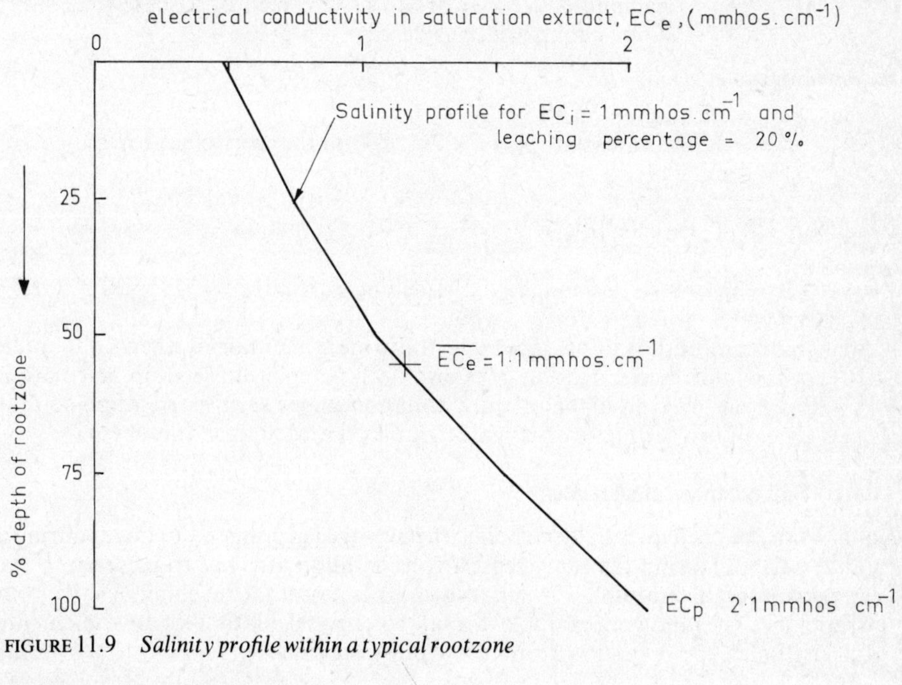

FIGURE 11.9 *Salinity profile within a typical rootzone*

(b) the rate of downward percolation and leaching varies with the soil moisture content. This rate is highest when the rootzone is above field capacity (FC), although this situation only prevails during the first 1 to 2 days after irrigation. Thereafter leaching continues at a decreasing rate. The average soil moisture content during leaching is about equal to field capacity. As indicated in figure 10.6, in many soils the soil solution at FC is roughly twice as concentrated as at saturation point, or $EC_{fc} \simeq 2 \times EC_e$;

(c) not all of the deep percolation is equally effective in contributing to the leaching of salts from the rootzone. The most effective leaching results from water which moves through the mass of the soil. The water that moves rapidly downwards through the larger pores, cracks, etc, picks up very little salt and has only a minimal leaching effect. EC_p, it should be emphasised, refers to the average salinity of the combined mass flow and crack flow, comprising the total deep percolation. This may be expressed by introducing the *leaching efficiency factor* (f), being the mass flow proportion of the total deep percolation, the remainder (1-f) being the crack flow. Assuming the salinity of the mass flow to be equal to EC_{fc} and that of the crack flow to be equal to EC_i, if follows that $EC_p = f.EC_{fc} + (1\text{-}f)EC_i$. As values of f are difficult to determine and to extrapolate, this formula is mainly of conceptual value only.

On the basis of these considerations, values for EC_p in eq. 11.6 are often taken as (RHOADES 1974):

- $EC_p = 2EC_e 25\%$ in cases where crops are mostly sensitive, or rather low leaching efficiencies are to be expected or where a high standard of salinity control is desirable
- $EC_p = 2EC_e 50\%$ in cases where crops are more tolerant to salinity, or high leaching efficiencies are to be expected or a somewhat lower salinity control is considered acceptable.

Examples of application

Given: $EC_i = 1{\cdot}2$ mmhos.cm^{-1}
$EC_p = 12{\cdot}0$ mmhos.cm^{-1} ($= 2 \times EC_e 50\%$ for the crop to be grown)
$I_c = 6$ mm.day^{-1}

In eq. 11.6: $LR = \dfrac{EC_i}{EC_p - EC_i} . I_c = \dfrac{1{\cdot}2}{12{\cdot}0 - 1{\cdot}2} . 6{\cdot}0 = 0{\cdot}7$ mm.day^{-1}

$I = I_c + LR = 6{\cdot}0 + 0{\cdot}7 = 6{\cdot}7$ mm.day^{-1}; leaching percentage $= (EC_i/EC_p) . 100\%$ $= (LR/I) . 100\% = (0{\cdot}7/6{\cdot}7) . 100\% = 10\%$.

So in this example the salt balance in the rootzone is maintained when a minimum of 10% of the infiltrated irrigation goes into deep percolation. A deep percolation loss of this order or even higher is quite common under surface irrigation so that there is generally no need to over-irrigate to satisfy the leaching requirement.

11.4.2 Salt balance calculations

Eq. 11.6 may be used directly to calculate the average leaching requirement during a rainfree period. During the rainy season consideration also has to be given to the deep percolation from rainfall which also contributes to the leaching of salts from the rootzone. The following example of a salt balance calculation for an arid climate in which rainfall and irrigation are involved illustrates this.

Data

Annual crop evapotranspiration	ET	= 1655 mm
Total annual rainfall	R	= 695 mm
Component of annual rainfall being used as deep percolation*	LR_r	= 140 mm
Component of annual irrigation being used as deep percolation*	LR_i	= to be established
Capillary rise from groundwater	G	= 0 mm
Irrigation water	EC_i	= 1·75 mmhos.cm^{-1}
Saturation extract at which 25% yield reduction occurs	EC_e	= 3·5 mmhos.cm^{-1}
Deep percolation ($EC_p = 2 \times EC_e - 25\%$)	EC_p	= 7·0 mmhos.cm^{-1}

Calculations

Net rainfall utilised by crop ($R_c = R - LR_r$)	R_c	= 555 mm
Net crop irrigation requirement ($I_c = ET - R_c$)	I_c	= 1100 mm
Influx of salts in irrigation:		
$I \times EC_i = (I_c + LR_i) \times EC_i = (1100 + LR_i) \times 1{\cdot}75$		$= 1925 + 1{\cdot}75\,LR_i$ units
Efflux of salts in deep percolation:		
$LR \times EC_p = (LR_i + LR_r) \times EC_p = (LR_i + 140) \times 7$		$= 7\,LR_i + 980$ units
For a salt balance to exist:		
Salt influx = salt efflux:		
$1925 + 1{\cdot}75\,LR_i = 7\,LR_i + 980$ from which	LR_i	= 180 mm
Total irrigation required $I = I_c + LR_i$	I	= 1100 + 180
		= 1280 mm
Leaching percentage	L%	$= \frac{180}{1280} \times 100\% = 14\%$

Had there been no rainfall, the leaching percentage would have been much higher, satisfying the following balance:

$$I \times EC_i = LR_i \times EC_p \text{ or } (1100 + LR_i) \times 1{\cdot}75 = LR_i \times 7$$

From which $LR_i = 367$ mm. It follows that $I = I_c + LR_i = 1100 + 367 = 1467$ mm and that the leaching percentage is: $\frac{367}{1467} \times 100\% = 25\%$.

11..4.3 Regional salt balances

The principle of the salt balance can also be applied regionally:

$$\left\{\begin{matrix}\text{salt influx}\\ \text{in a region}\end{matrix}\right\} - \left\{\begin{matrix}\text{salt efflux}\\ \text{from the region}\end{matrix}\right\} = \begin{matrix}\text{change in salt}\\ \text{storage in the region}\end{matrix}$$

The salt influx or efflux can be calculated as: (waterflux) × (salt concentration) × (time). Both the flux and the concentration are liable to vary during the year so that annual fluxes have to be determined by summing the assumed uniform

* A soil moisture balance may be made using the procedures in Section 15.1 assuming that all excess goes to deep percolation that S = 90 mm (soil moisture storage capacity) and I_c = 50 mm/10–14 days.

salt fluxes occurring during shorter periods of time. Various dimensional units may be used; $m^3.sec^{-1}$ or $mm.day^{-1}$ for flux and $gr.L^{-1}$, ppm or $mmhos.cm^{-1}$ for concentration.

The waterfluxes in a region, especially those due to groundwater, are quite difficult to quantify and one frequently has to rely on estimates. An example of a simple regional salt balance is illustrated in figure 11.10. In this case there is one source of salt influx (irrigation water abstracted from the river), while all drainage water passes through one outlet point.

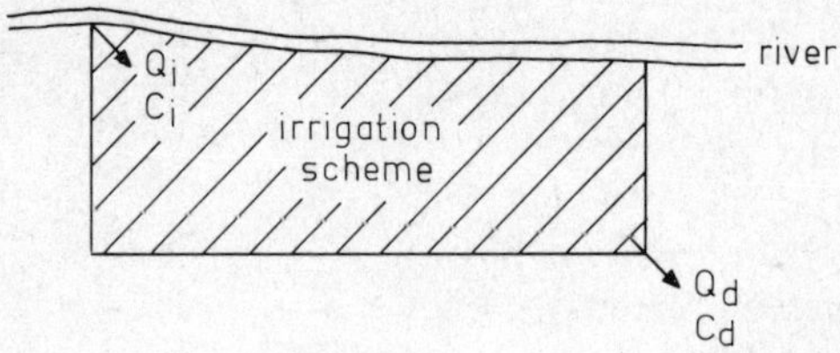

FIGURE 11.10 *Salt balance for an irrigation scheme*

The salt balance reads:

$$\Sigma Q_i \times C_i - \Sigma Q_d \times C_d = \Delta S$$

In this way it can be checked whether in the long term salts are liable to accumulate in the irrigated area.

11.5 Irrigation water quality

The salinisation/sodification hazards posed by irrigation water can be readily predicted on the basis of the amount and types of salts contained in the water. Irrigation development should not therefore be undertaken without prior analysis and appraisal of the water to be used for irrigation. Three different hazards are distinguished, respectively corresponding to the three types of salinity problems identified in section 10.2:

- salinity hazard: refers to the danger that the use of the irrigation water will lead to osmotic problems in the soil/plants;
- sodicity hazard: the same with respect to sodic problems;
- toxicity hazard: the same with respect to toxic problems.

A number of different irrigation water quality appraisal systems have been developed. Until recently, the system developed by the US Salinity Laboratory (1954) was the most widely used. The system proposed by FAO (1976) incorporates more recent research findings and experiences.

11.5.1 Salinity hazard

This hazard may be diagnosed on the basis of the EC-value of the irrigation water (EC_i in $mmhos.cm^{-1}$). The relevance of this parameter stems from the fact that the EC_e-value of a soil correlates rather well with the EC-value of the water with which it has been irrigated for some time. Generally it can be stated that:

$$EC_e \simeq \beta\, EC_i \qquad \text{......(eq. 11.7)}$$

where the value of β varies with the applied leaching percentage:

for leaching percentage 5–10% $\beta = 2{\cdot}0$–$2{\cdot}5$
for leaching percentage 15–20% $\beta = 1{\cdot}5$
for leaching percentage 25–30% $\beta = 1{\cdot}0$

The EC_e-value in these relationships represents the weighted average EC_e over the main rootzone, taking into account the normal decrease in water uptake by the roots with depth (parallel with the decrease in root density).

Few crops experience osmotic problems when the EC_e in the rootzone is less than 2 mmhos.cm^{-1}. This suggests that if one assumes a conservatively low leaching percentage of 5–10%, a safe limit for the salinity of the irrigation would be around 1 mmhos.cm^{-1}. The US Salinity Laboratory system recommends the following limits (see also figure 11.11):

	salinity hazard	
$EC_i < 0{\cdot}250$ mmhos.cm^{-1}	low	(class C-1)
$EC_i = 0{\cdot}250$–$0{\cdot}750$ mmhos.cm^{-1}	medium	(class C-2)
$EC_i = 0{\cdot}750$–$2{\cdot}250$ mmhos.cm^{-1}	high	(class C-3)
$EC_i > 2{\cdot}250$ mmhos.cm^{-1}	very high	(class C-4)

The US Salinity Laboratory limits appear in the light of the preceding discussion to be well on the safe side. In the FAO system somewhat higher limits apply than in the US system (see table 11.1; the FAO system is based upon an assumed irrigation regime with a leaching percentage of 15%, so $EC_e \sim 1{\cdot}5\, EC_i$).

11.5.2 Sodicity hazard

This hazard may be appraised on the basis of the two main diagnostic parameters (see table 11.1):

(a) *the EC_i-value*: used to appraise the possible harm likely to result from using water which is too fresh for irrigation. Fresh water may dilute the soil solution too much and enhance soil dispersion (see section 10.2.3). Also, soils may be 'over-leached' (in the sense that too much of the semi-soluble Ca and Mg-salts are leached).

(b) *the SAR_i-value* (the SAR-value of the irrigation water):

$$SAR_i = \frac{Na^+}{\sqrt{\dfrac{Ca^{++} + Mg^{++}}{2}}}$$ where Na^+, Ca^{++} and Mg^{++} represent the concentrations of these elements in the irrigation water (meq.L^{-1}).

The relevance of this parameter stems from the close correlation between

$$SAR_i \text{--------}\rightarrow SAR_e \begin{bmatrix}\text{of the soil solution}\\ \text{at saturation point}\end{bmatrix} \text{--------}\rightarrow \text{ESP of the soil}$$

The relationship between the SAR_e and the corresponding ESP of the soil has been discussed in section 10.1.3, and is expressed in figure 10.3. At a low leaching percentage of 5–10%, the soil solution in the saturated paste (= saturation extract) would be about 2·0–2·5 times as concentrated as the irrigation (see section 11.5.1). On the basis of the Gapon equation, also discussed in section 10.1.3, it follows that:

$$SAR_e = \sqrt{2{\cdot}0 \text{ to } 2{\cdot}5}\,.SAR_i \quad \text{or} \quad SAR_e \simeq 1{\cdot}5\,SAR_i.$$

This relationship and those equations discussed in section 10.1.3 enable both the SAR_e value of the saturation extract and the ESP-value of the soil to be predicted for a certain irrigation regime. For example: irrigation water with $Na^+ = 30$, $Ca^{++} = 6$ and $Mg^{++} = 2$ meq.L^{-1}, leaching percentage = 15–20% ($\beta \simeq 1{\cdot}5$).

$$SAR_i = \frac{30}{\sqrt{\dfrac{6+2}{2}}} = 15;\, SAR_e = \sqrt{1{\cdot}5}.\,SAR_i \simeq 1{\cdot}2 \times 15 = 18.$$

The corresponding ESP-value (read from figure 10.3) is 20%. For higher leaching percentages $SAR_e \simeq SAR_i$, and so ESP may be determined without significant error from figure 10.3 by using the SAR_i of the irrigation water directly.

The relationship between SAR and ESP is complicated by the semi-soluble nature of the Ca and Mg-salts in the soil. When the $(HCO_3^- + CO_3^{--})$ concentration of the irrigation water is high, Ca^{++} and Mg^{++} may precipitate as carbonates (especially in between irrigations when the soil dries out and the solubility limits of the Ca/Mg-carbonates are easily exceeded). Alternatively, Ca^{++} and Mg^{++} may come into solution, eg when the soil solution becomes more diluted. This particular aspect of the sodicity hazard posed by irrigation may be diagnosed by what is termed the

TABLE 11.1 *FAO guidelines for irrigation water quality appraisal* (abstracted from FAO 1976)

Hazard	*Diagnostic parameter*	*Criteria* *No problem*	*Moderate problem*	*Severe problem*
Salinity	*EC-irrigation water* (EC_i in mmhos.cm^{-1})	<0·75	0·75–3·0	>3·0
Sodicity	EC-irrigation water (EC_i in mmhos.cm^{-1})	>0·5	0·5–0·2	<0·2
	SAR_i adj (irrigation water)			
	– montmorillonite (2:1 clays)	<6	6–9	>9
	– illite/vermiculite (2:1 clays)	<8	8–16	>16
	– kaolinite/sesquioxides (1:1 clays)	<16	16–24	>24
Toxicity (for sensitive crops only)				
– sodium: SAR_i adj (irrigation water)		<3	3–9	>9
– chloride: Cl concentration irr water (meq.L^{-1})		<4	4–10	>10
– boron: B concentration irr water (mg.L^{-1})		<0·75	0·75–2·0	>2·0

Calculation of SAR_i adj: $SAR_i \text{ adj} = \dfrac{Na^+}{\sqrt{(Ca^{++}+Mg^{++})/2}}\,[1+(8{\cdot}4-pH_c)]$

$pH_c = (pK_2' - pK_c') + p(Ca^{++} + Mg^{++}) + p(Alk)$, table 11.2.

TABLE 11.2 *Basic data for the calculation of pH_c*
$[pH_c = (pK'_2 - pK'_c) + p(Ca^{++} + Mg^{++}) + p(Alk)]$

ΣConcentration (meq.L^{-1})	$pK'_2 - pK'_c$	$p(Ca^{++} + Mg^{++})$	p(Alk)
0·05	2·0	4·6	4·3
0·10	2·0	4·3	4·0
0·15	2·0	4·1	3·8
0·20	2·0	4·0	3·7
0·25	2·0	3·9	3·6
0·30	2·0	3·8	3·5
0·40	2·0	3·7	3·4
0·50	2·1	3·6	3·3
0·75	2·1	3·4	3·1
1·00	2·1	3·3	3·0
1·25	2·1	3·2	2·9
1·50	2·1	3·1	2·8
2·00	2·2	3·0	2·7
2·5	2·2	2·9	2·6
3·0	2·2	2·8	2·5
4·0	2·2	2·7	2·4
5·0	2·2	2·6	2·3
6·0	2·2	2·5	2·2
8·0	2·3	2·4	2·1
10·0	2·3	2·3	2·0
12·5	2·3	2·2	1·9
15·0	2·3	2·1	1·8
20·0	2·4	2·0	1·7
30·0	2·4	1·8	1·5
50·0	2·5	1·6	1·3
80·0	2·5	1·4	1·1

$(pK'_2 - pK'_c)$ is obtained from using the sum of Ca + Mg + Na in meq.L^{-1}
p(Ca + Mg) is obtained from using the sum of Ca + Mg in meq.L^{-1}
p(Alk) is obtained from using the sum of $CO_3 + HCO_3$ in meq.L^{-1}
} Obtained from water analysis

Example calculations

	Sample No. 1	2	3
$Ca^{++} + Mg^{++} + Na^{+}$	6·2	38.5	38·0
$Ca^{++} + Mg^{++}$	4·8	6·5	26·5
$HCO_3^{-} + CO_3^{--}$	2·9	4·5	3·2
$pk'_2 - pk'_c$	2·2	2·45	2·45
$p(Ca^{++} + Mg^{++})$	2·6	2·5	1·9
p(Alk)	2·5	2·35	2·5
pH_c	7·3	7·3	6·85
$1 + (8{\cdot}4 - pH_c)$	2·1	2·1	2·55
$(SAR)_i$	0·9	17·8	3·2
$SAR_i\,[1 + (8{\cdot}4 - pH_c)]$	1·9	37·4	8·2

'*residual sodium carbonate*' parameter (RSC) of the irrigation water, where:

$$RSC = (HCO_3^- + CO_3^{--}) - (Ca^{++} + Mg^{++})$$

with all concentrations in $meq.L^{-1}$ (Eaton 1950).

Example: irrigation water with $HCO_3^- = 5{\cdot}2$, $CO_3^{--} = 0{\cdot}2$, $Ca^{++} = 3{\cdot}3$ and $Mg^{++} = 1{\cdot}6\ meq.L^{-1}$; $RSC = (5{\cdot}2 + 0{\cdot}2) - (3{\cdot}3 + 1{\cdot}6) = 0{\cdot}5\ meq.L^{-1}$.

Criteria: $RSC < 1{\cdot}25\ meq.L^{-1}$: no problems to be expected
$RSC = 1{\cdot}25$–$2{\cdot}50\ meq.L^{-1}$: moderate problems to be expected
$RSC > 2{\cdot}50\ meq.L^{-1}$: severe problems to be expected.

In the FAO system this aspect of the sodicity hazard has been integrated with the general diagnosis on the basis of the SAR_i value. This parameter has been adjusted so as to take the carbonate precipitation hazard into account (SAR_i adj). The US Salinity Laboratory System employs the (unadjusted) SAR_i as the sole diagnostic parameter of the sodicity hazard (the carbonate problem is then often appraised separately by the RSC parameter).

In the US Salinity Laboratory system, three sodicity classes are distinguished:

- low sodicity hazard Class S-1
- moderate sodicity hazard Class S-2
- high sodicity hazard Class S-3

(Classes S-1 to S-3: see figure 11.11)

These classes are bounded by SAR_i limits which also vary with the EC_i value: the higher the salinity of the irrigation water (EC_i-value), the lower the SAR_i limit for a certain sodicity class. There is no sound theoretical base for this lowering of the SAR_i limits with increasing EC_i value, although it does express the general experience that the sodicity hazard of the irrigation water increases somewhat with the salinity hazard of the water. For water of low salinity (class C-1), the upper limit for sodicity class S-1 is at $SAR_i = 8$–10, while at very high salinity the limit for the S-1 classification is at $SAR_i = 2$–4 (see figure 11.11).

11.5.3 Toxicity hazards

Toxicity hazards are not explicitely appraised within the US Salinity Laboratory system. However, as discussed in section 10.2.2, the diagnosis of general toxicity caused by high salinity is covered indirectly by the EC_e parameter (and the corresponding irrigation water appraisal is thus covered by the EC_i parameter). Similarly, the SAR_i parameter covers the appraisal of Na-toxicity.

In the FAO system the toxicity is appraised as follows (see table 11.1):

- Na-toxicity: diagnosed by the adj. SAR_i parameter; given criteria apply to sensitive crops only (woody perennials).
- Cl-toxiciy: diagnosed by the Cl concentration of the irrigation water; given criteria apply to sensitive crops only (woody perennials). Water which poses a Cl-toxicity hazard, normally also far exceeds the limits of the salinity hazard discussed.
- B-toxicity: diagnosed by the B-concentration of the irrigation water; given criteria apply mostly to sensitive crops only.

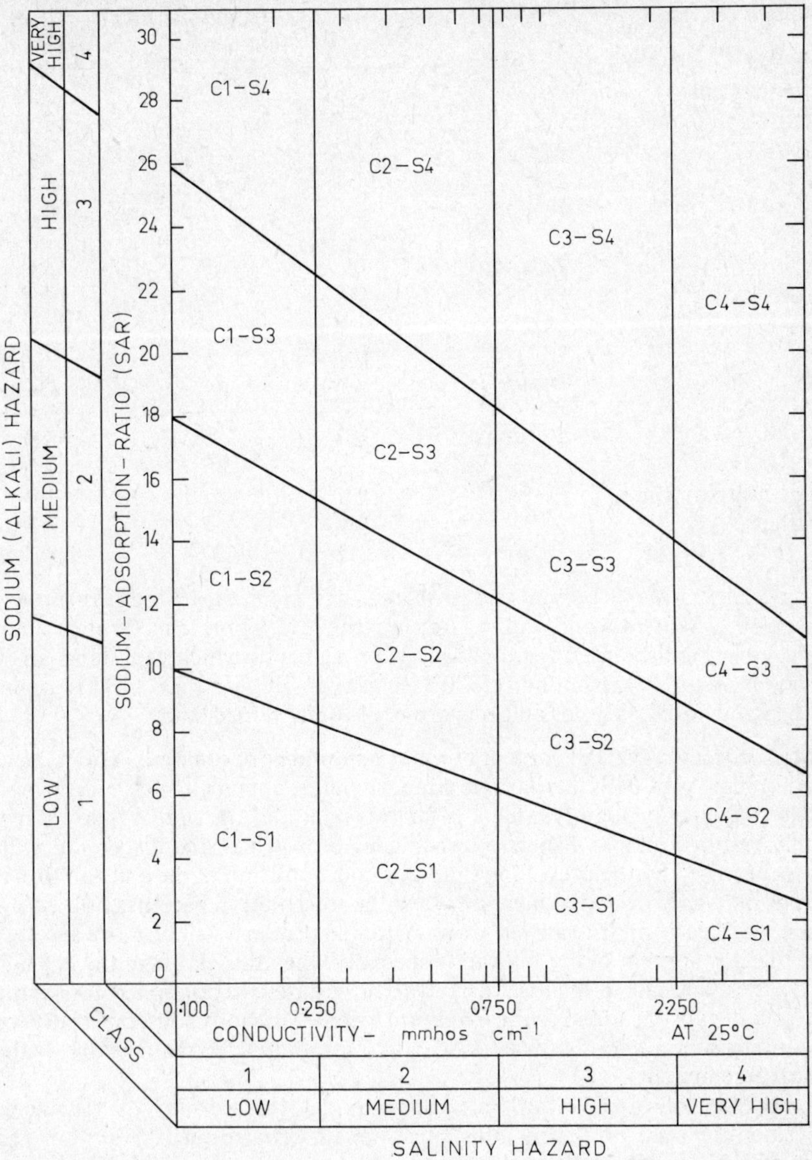

FIGURE 11.11 *Diagram for the classification or irrigation water according to US Salinity Laboratory* (USDA 1954)

11.5.4 Examples of irrigation water quality appraisal

Three different types of irrigation water will be appraised, respectively characterised by the following results of analysis:

		Irrigation water 1	2	3
EC_i (mmhos.cm^{-1})		0·51	3·60	3·21
concentration in meq.L^{-1}:	Ca^{++}	2·6	2·5	17·3
	Mg^{++}	2·2	4·0	9·2
	Na^{+}	1·4	32·0	11·5
	K^{+}	0·1	1·2	0·6
	Σ cations	6·3	39·7	38·6
	Cl^{-}	1·5	25·8	12·2
	SO_4^{--}	1·6	9·0	23·1
	CO_3^{--}	0·3	0·0	0·0
	HCO_3^{-}	2·6	4·5	3·2
	Σ anions	6·0	39·3	38·5
pH		7·8	7·7	7·7
concentration of B (meq.L^{-1})		nil	nil	1·0
calculated: SAR_i		0·9	17·8	3·2
SAR_i adj		1·9	37·4	8·2

Salinity hazard: water 1 presents no problems, not even at leaching percentages as low as 5–10%. Waters 2 and 3 may both present severe problems. Waters 2 and 3 may, however, still be used for tolerant crops, eg crops which can stand an EC_e = 6–7 mmhos.cm^{-1} (assuming 15–20% leaching: $EC_e \sim 2 \times EC_i$). This scale of leaching is, however, only possible on permeable well drained land.

Sodicity hazard: the low salt content of water 1 should be of some concern in the long run although in view of the favourable adj. SAR_i value, not much harm is anticipated.

The adj. SAR_i value of water 2 is far too high and the use of this water will certainly result in major dispersion problems, even where the dispersion will be partly suppressed by the prevailing high salt concentration of the soil solution (the problems may be expected to increase when the soil solution becomes diluted as for example, may occur in the rainy season). Also, the problem will be more severe, the higher the clay content of the soils and especially when this clay is of the 2:1 lattice type. On these soils amendments are likely to be required (to be applied, for example, in the form of gypsum added to the irrigation water). On sandy soils, especially when they contain some lime, the dispersion problems should be manageable without chemical amendments.

The adj. SAR_i of water 3 is fairly high but should not create much of a problem on sandy soils nor even on loamy soils since the EC_i is also fairly high. On heavy montmorillonitic soils, however, more problems are anticipated. These could, however, probably be solved without the use of amendments by providing good surface drainage to the land and growing the more tolerant crops only.

Toxicity hazard: there should be no problems with water 1. The Na and the Cl concentrations of water 2 are far too high for sensitive crops which should not therefore be grown. For the more tolerant crops, conditions which aggravate the problems (such as the adoption of long irrigation intervals leading to advanced soil drying; under-irrigation, etc) should be avoided.

Water 3 has a high Cl-concentration while in addition the adj. SAR_i-value is also such that severe toxicity is likely to occur under unfavourable conditions as indicated above when sensitive crops are grown. Most crops, however, will experience no problems.

The boron concentration of water 3 is of some concern for the more sensitive crops.

References and further reading

Boumans, JH 1976: 'Wasserbedarf für Salzauswaschung und Entwasserungsmassnahmen zum Zwecke der Salzbodenmelioration', Zeitschrift für Bewässerungswirtschaft *10*: 7–24 (in German)

Eaton, FM 1950: 'Significance of carbonates in irrigation water', *Soil Science 69*: 123–133

FAO 1976: 'Water quality for irrigation', Irrigation and Drainage Paper no 29, Food and Agriculture Organisation of the United Nations, Rome

van Hoorn, JW 1979: 'The effect of capillary flow on salinisation and the concept of critical depth for determining drain depth'. In: *Proceedings of the International Drainage Workshop*, J Wesseling (ed), publication 25, International Institute for Land Reclamation and Improvement, Wageningen

Dieleman, PJ (ed) 1963: *Reclamation of Salt Affected Soils in Iraq*, publication 11, International Institute for Land Reclamation and Improvement, Wageningen

van der Molen, WH 1973: 'Salt balance and leaching requirement'. In: *Drainage Principles and Applications*, volume II, publication 16, International Institute for Land Reclamation and Improvement, Wageningen

Rhoades, JD 1974: 'Drainage for salinity control'. In: *Drainage for Agriculture*, J van Schilfgaarde (ed), Agronomy Monograph 17, American Society of Agronomy, Madison, Wisconsin

Rijtema, PE 1963: 'Soil moisture forecasting', Nota 513, Institute for Land and Water Management, Wageningen

UNESCO 1970: 'Research and training on irrigation with saline waters', Technical Report of the UNDP/UNESCO project Tunisia 5, UNESCO, Paris

USDA 1954: Diagnosis and improvement of saline and alkali soils, Agricultural Handbook no 60, US Department of Agriculture, Washington DC

Chapter 12 Drainage of irrigated land

Although, by its nature, irrigated land will generally feature a more arid climatic/physiographic setting than land used for rainfed agriculture, the occurrence of excess rainfall during some period of the year is still quite common.

Irrigation water losses constitute an additional source of water in the drainage of irrigated land. In total these losses may add up to more than 50% of the irrigation water intake at source, half of which would typically be removed by surface drainage, the other half by groundwater drainage. With intake rates normally ranging from 1·0–1·5 $1.sec^{-1}.ha^{-1}$ (8–13 $mm.day^{-1}$), drainage loads due to irrigation losses for each type of drainage would generally be less than 0·25–0·40 $1.sec^{-1}.ha^{-1}$ (2–3·5 $mm.day^{-1}$).

Rainfall in arid climates falls typically as intense storms, creating surface drainage requirements which are easily 5 to 10 times greater than those due to the irrigation losses. For this reason design discharges for surface drainage systems for irrigated land may, in most cases, be based on rainfall alone. This also applies to groundwater drainage when the drainage problems in the soil are exclusively of the wetland type, caused by excess rainfall (as may for example be the case with supplemental irrigation in semi-humid climates). Under the typical arid to semi-arid setting of most irrigated land where groundwater drainage is mainly needed for salinity control, different design criteria apply.

12.1 Pipe drainage systems

Groundwater drainage systems for salinity control in irrigated land should be capable of draining the recharge to the groundwater whilst maintaining the watertable at a level that is not harmful to the crop and that minimises capillary salinisation of the rootzone.

Horizontal drainage using pipes or ditches as discussed in chapter 3 is most common. Vertical drainage using pumped wells can also be used under certain circumstances (see discussion in section 12.2).

The design of a pipe drainage system, discussed in section 7.1, usually involves the following steps:

(a) Selection of the type of system and layout of the field drains.
(b) Determination of the field drainage base depth (W = drain depth below the soil surface).
(c) Determination of the basic design criteria, ie the design discharge (q) in combination with the watertable depth to be maintained (H).

This information, fed into an appropriate drain spacing formula, yields the required drain spacing (L), see figures 7.1 and 7.2.

The reader is referred to chapter 3 in general since most of that discussion is also relevant to groundwater drainage for salinity control in irrigated land. Specific aspects have been included in this section.

12.1.1 Drain depth

Most factors influencing the drain depth (W) listed in section 7.6 apply also to groundwater drainage for salinity control. In contrast to humid areas, the local/regional drainage base will seldom be a limitation. Instead, the drain depth W may be limited either by soil conditions (unfavourable soil hydrological/mechanical conditions might occur at some depth), the available machinery (few machines go beyond 2·5 m) or the cost/benefit ratio.

To prevent capillary salinisation, the depth W should adhere to certain minimum depth criteria. The critical period for capillary salinisation occurs once the downward percolation of excess irrigation has stopped. This often coincides with the period just preceding harvest time when irrigation is discontinued or it could occur towards the end of a particularly lengthy irrigation interval. The watertable will then have fallen back to near the drain depth W or will have remained some 10–20 cm above the drainage base, constituting the required head to drain off seepage inflow to the area. This level should be no higher than the critical watertable depth D_c discussed in section 11.2.1. This of course depends to some extent upon the depth of the evaporation zone, but for the most critical conditions during the fallow season this zone is very shallow so that $D_c \sim H_c$. This leads to the minimum requirements:

$W \geqslant H_c$ when there is no seepage
$W \geqslant H_c + 10$–20 cm where there is seepage inflow.

H_c-values for most soils are of the order of 100–150 cm rising to over 200 cm in very fine sandy or silty soil profiles. Consequently the values for W range between 150 and 250 cm with a bias resulting from the optimisation of costs towards the deeper values, 200–250 cm. These depths are considerably deeper than those normally encountered in groundwater drainage systems for wetland conditions.

12.1.2 Drain spacing criteria

The q-H package needed for calculating drain spacings may be formulated in a steady state or a non-steady state form. The latter formulation is closer to reality although both represent rather fictitious situations, the significance of which is statistical rather than physical. The use of these criteria is validated only by their good correlation with such response indicators as crop yield or rootzone salinity level. Groundwater drainage design for salinity control in irrigated land can normally be carried out quite adequately using a steady state formulation which reflects the average discharge and watertable level during (part of) the season. This formulation allows the use of the simple steady state drain spacing formulae.

Watertable depth (H)

Figure 12.1 illustrates the common case of a permitted general rise of the watertable during the season with, superimposed on it, the short term fluctuations due to the periodic irrigations. Capillary salinisation is of little concern since the predominant water movement in the soil will be downward due to the percolation of excess irrigation (and rain). The watertable depth may therefore be considerably less than H_c. The average watertable should, however, remain deep enough to assure sufficient rootzone aeration. Recommended criteria are set out in table 12.1.

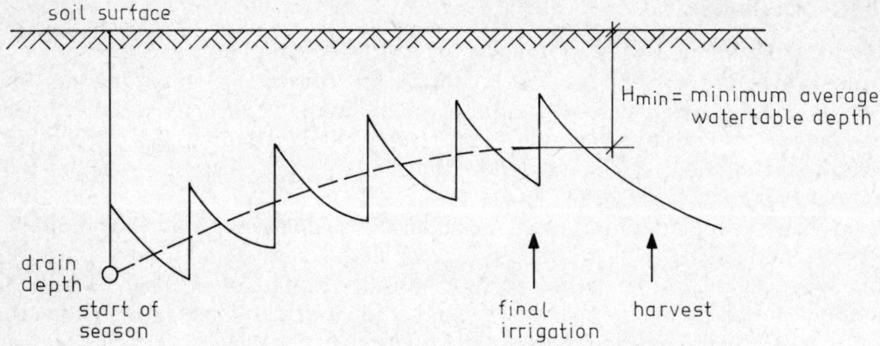

FIGURE 12.1 *Normal watertable regime in irrigated land with groundwater drainage*

TABLE 12.1 *Minimum average watertable depths H_{min} for irrigated land* (FAO 1980)

	Fine textured, moderately permeable soils	*Light textured soils*
Field crops	1·20 m	1·00 m
Vegetables	1·10 m	1·00 m
Tree crops	1·60 m	1·20 m

Given the drain depth W, the design watertable head (h) to be used in the steady state drain spacing equation may be determined as h = W–H with H to be taken from table 12.1.

Design discharge (*q*)

This rate may in principle be determined from the following groundwater balance:

$$q = \frac{R + S_c + S_i - D_n}{T}$$

where: q = drainage discharge mm.day^{-1}
R = estimated deep percolation mm.period^{-1}
S_c = seepage from canals mm.period^{-1}
S_i = seepage inflow to the area mm.period^{-1}
D_n = natural groundwater drainage from the area mm.period^{-1}
T = time period over which the excess is removed. days

To arrive at matching q and H values, this balance should be applied to the same period to which H applies, ie the period during which the watertable has risen to its maximum permitted level (generally the last part of the irrigation season).

The amount of deep percolation (R) from rainfall or irrigation is difficult to predict with any great accuracy although the minimum is dictated by the leaching requirement. Values that can be applied to irrigation should be determined using locally available data or where unavailable, using values such as those set out in table 12.2.

TABLE 12.2 *Guidelines for estimating deep percolation from field irrigation* (FAO, 1980)

Irrigation method	*Conditions influencing percolation*	*Percolation as a percentage of the water delivered to the field*	
		fine textured moderately permeable soils	*light textured soils*
Sprinkler	Daytime application, moderately strong wind	30	30
	Night-time application	25	25
Trickle		15	15
Basin	Poorly levelled and shaped	30	40
	Well levelled and shaped	20	30
Furrow, border	Poorly graded and sized	30	40
	Well graded and sized	25	35

Canal seepage (S_c) varies with the nature of the soil in which the canal is formed and with the depth, velocity and sediment load of the water in the canal. Its effect depends upon the depth of permeable soil below the canal as well as the proximity of deep intercepting ditches. The seepage rate may be estimated by standard methods (FAO, 1977) or more simply may be estimated to be 5% of the flow in lined canals rising to 10–20% for unlined canals.

The movement of groundwater into (seepage, S_i) and out of (natural drainage, D_n) the project area may be estimated on the basis of the regional groundwater gradients and corresponding transmissivity values, using Darcy's Law. The (relative) strengths of the seepage inflow and natural drainage in an area may also be evaluated on the basis of the fall of the watertable during the dry season in combination with the salt concentration of the groundwater whereby the following interpretation scheme may be used:

End of dry season		
Watertable	*Groundwater Salinity*	*Interpretation*
Falling to drain base depth	Low	Natural drainage
Remaining well above drain base depth	High	Seepage inflow

All of this recharge to the groundwater has to be discharged without delay as no further rise of the watertable can be permitted.

The discharge period T may be taken equal to the one or two irrigation interval periods over which the critical recharge is determined. The drain outflow will of course not be steady during an irrigation interval. It will rise sharply to a peak immediately after the irrigation and then recede gradually, in line with the fluctuation of the watertable. The outflow recession will occur more rapidly for shallow than for deeply installed drains since the watertable in the former case will fall relatively that much faster. The drains may even stop discharging in the case of shallow drains due to the fact that the watertable has fallen back to drain depth well before the next irrigation is due. This influence of drain depth on discharge may be taken into account by introducing the notion that the effective discharge period T varies with the drain depth as provisionally indicated below (values applicable for values of H taken according to table 12.1):

	Effective discharge period as a percentage of the total available discharge period
Drain depth W = 150 cm	50%
W = 200 cm	75%
W = 250 cm	100%

Design discharges for field drainage of irrigated land in a typical arid setting are generally much lower than those used for field drainage design under the typical wetland conditions. The following values are commonly encountered in practice in situations with negligible seepage inflow and negligible natural drainage (FAO, 1980):

– soils with low infiltration rate	$q = 1{\cdot}5$ mm.day^{-1}
– typical of most situations; use the higher value on intensively farmed permeable soils	$q = 1{\cdot}5$–$3{\cdot}0$ mm.day^{-1}
– extreme conditions with high irrigation losses or high water salinity	$q = 3{\cdot}0$–$4{\cdot}5$ mm.day^{-1}

Example

Fine textured soils; furrow irrigation, poorly graded; irrigation interval 12 days; field application per irrigation 75 mm (gross); rainfall during critical period nil; drain depth 200 cm; field crops.

	During critical period (near the end of the irrigation season when watertables have risen to their highest permitted level)
– deep percolation due to irrigation ..... (from table 12.2)	30% of 75 mm = 22·5 mm per 12 days = 1·9 mm.day^{-1}
– seepage from canals (estimate).........	= 0·3 mm.day^{-1}
– seepage inflow minus natural drainage.............................	= —
	2·2 mm.day^{-1}
– effective discharge time 75%...........	$q = 2{\cdot}9$ mm.day^{-1}
	$H = 1{\cdot}20$ m; $h = W - H = 0{\cdot}80$ m

12.1.3 Layout patterns

Figure 12.2 shows a typical layout of irrigated land (surface irrigation; the layout of sprinkler or trickle irrigated land would generally be similar to that of rainfed land). The layout is designed to promote controlled overland flow of irrigation water. Excess irrigation water and rainfall is collected at the lower end of the field by a surface drain, typically consisting of a shallow V-shaped passable ditch. Length of irrigation runs are in the first place based on the needs of irrigation but they should also conform to the following requirements for surface drainage:

- minimum slope from 0·10–0·20% for clean furrows, to 0·20–0·25% for rough trashy furrows (see 'Row drainage', section 4.2.2).
- drainage discharge in the furrow due to rainfall should not exceed its safe carrying capacity (which may lead to erosion and/or overtopping); this may pose the critical limit on furrow length.

A field which has been graded/smoothed and properly laid out for good surface irrigation will normally also provide good surface drainage (no ponding, no erosion).

The comparison between the use of pipes and ditches for groundwater drainage in section 3.2 also largely applies to their use in irrigated land. The dense system of irrigation canals inevitably leads to a great many crossings of ditches that are avoided in a pipe drainage system installed underground. Ditches within the field, especially those aligned along the contours (as they should be in sloping land) interfere quite severely with the field irrigation so that in general they should only be used as widely spaced boundary ditches for large fields (spacings of 200–300 m). The maintenance of ditches should also be a major consideration as weed growth in the ditches is often rapid and abundant (especially various reed species).

The present trend is clearly in favour of pipe drainage using layouts such as those detailed in figure 12.3. Both singular and composite pipe systems may be used, the choice depending on the weighting given to the factors listed in section 3.1.2. The major advantage of composite systems is that the number of open drains, and in consequence the crossings with irrigation canals, are reduced. Areas of land up to 100 ha or more may be drained wholly by a subsurface system enabling the main canals to be spaced 1000–2000 m apart (see figure 12.3-B).

The extended singular system shown in figure 12.3-C is an interesting alternative layout. This also permits draining relatively large areas of land wholly by a subsurface system while it is often less costly than composite drainage. The drains are best aligned with the natural slope but the system is still applicable in nearly flat land, using minimal drain slopes of not less than 0·05%. Drain length between two access points should generally be no more than 250–300 m.

Of the two alternatives shown for the composite system (figure 12.3-B), the layout with the laterals running down the slope, parallel to the direction of the irrigation runs, is often preferable. This is because irrigation water running over a recently backfilled trench easily results in piping. The solutions to this problem are discussed in section 3.3 and essentially consist of preventing water from coming into direct contact with the trench backfill either by using bunds bordering the trench or by constructing a broad based ridge over the trench. These should be retained until the trench fill has stabilised. In the layout where the laterals runs along the contour intersecting the irrigation runs, the avoidance of direct contact between the trench fill and the irrigation water is not possible. In these circumstances the backfill should be thoroughly compacted but even this might not be sufficient to eliminate piping altogether.

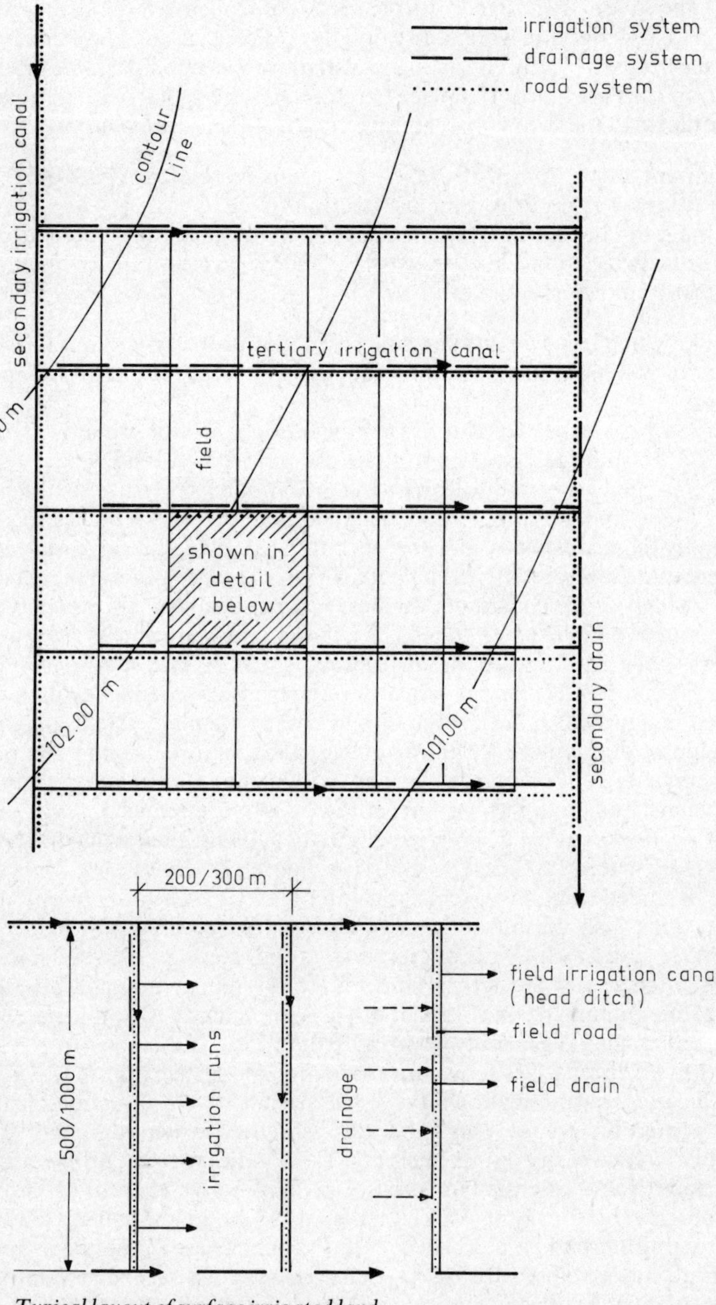

FIGURE 12.2 *Typical layout of surface irrigated land*

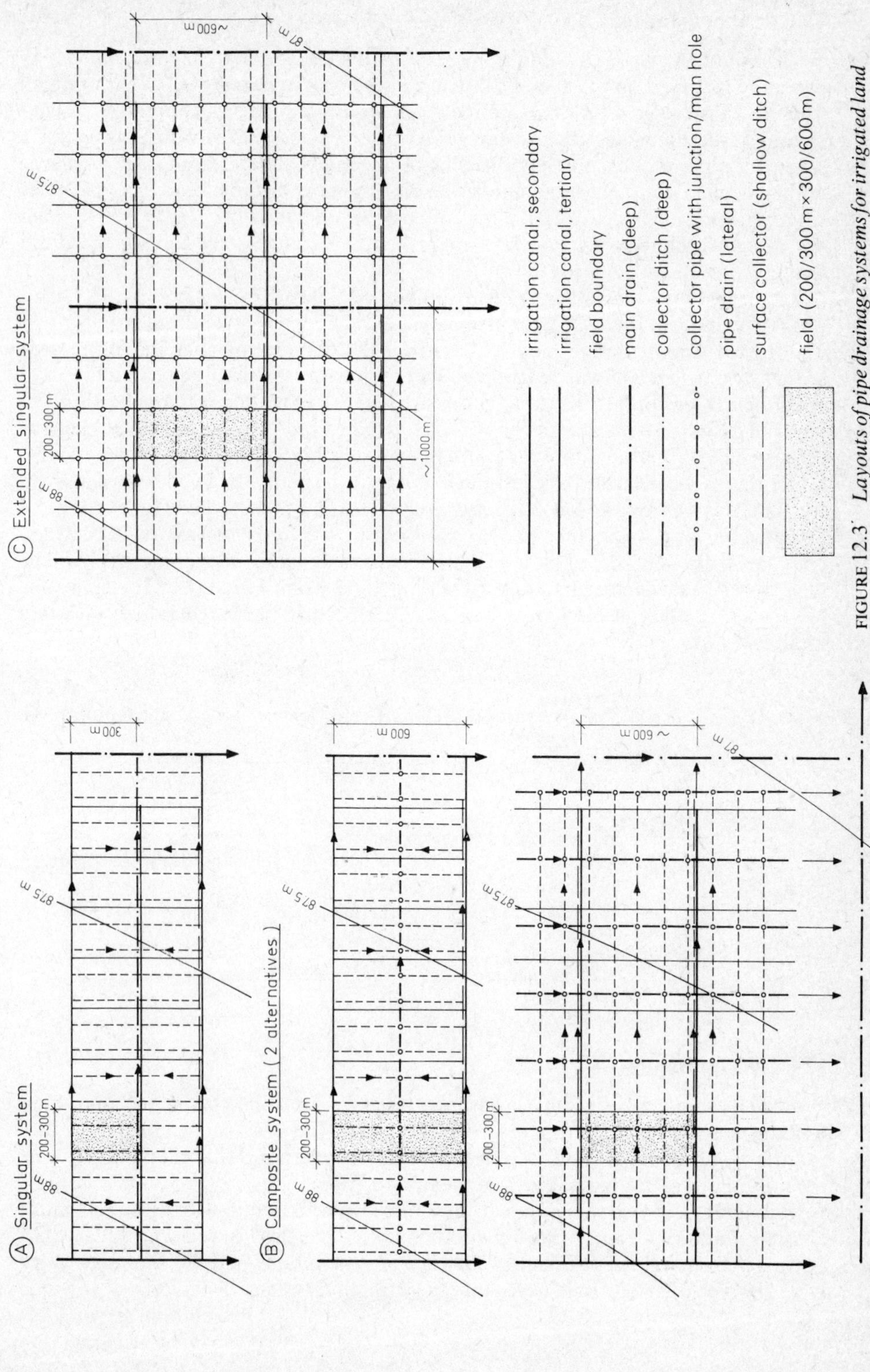

FIGURE 12.3 *Layouts of pipe drainage systems for irrigated land*

12.1.4 Pipe diameter

Whereas under wetland conditions, design discharges for the pipe flow may generally be based on the average discharge resulting from a period of high rainfall (the same discharge used to determine drain spacing), pipe drains in irrigated land should normally be designed to carry safely the peak discharge which occurs for a short duration after each irrigation. This value may be determined as the discharge of the installed drainage system under its maximum watertable head (h_{max}). For the situation depicted in figure 12.1, it may often be assumed that $h_{max} = W - \frac{1}{2} H_{min}$ and the design discharge for pipe flow may then be estimated by solving the drain spacing formula for q, taking $h = h_{max}$.

For drain depths W = 2·0 to 2·5 m and H_{min} = 1·0 to 1·2 m (as in table 12.1), this roughly works out that q_{design} for pipe diameter = $1{\cdot}5 \times q_{design}$ for drain spacing.

For field drains, design may be based on the assumption that the entire area served by a field drain will be irrigated within one day (so that the peak inflow will occur nearly simultaneously along its entire length). However, this assumption will generally lead to over-design for collector drains (and also for extended singular drains). The irrigation supply is normally rotated among fields or farms and the larger the area considered, the less likely it is that the entire area will be irrigated on the same day. An area of 100 ha or more served by one collector may for example be irrigated in five days at 20 ha per day. So the inflow into the collector will be at the peak rate from only 20% of the area while the inflow from the remaining areas will be at a lower rate, decreasing with increasing time from irrigation. Guidelines for estimating the design discharge rate for collectors are presented in the following table.

TABLE 12.3 *Area reduction factors for collector drains in irrigated land* (adapted from FAO, 1980)

% of area served by one collector irrigated per day	q_{design}
50–100%	100% (= q_{design} for field drains)
30– 50%	90%
20– 30%	85%
10– 20%	80%
< 10%	70%

12.2 WELL DRAINAGE

With well drainage, watertable levels in an area are controlled by the installation of wells in a suitable pattern across the area such that their 'cones of depression' (= drawdown zones around each well) adequately overlap. Suitable layout patterns are the unilateral triangular grid and the square grid.

In general these wells are pumped, intermittantly or continuously, although under artisian conditions free flowing wells could in principle also be used. The wells would normally discharge into the (surface) drainage system but where the drainage water is of suitable quality, it could be fed into the irrigation system.

12.2.1 Types of aquifers

The applicability of well drainage depends especially on the geohydrological situation within the area. Two types of aquifer may be encountered:

Confined/Semi-confined aquifers

Where an area is underlain by artesian water as in figure 12.4, well drainage may be used to take some of the pressure off this water and as such help to reduce the upward seepage of this artesian water through the confining layer (*aquitard*) up into the overlying soil layer.

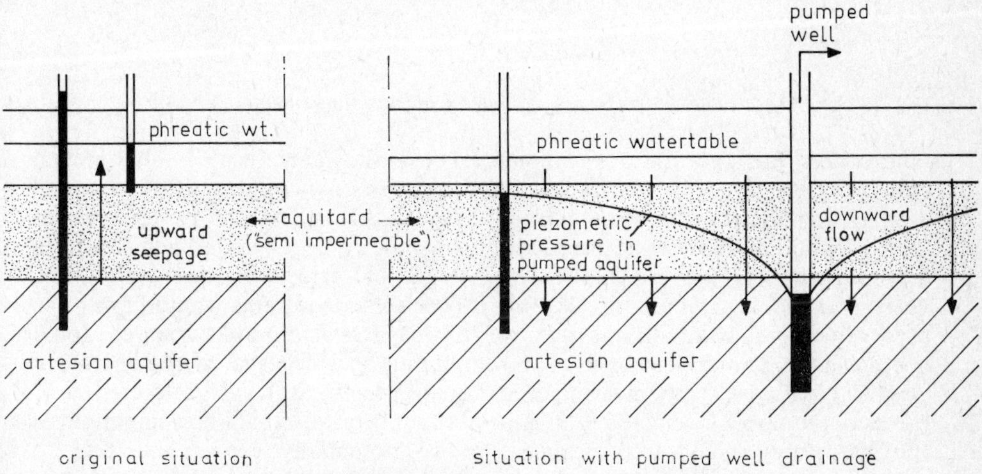

FIGURE 12.4 *Relief of pressure on artesian water by means of well drainage*

Free flowing wells may provide some relief of pressure and reduction in seepage in cases where the artesian pressure reaches up to the surface, although an effective solution of the seepage problem will generally require pumped wells. Through pumping the pressure in the aquifer may be reduced to the extent that sufficient downward flow through the aquitard occurs to control the phreatic watertable (thereby eliminating the need for watertable control by means of horizontal drainage).

Phreatic groundwater

The functioning of well drainage in this case (depicted in figure 12.5) is similar to that of horizontal drainage and both systems could generally be considered.

Under wetland drainage conditions, a large quantity of water needs to be discharged and operational costs of well drainage would generally be too high to be competitive with horizontal drainage. Also, well drainage is most suitable when the watertable can be drawn down deep (allows wide spacings), which is generally not necessary, or may even be undesirable, under rainfed cropping. Enclosed depressions might be drained by well drainage (with connecting pipe outlet) as such areas are difficult to drain by horizontal systems.

Discharge requirements are much lower for groundwater drainage for salinity control of irrigated land and in addition, the control of deep watertables is desirable.

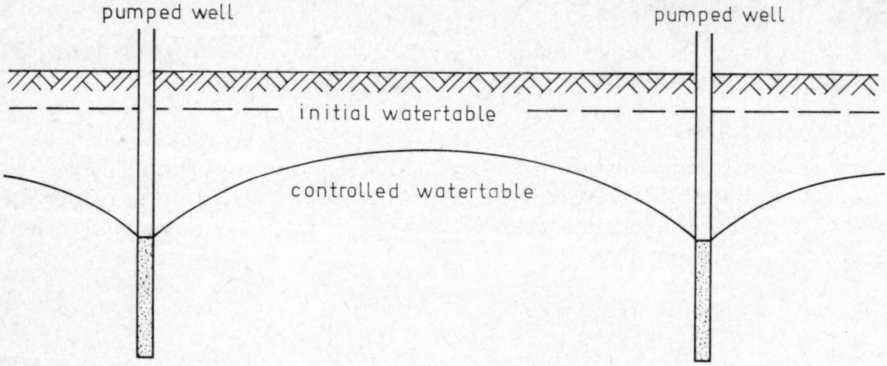

FIGURE 12.5 *Watertable control by well drainage in an unconfined aquifer*

Investments, fuel and other operational expenses, however, result in high annual costs of well drainage. In general, well drainage can only compete with horizontal drainage in situations where the upper soil layers, through which the flow towards the horizontal drains occurs, are poorly permeable while the underlying substrata are deep and highly permeable (have a high KD-value). In such a situation, horizontal drainage would be relatively expensive as the drains would have to be narrowly spaced, while conditions for well drainage are favourable. The wells could operate in the permeable substrata, and could thus be widely spaced and operate at low heads (relatively low investment and energy costs). Still, even at quite low K values in the upper layers, the KD-value of the substrata, should be a minimum of 100–200 $m^2.day^{-1}$ for well drainage to be potentially economic, rising to $> 500\ m^2.day^{-1}$ for the case of moderately permeable conditions in the upper layers. In all cases it holds that the economic feasibility of well drainage is considerably enhanced when the quality of the pumped water is good enough to be used for irrigation.

12.2.2 Design

Design of a well field for drainage purposes requires considerable knowledge about geohydrology and well technology, which is beyond the scope of this book. However, a simple well flow equation is presented enabling the required well spacing and drawdown to be approximated for a first comparison between well drainage and horizontal drainage. The equation applies to steady state well flow in a situation as depicted in figures 12.4 and 12.6. This is based upon the assumption that the deep percolation through the aquitard is uniform whereas in reality, of course, it decreases with increasing distance from the well (in practice, the variation in deep percolation rate may well be neglected):

$$H - c\,q = \frac{\pi r_e^2 q}{2\pi KD}\left[\ln\frac{r_e}{r_w} - 0{\cdot}5\right] \qquad \text{......(eq. 12.1)}$$

where:

H = the head = the difference in level between the watertable and the piezometric level adjacent to the well (m) c = the hydraulic resistance of the aquitard calculated as $c = \frac{D'}{K_v'}$ where D′ is measured from the watertable to the base of the aquitard and K_v' is the vertical component of hydraulic conductivity in the aquitard (days)

q = the steady uniform rate of percolation of phreatic groundwater through the aquitard during pumping (m.day^{-1})

r_e = the radius of influence of the well (m) ie the radius over which the well exerts a drainage influence

r_w = the radius of the well plus the thickness of any gravel filter (m)

KD = the transmissivity of the aquifer having a thickness D and hydraulic conductivity K (m^2.day^{-1}).

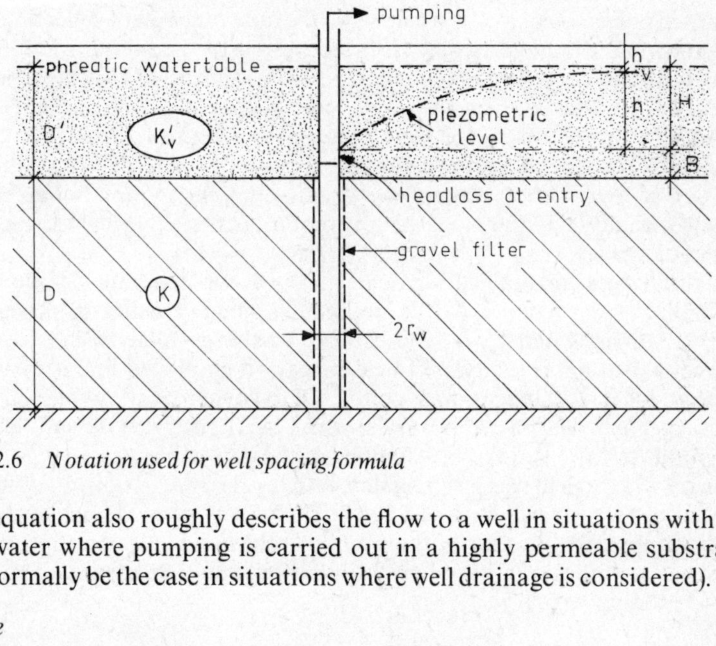

FIGURE 12.6 *Notation used for well spacing formula*

This equation also roughly describes the flow to a well in situations with phreatic groundwater where pumping is carried out in a highly permeable substratum (as would normally be the case in situations where well drainage is considered).

Example

Annual drainable surplus 320 mm
Annual operating time 290 days } $q = 1{\cdot}1$ mm.day^{-1}
(accounting for repairs, maintenance, etc)

Aquitard: $D' = 13$ m
$K_v' = 0{\cdot}033$ m.day^{-1} } $c = D'/K_v' = 394$ days

Head lost in vertical flow through aquitard: $h_v = c.q = 0{\cdot}43$ m
Drawdown adjacent to well: $h = D' - h_v - B = 13 - 0{\cdot}43 - 3 = 9{\cdot}57$ m (for B = 3 m, see figure 12.6).

Aquifer: $K = 8$ m.day^{-1}
$D = 25$ m } $KD = 200$ m^2.day^{-1}

Diameter of well, including filter: $2\,r_w = 0{\cdot}4$ m

Equation 12.1: $h = H - h_v = \frac{\pi r_e^2 q}{2\pi KD}\left[\ln\frac{r_e}{r_w} - 0{\cdot}5\right]$

$$9{\cdot}57 = \frac{r_e^2(0{\cdot}0011)}{2(200)}\left[\ln\frac{r_e}{0{\cdot}2} - 0{\cdot}5\right] \quad \text{so } r_e = 670\,\text{m}$$

Well discharge: $Q = \pi r_e^2.q = 1551\,m^3.day^{-1} = 18\,l.sec^{-1}$
Drained area: $A = \pi r_e^2 = 141$ ha

12.3 MAIN DRAINAGE

Main drains in an irrigated area as a rule serve the dual function of collecting/disposing of groundwater drainage as well as surface drainage. To fulfil the first function, the drains must be deep (levels often 2·5–3·5 m below the soil surface), although groundwater discharges are usually very low compared to the surface drainage discharges. The storm discharge requirements in irrigated areas (even in arid climates due to the intensity of the storms) far exceed those due to groundwater drainage or due to irrigation waste. These relatively minor sources of excess water can normally be neglected, main drainage requirements being based solely upon storm water discharge.

Design discharges for main drains can generally be determined following the procedures discussed in chapter 8. The method discussed in section 8.4.2 is generally most relevant to irrigated land. The rainfall–discharge relationship to be used depends very much upon the type of field irrigation employed and is for example quite different for furrow irrigation as compared to basin irrigation (basin irrigation can provide a great deal of temporary storage, see figure 8.9). It should also be borne in mind that the storage possibilities of the soil will normally vary greatly, depending on when a field was last irrigated. At a ten days irrigation schedule, at one extreme 10% of the area will have just been irrigated when the design storm occurs and can therefore store virtually nothing while at the other extreme another 10% of the area has a maximum storage potential (just about to be irrigated).

References and further reading

BOUMANS, J H 1979: 'Extended singular drainage: an interesting alternative lay-out for drainage of irrigated land'. In: *Proceedings of the International Drainage Workshop*, Publication 25, International Institute for Land Reclamation and Improvement, Wageningen.

FAO 1977: 'Irrigation canal lining', Irrigation and Drainage Paper no 2 (new edition), Food and Agriculture Organisation of the United Nations, Rome

FAO 1980: 'Drainage design factors', Irrigation and Drainage Paper no 38, Food and Agriculture Organisation of the United Nations, Rome

ICID 1978: 'Drainage construction techniques for vertical/tubewell drainage', International Commission on Irrigation and Drainage, New Dehli

KESSLER, J et al 1974: 'Field drainage criteria'. In: *Drainage Principles and Applications*, volume III, publication 16, International Institute for Land Reclamation and Improvement, Wageningen

NAZIR AHMAD 1979: *Tubewells, theory and practise*, Monograph 4, Pakistan Academy of Science, Islamabad

OCHS, W J et al 1980: 'Drainage requirements and systems'. In: *Design and Operation of Farm Irrigation Systems*, M E Jensen (ed), Monograph 3, American Society of Agricultural Engineers, St Joseph, Michigan

DE RIDDER, N A 1973: 'Drainage by means of pumping from wells'. In: *Drainage Principles and Applications*, volume II, publication 16, International Institute for Land Reclamation and Improvement, Wageningen

PART V
MISCELLANEOUS DRAINAGE PROBLEMS

Chapter 13 Seepage and related drainage problems

Groundwater flow into an area frequently creates local problems of excess water which need to be controlled by drainage measures. Such flow is generally referred to as *seepage* while the affected areas are termed *seep areas*. The required drainage measures may involve the collection and disposal of the seepage in the seep area or the seepage flow may be intercepted at a convenient point somewhere in between the source and the seep area.

The source may be remote (10–100 km and more) and the seepage may pass through deep strata. This will generally be the case with seepage from higher surrounding hills into a plain or valley. Diagnosis of such cases requires considerable insight into the geohydrological conditions in the area, up to great depths (50–500 m and deeper). An example of a more local problem is the seepage from a canal or reservoir to lower lying adjacent land (see also figure 11.7). A number of common seepage problems having relevance to land drainage, together with their possible solutions, are discussed in this chapter.

13.1 GROUNDWATER DRAINAGE FOR SLOPING LAND

The hydraulic gradients that are associated with natural groundwater flow are generally fairly small (<1%). Those associated with artificial drainage are often somewhat higher although they seldom exceed 2–5% (this would apply to a parallel groundwater drainage system with a 20–50 m spacing and a mid-spacing watertable head above the field drainage base of 0·5 m). The downslope directed gravitational gradients generated by the elevation differences in sloping land are often also of this order and therefore may be expected to have a significant influence on the groundwater flow pattern within such land.

13.1.1 Interception of seepage down the slope

This situation is depicted in figure 13.1. A seepage flow (Q in $m^3.day^{-1}$ per m slope width), for example generated by excess water on the upper part of the slope, moves downslope through a permeable upper soil layer overlying an impermeable substratum. To protect downslope land from becoming waterlogged, an *interceptor drain* may be installed across the slope as indicated in figure 13.1. This may either be a ditch or a pipe drain.

Under the assumption of parallel flow (streamlines parallel with the impermeable substratum), the downslope watertable establishes itself at the water level in the interceptor. It then holds that:

upslope of interceptor: $$Q_1 = K H_1 \operatorname{tg} \alpha \quad \text{......(eq. 13.1)}$$

downslope of interceptor: $$Q_2 = K H_2 \operatorname{tg} \alpha \quad \text{......(eq. 13.2)}$$

intercepted flow: $$Q_1 - Q_2 = K (H_1 - H_2) \operatorname{tg} \alpha$$

$$= \frac{H_1 - H_2}{H_1} Q_1 \quad \text{......(eq. 13.3)}$$

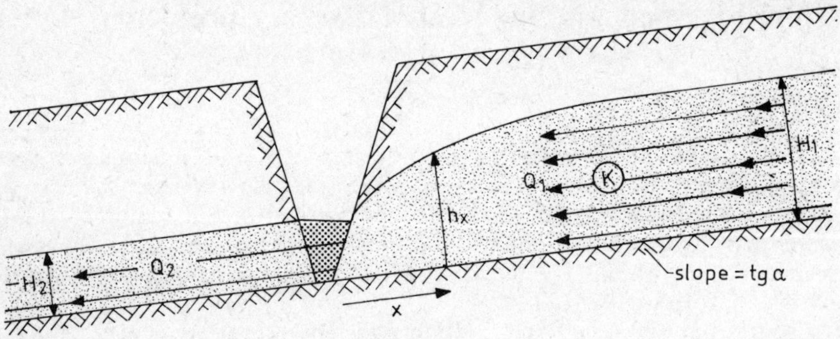

FIGURE 13.1 *Interception of seepage flow down the slope*

The difference in level between the upstream flow depth (H_1) and the water-level in the interceptor (H_2) thus determines how much of the seepage will be intercepted. When the water-level in the interceptor is maintained at 0·75 H_1, 25% of the flow will be intercepted; when at 0·5 H_1, 50% will be intercepted; when at 0·25 H_1, 75% and so on. For full interception, the drainage base should be at or in the impermeable substratum, meaning in practise excavating the drain well into this substratum. When pipe drains are installed in a trench excavated into the impermeable substratum, there is a danger, that considerable flow may pass above the drain without being intercepted, especially when the trench is backfilled with soil. This so-called 'bridging' should be minimised either by backfilling the trench with gravel or by installing the pipes just on top of the impermeable base rather than into it. In general, bridging is liable to become a problem on slopes steeper than 3–4%.

The flow in the drawdown zone immediately upslope of the interceptor, may be described as (Dupuit-Forcheimer assumptions):

$$Q_1 = K\,h_x\left(tg\,\alpha + \frac{dh_x}{dx}\right) \qquad \text{......(eq. 13.4)}$$

which, when combined with eq. 13.1, yields:

$$K\,H_1\,tg\,\alpha = K\,h_x\left(tg\,\alpha + \frac{dh_x}{dx}\right)$$

$$dx = \frac{h_x}{(H_1 - h)\,tg\,\alpha}\,dh_x \qquad \text{......(eq. 13.5)}$$

Integration of eq. 13.5 gives:

$$x = -\frac{H_1}{tg\,\alpha}\ln(H_1 - h_x) + \frac{H_1 - h_x}{tg\,\alpha} + C$$

For $x = 0, h_x = H_2$ so $C = \dfrac{H_1}{tg\,\alpha}\ln(H_1 - H_2) - \dfrac{H_1 - H_2}{tg\,\alpha}$

$$x = \frac{1}{\operatorname{tg}\alpha}\left[H_1 \ln\frac{H_1-H_2}{H_1-h_x} + (H_2-h_x)\right] \qquad \text{......(eq. 13.6)}$$

This shows that the watertable upslope of the interceptor approaches the undisturbed watertable asymptotically. Defining the effective drawdown zone as the zone in which the drawdown is at least 10%, the upslope extent of this zone (x_{eff}) may be found by solving eq. 13.6 for $h = 0{\cdot}9\,H_1$:

$$x_{eff} = \frac{1}{\operatorname{tg}\alpha}\left[H_1 \ln\frac{H_1-H_2}{0{\cdot}1H_1} + (H_2-0{\cdot}9\,H_1)\right]$$

For $H_2 = 0$ (100% interception):

$$x_{eff} = \frac{1{\cdot}4\,H_1}{\operatorname{tg}\alpha}$$

$H_2 = 0{\cdot}5\,H_1$ (50% interception):

$$x_{eff} = \frac{1{\cdot}2\,H_1}{\operatorname{tg}\alpha}$$

The parallel flow assumption underlying all these relationships is most realistic for an interceptor ditch cutting deep into the flow zone (with the ditch bottom at or close above the impermeable substratum as in figure 13.1). In situations where a pipe interceptor is installed well above the impermeable substratum as in figure 13.2, additional head will be lost due to radial flow towards the pipe, rendering these equations less accurate. Under these circumstances some water will also flow towards the pipe from the downslope side (to provide head for this, the watertable just downslope of the interceptor must be somewhat above the water level in the interceptor).

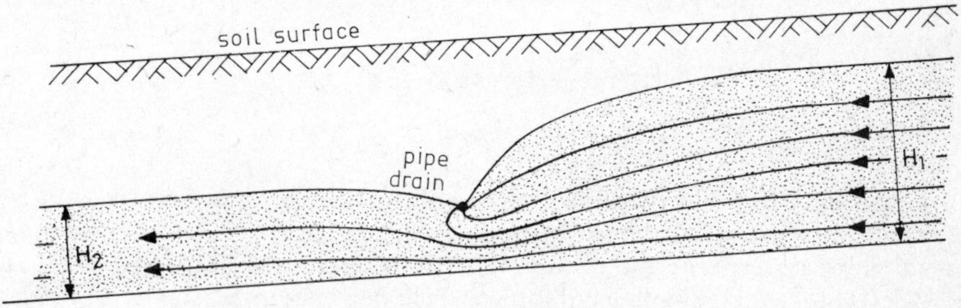

FIGURE 13.2 *Pipe interceptor with pronounced radial flow*

13.1.2 Natural drainage of river valleys

Figure 13.3 depicts a situation where seepage flow (Q in $m^3.day^{-1}$ per m width) from the hills bordering a valley, enters directly into the water bearing strata underlying the valley floor. The latter slope towards the river. In the valley the groundwater is recharged by irrigation water losses and/or by rainfall (recharge q in $m.day^{-1}$). The groundwater in the valley drains naturally towards the incised river.

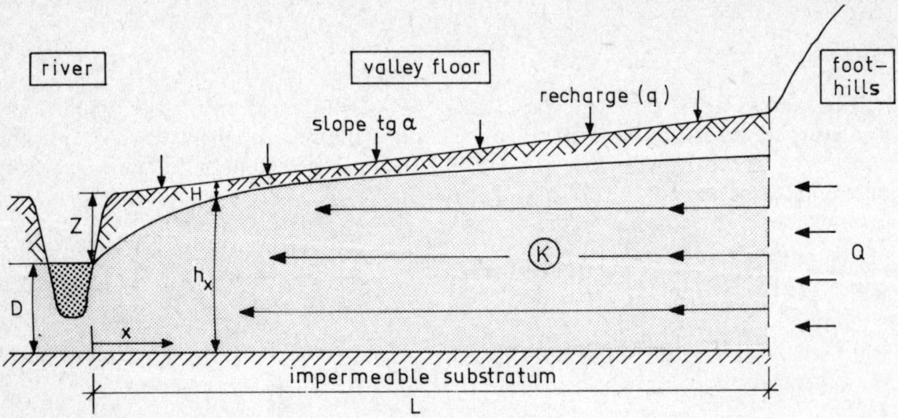

FIGURE 13.3 *Natural groundwater flow in a river valley*

Assuming horizontal flow (all streamlines parallel with the top of the impermeable substratum), it holds that:

$$(L-x)q+Q = K h_x \frac{dh_x}{dx} \qquad \text{......(eq. 13.7)}$$

Integration over the limits $x = 0, h_x = D; x = x, h_x = h_x$ yields

$$\left[-\tfrac{1}{2}q(L-x)^2+Qx\right]\Bigg|_{x=0}^{x=x} = \tfrac{1}{2}K h_x^2\Bigg|_{h_x=D}^{h_x=h_x}$$

$$-\tfrac{1}{2}q(L-x)^2+Qx+\tfrac{1}{2}qL^2 = \tfrac{1}{2}K h_x^2-\tfrac{1}{2}K D^2$$

$$h_x^2 = \frac{q}{K}(2L-x)x+\frac{2Q}{K}x+D^2 \qquad \text{......(eq. 13.8)}$$

Eq. 13.8 indicates that two flows occur which are super-imposed *viz* a uniform seepage flow from the hills (represented by the linear term) and a non-uniform flow generated by the recharge in the valley (quadratic term). Also according to eq. 13.8, the watertable is curved with a curvature which becomes increasingly pronounced towards the river (the more so when radial resistance is taken into account as it should be; this resistance is of course not included in the watertable curvature described by eq. 13.8).

At minimum watertable depth (relative to the valley floor surface), it holds that $dh_x/dx = \text{tg}\ \alpha$ which, applied to eq. 13.8, enables the location of the minimum position to be determined. Critical zones in this respect occur typically in the middle to lower part of the slope (see figure 13.3), their exact location depending on the relative values of D, Z, tg α, K, q and Q. Groundwater drainage may be required in this zone (interception type) while the areas above and below would normally be adequately drained naturally.

13.1.3 Drainage systems for sloping land

Field drains on sloping land may be aligned down the slopes (*longitudinal drainage*) or across the slope (*transverse drainage*). In fact these are the extremes in between which there are an endless variety of oblique alignments of the drains relative to the contours. The systems may be of the singular or of the composite type, usually of the latter (see figure 13.4).

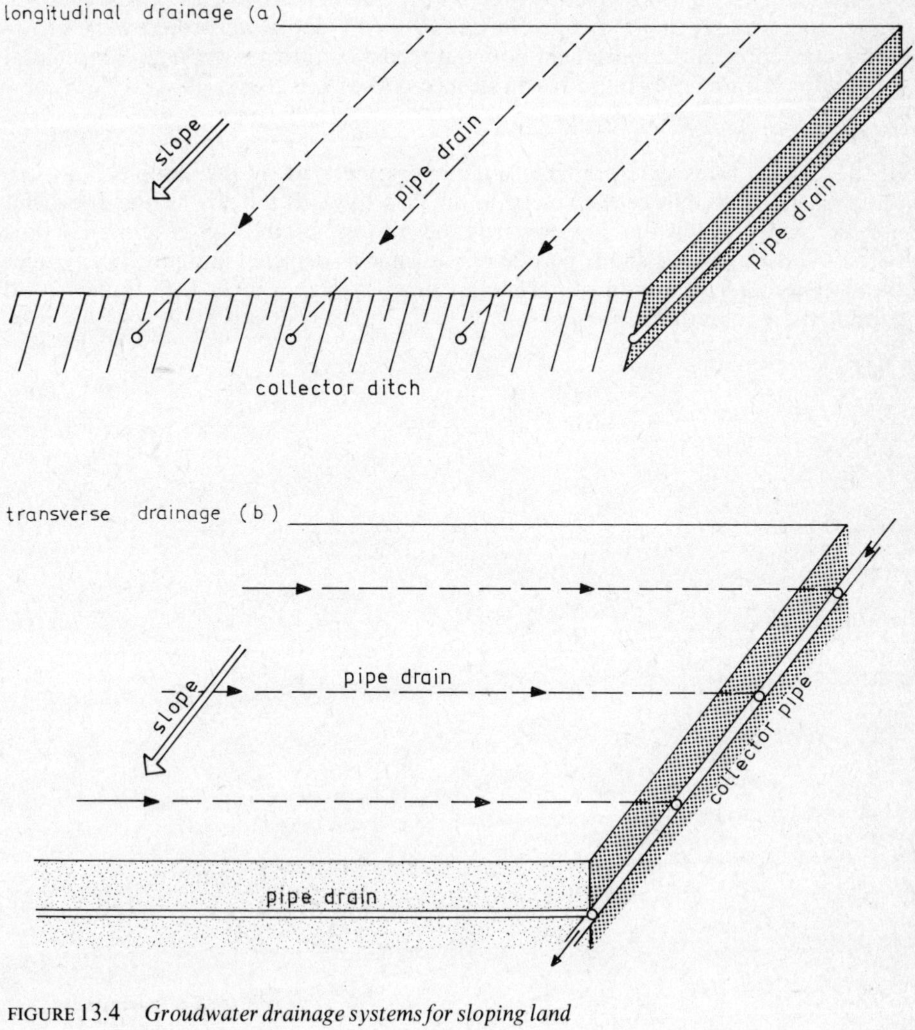

FIGURE 13.4 *Groudwater drainage systems for sloping land*

Longitudinal drainage

In the case of longitudinal drainage, the transverse gradient generated by the pipe drains acts in a direction parallel to the contours while the gravitational gradient generated by the slope of the land acts in a direction perpendicular to the contours.

The water will thus flow at an angle towards the drains, the degree of which depends on the relative strength of these two gradients. The performance of the system (measured in terms of drain discharge and mid-spacing watertable control) is however virtually the same as that of a similar system in a horizontal field. Formulae discussed in chapter 7 may be used to calculate the required drain spacing, at least on mild slopes. When the gradient generated by the slope is of the same order or exceeds those generated by the drainage system, the flow direction becomes strongly downslope oriented. Consequently much of the flow in the lower part of the field will either enter directly into the collector (see figure 13.4-a) or when there is no cross-slope interceptor at the lower field boundary, pass on into a lower field. This makes longitudinal drainage less effective on steeper slopes ($\sim > 2$–5%).

Transverse drainage

In the case of transverse drainage, gradients generated by the slope and by the drainage system reinforce each other in the flow towards the downslope drain, but oppose each other in the flow towards the upslope drain. This results in a flow pattern and watertable shape between the drains as depicted in figure 13.5. Under the simplifying assumption of horizontal flow through a layer with uniform K d conditions, it holds that:

$$\left.\begin{aligned} Q_x &= -Kd\frac{dh_x}{dx} \\ \frac{dQ_x}{dx} &= q \end{aligned}\right\} \frac{d^2h_x}{dx^2} = \frac{-q}{Kd} \qquad \text{......(eq. 13.9)}$$

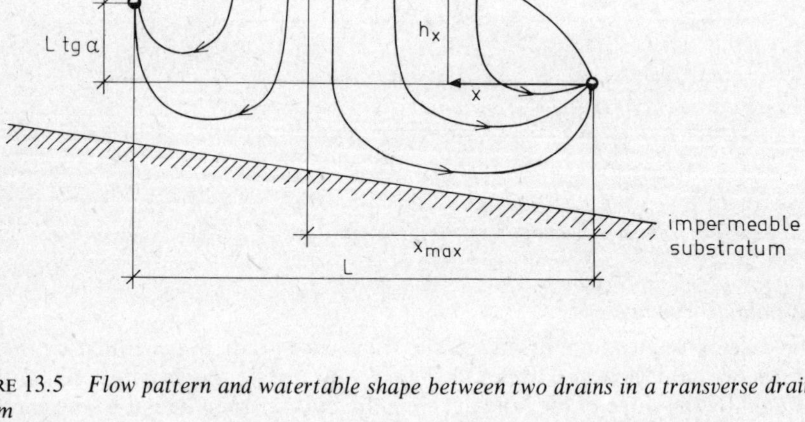

FIGURE 13.5 *Flow pattern and watertable shape between two drains in a transverse drainage system*

Integration of eq. 13.9 gives:

$$h_x = \frac{-q}{2\,Kd}x^2 + Ax + B \qquad \text{......(eq. 13.10)}$$

For $x = 0, h_x = 0$ and $x = L, h_x = L\,tg\,\alpha$

it follows that: $B = 0$ and $A = \frac{qL}{2\,Kd} + tg\,\alpha$

and $$h_x = \frac{q}{2\,Kd}(L-x)x + tg\alpha\,x \qquad \text{......(eq. 13.11)}$$

For x_{max} where h_x attains its maximum value (h_{max}), it holds that $dh_x/dx = 0$:

$$\frac{dh_x}{dx} = \frac{-q}{Kd}x_{max} + \frac{q}{2\,Kd}L + tg\,\alpha = 0$$

$$x_{max} = \frac{L}{2} + \frac{Kd}{q}tg\,\alpha \qquad \text{......(eq. 13.12)}$$

The maximum is always located closer to the upslope drain, the more so the smaller q is in relation to the Kd-value and the steeper the slope. The value of h_{max} may be found by inserting x_{max} as given by eq. 13.12 into eq. 13.11. A limit posed on the watertable depth (H) sets a limit on h_{max} since $h_{max} = W - H$. The required spacing may then be found by solving eq. 13.11 for L.*

Eq. 13.12 defines the extent of the upslope drained area from which it follows that:

- flow towards the drain from upslope: $Q_u = \frac{qL}{2} + Kd\,tg\,\alpha$
- flow towards the drain from downslope: $Q_d = \frac{qL}{2} - Kd\,tg\,\alpha$

Actually, the results obtained with these equations are often not very different from those for flat land (using the equations discussed in chapter 7), even for slopes up to 20%.

Guidelines

For small land slopes (<1–2%), longitudinal and transverse drainage should be nearly equally efficient in controlling the groundwater. On steeper slopes there will always be a significant downslope flow component and here transverse drainage should be used as the downslope component is most efficiently intercepted by cross-slope drains. Field drains in a transverse system may be aligned at a slight angle to

* The minimum depth to the watertable is actually less than H and occurs at $\frac{1}{2}L$ where $dh_x/dx = tg\alpha$; for gentle slopes the difference is negligible.

the contours ($\sim 10°$) to provide gradient for flow in the drain lines. The collector, usually a pipe as in figure 13.4, runs down the slope at maximum gradient.

13.2 SEEPAGE INTO A POLDER

By their very nature, involving a difference between the inner and outer drainage base (see figure 5.7), polders are prone to seepage inflow.

13.2.1 Semi-confined flow

Figure 13.6 illustrates a fairly common situation where a polder is underlain by a semi-confined aquifer which is recharged from outside. Inside the polder the water in the aquifer is under over-pressure (as indicated in figure 13.6 by the piezometer readings), creating a head difference for upward flow through the overlying semi-permeable layer to the phreatic groundwater.

Under steady state conditions, the head of the semi-confined water (h_x in figure 13.6) decreases in the flow direction in a manner as depicted by the piezometric grade line in figure 13.6. The upward flow from the semi-confined aquifer to the phreatic groundwater is given by:

$$q_x K' \frac{h_x - H_2}{D'} = \frac{h_x - H_2}{c} \qquad \text{......(eq. 13.13)}$$

where: q_x = upward seepage flow ($m.day^{-1}$)
h_x and H_2 = head (m)
K' and D' = vertical hydraulic conductivity ($m.day^{-1}$) and thickness (m) of the semi-permeable layer (often combined into $c = D'/K'$, *the hydraulic resistance* of the semi-permeable layer, in days).

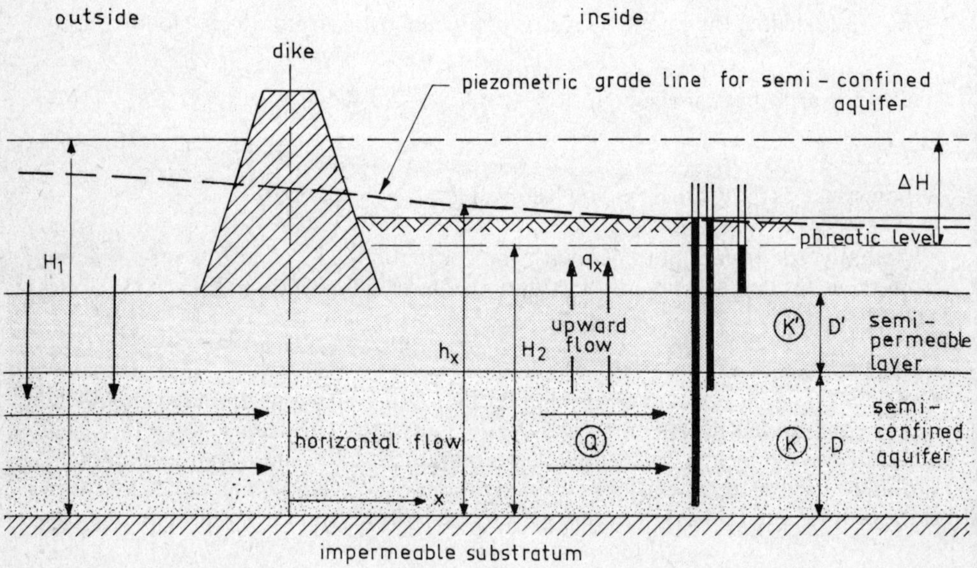

FIGURE 13.6 *Seepage into a polder from a semi-confined aquifer*

The horizontal seepage flow through the aquifer (Q_x in $m^3.day^{-1}$ per m width) is given by:

$$Q_x = -KD\frac{dh_x}{dx} \qquad \text{......(eq. 13.14)}$$

This flow decreases in the flow direction, the rate of decrease being equal to the rate of upward seepage:

$$\frac{dQ_x}{dx} = -q_x \qquad \text{......(eq. 13.15)}$$

Combining eqs. 13.13, 13.14 and 13.15 results in:

$$Q_x = cKD\frac{d^2Q_x}{dx^2} \qquad \text{......(eq. 13.16)}$$

Where the semi-confined aquifer extends infinitely both outside and inside the polder, the following boundary conditions apply:

at $\quad x = o \quad Q_x = Q_o \quad h_x = H_1 - \frac{\Delta H}{2} \quad q_x = q_o$

and $\quad x = \infty \quad Q_x = o \quad h_x = H_2 \quad q_x = o$

Assuming that Q_x varies exponentially with x, eq. 13.16 may be solved to show that (ILRI 1972):

$$Q_x = Q_o e^{-x/\sqrt{cKD}} \qquad \text{......(eq. 13.17)}$$

$$Q_o = \frac{\Delta H}{2}\sqrt{\frac{KD}{c}} \qquad \text{......(eq. 13.18)}$$

$$q_x = q_o e^{-x/\sqrt{cKD}} \qquad \text{......(eq. 13.19)}$$

$$q_o = \frac{\Delta H}{2c} \qquad \text{......(eq. 13.20)}$$

$$h_x = H_2 + \frac{\Delta H}{2}e^{-x/\sqrt{cKD}} \qquad \text{......(eq. 13.21)}$$

Defining the effective extent (x_{eff}) of the seepage zone as the zone in which $q_x > 0{\cdot}1\, q_o$, it follows from eq. 13.19 that:

$$x_{eff} = -\ln(0{\cdot}1).\sqrt{cKD} = 2{\cdot}3\sqrt{cKD} \qquad \text{......(eq. 13.22)}$$

The piezometric grade line approaches asymptotically both the levels H_1 (outside the polder) and H_2 (inside). Half of the driving head ΔH is used outside the polder, half inside.

It may be shown that in situations where the semi-permeable layer does not extend outside the polder, the full head applies to Q_o and q_o, eq. 13.18 and 13.20 becoming:

$$Q_o = \Delta H \sqrt{\frac{KD}{c}} \qquad \text{......(eq. 13.23)}$$

$$q_o = \frac{\Delta H}{c} \qquad \text{......(eq. 13.24)}$$

whilst the piezometric grade line in this case is described by:

$$h_x = H_2 + \Delta H\, e^{-x/\sqrt{cKD}} \qquad \text{......(eq. 13.25)}$$

13.2.2 Phreatic flow

In this case seepage takes the form of phreatic groundwater flow as shown in figure 13.7. The inflow into the polder may in this case be estimated by the flow-net method,* which allows the different actual geometrical situations to be taken into account. A general solution has also been derived by conformal transformation which uses the B/D-ratio to characterise the geometrical situation. The seepage inflow is given by:

$$Q_o = K.\Delta H.\beta \qquad \text{......(eq. 13.26)}$$

where: Q_o = seepage flow passing beneath the dike into the polder ($m^3.day^{-1}$ per m length of dike).
K = hydraulic conductivity of the flow zone ($m.day^{-1}$)
ΔH = head (m)
β = geometry factor, function of $\frac{B}{D}$ (see figure 13.8)

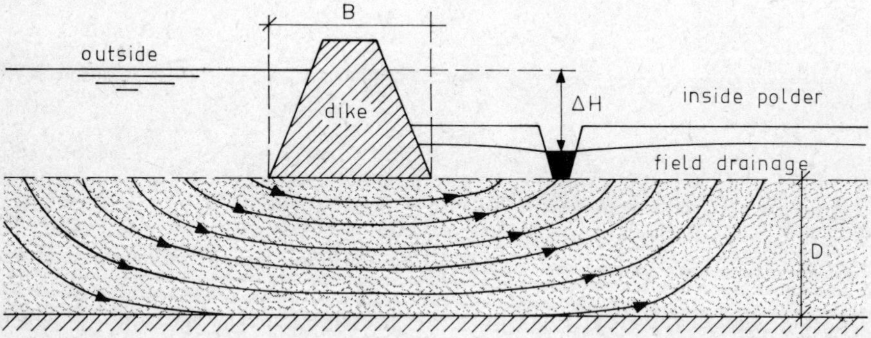

FIGURE 13.7 *Seepage into a polder by phreatic flow*

* For a description of this method see the handbooks on Groundwater Hydraulics detailed at the end of this chapter.

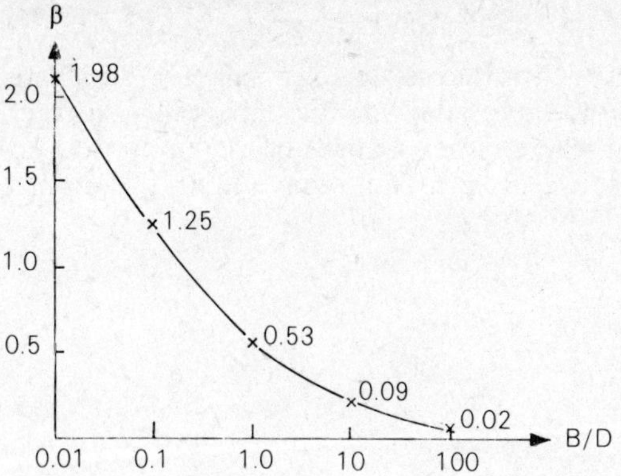

$$\beta = \frac{\int_0^{\frac{\pi}{2}} \frac{dx}{\sqrt{1-m'\sin^2 x}}}{\int_0^{\frac{\pi}{2}} \frac{dx}{\sqrt{1-m\sin^2 x}}}$$

where $m^1 = e^{-\pi B/D}$
$m = 1 - m^1$

FIGURE 13.8 *The geometry factor β in eq. 13.26 as a function of B/D* (ILRI 1972)

Examples

Semi-confined flow (with semi-permeable layer not extending outside the polder)
$D' = 5$ m, $K' = 0{\cdot}01$ m.day^{-1} which gives $c = 500$ days; $D = 200$ m, $K = 25$ m.day^{-1}; $KD = 5\,000$ m^2.day^{-1}; $\Delta H = 2$ m.

Eq. 13.23: $Q_o = 2\sqrt{\frac{5\,000}{500}} = 6{\cdot}3$ m^3.day^{-1} per metre length of dike.

Eq. 13.24: $q_o = \frac{2}{500} = 0{\cdot}004$ m.day^{-1} $= 4$ mm.day^{-1}.

Eq. 13.22: $x_{eff} = 2{\cdot}3\sqrt{500 \times 5\,000} = 3\,636$ m.

Phreatic flow
$B = 50$ m, $D = 200$ m, $B/D = 0{\cdot}25$ which gives $\beta = 0{\cdot}87$;
$K = 25$ m.day^{-1}; $\Delta H = 2$ m.
Eq. 13.26: $Q_o = 25 \times 2 \times 0{\cdot}87 = 43{\cdot}5$ m^3.day^{-1} per metre length of dike.

Polder

Square polder of 10 × 10 km; area 10 000 ha, dike length 40 km.
- semi-confined: total inflow: $40\,000 \times 6{\cdot}3 = 252\,000$ m^3.day^{-1}
  $= 2{\cdot}5$ mm.day^{-1}
- phreatic: total inflow: $40\,000 \times 43{\cdot}5 = 1\,740\,000$ m^3.day^{-1}
  $= 17{\cdot}4$ mm.day^{-1}

The seep area in the semi-confined case would essentially be restricted to a 3 600 m wide strip inside and adjacent to the dike. In the phreatic flow case the seep area may be determined by the flow-net method (usually restricted to a narrower strip, 2 to 3 times D or in this case 400–600 m).

In a polder there will obviously be overlap and inter-action of seepage flow in the corners where the seepage flow will be from different directions. In these simple calculations, this is neglected.

13.3 Seep zones and springs

Seep zones typically occur in situations such as the one depicted in figure 13.9 where at the foot of a hill slope the hydraulic grade line of the phreatic seepage flow emerges onto the surface. This is likely to occur where slopes change abruptly from steep to flat, although much depends upon the soil, geological conditions and stratification.

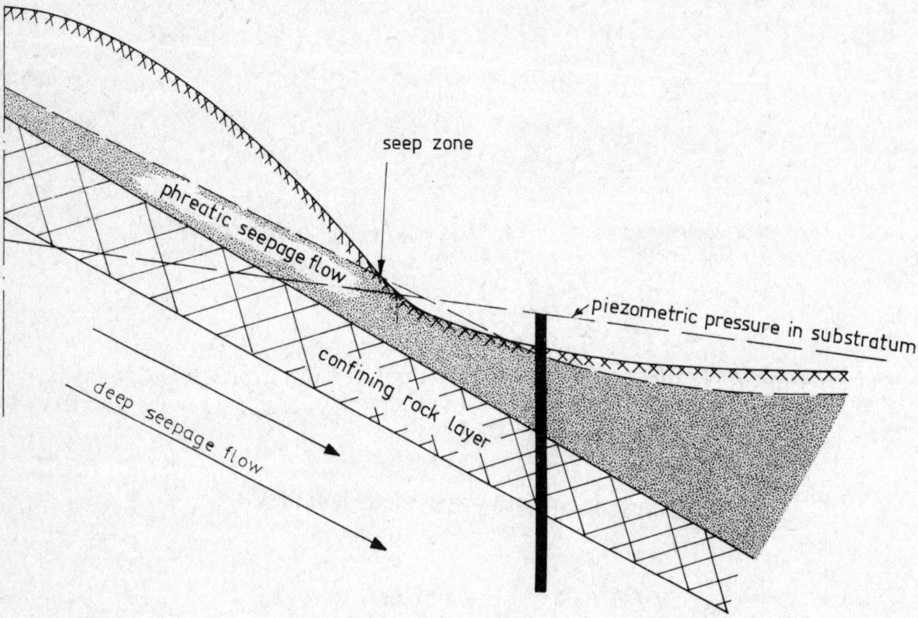

FIGURE 13.9 *Seep zones and springs at the foot of hill slopes*

In addition, there may be a seepage flow through deeper strata from the hills emerging in the valley. Springs arise when this seepage creates (semi)- confined groundwater conditions under the valley from which water spills to the valley floor through leaks such as fissures or faults in the overlying layers. Typically, a fault line occurs along the foot of hills, giving rise to a spring line. Seep zones are also commonly due to upward sloping and/or daylighting rock layers which force moving groundwater to the surface (see figure 13.10).

Naturally, seep areas often abound with reeds, rushes and similar wetland indicator plants.

Drainage measures

Seepage zones as in figures 13.9 and 13.10 may in principle be drained by intercepting the casual seepage flow 'above' the wet spot, just about where the watertable emerges onto the surface of the land. The interceptor drain (ditch or pipe drain) should be deep enough, cutting well into the layer transporting water. As discussed in section 13.1.1 the deeper the drain is installed in this layer, the more water it will intercept and for full interception the drain should reach to the base of the water transporting layer.

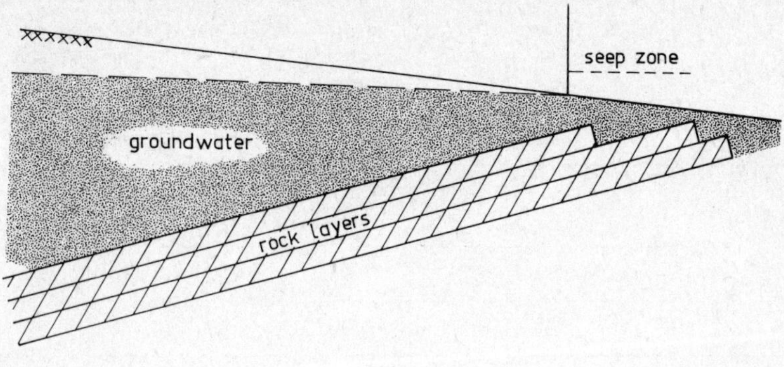

FIGURE 13.10 *Seep zone due to daylighting of rock layers*

Springs may be neutralised by an encircling ditch or be otherwise directly connected with a drainage ditch leading to an outlet.

The required discharge capacity for a seep area may in principle be determined by estimating the amount of deep percolation in the source area which feeds this seepage flow. As an example consider the case shown in figure 13.11. The irrigated area is underlain by rock which is weathered/fractured in its upper layers. Deep percolation losses of irrigation, and during the rainy season also of rain, sink into this layer and move down slope towards the lower land. In the transition zone the flow runs up against the 'impermeable' clay soil of the lower land and partly also

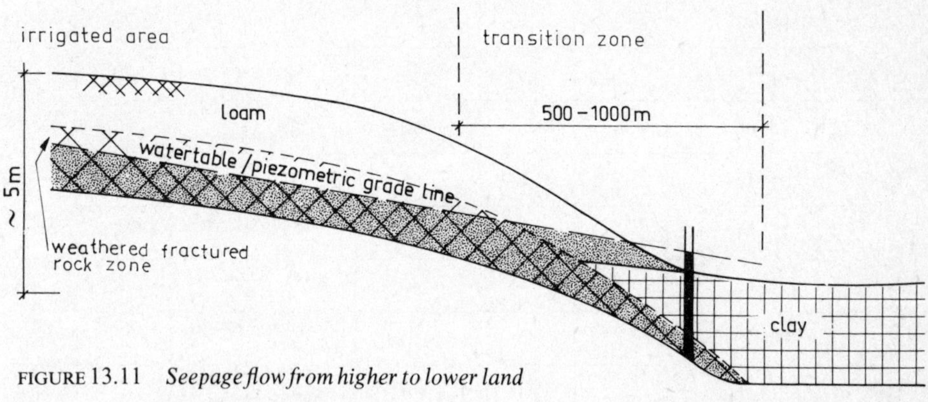

FIGURE 13.11 *Seepage flow from higher to lower land*

becomes confined (see piezometer reading). Some seepage emerges directly at the surface in a seep area at the toe of the slope and some appears through leaks in the clay layers which provide an escape for the underlying confined water. The total seepage load in the transition zone should be equal to the total deep percolation losses (= groundwater recharge) in the irrigated area. These losses may be roughly estimated from soil moisture balance calculations as described in section 15.1 (a). Calculations for two-week periods as in table 13.1 estimate average deep percolation losses throughout a year at 1·2 mm.day^{-1}. The highest losses occur in the rainy season when they average about 4 mm.day^{-1} (from early February till mid-April).

TABLE 13.1 *Soil moisture calculations for the determination of deep percolation losses*

		Rain-fall (*mm*)	*Irriga-tion* (*mm*)	*Evapo-transpira-tion* (*mm*)	*Storage* (*mm*)	*Defi-cit* (*mm*)	*Excess (deep percola-tion)* (*mm*)	
March	31				150			Crop I
April	1–15	160		95	150		65	
	16–30	20		90	80			
May	1–15	90		85	85			
	16–31	15		40	60			
June	1–15	15		25	50			
	16–30	—		20	30			
July	1–15	5	180	75	140			Crop II
	16–31	—	90	70	150		10	
Aug.	1–15	—	90	70	150		20	
	16–31	—	90	70	150		20	
Sept.	1–15	5	90	75	150		20	
	16–30	—	90	80	150		10	
Oct.	1–15	—	90	85	150		5	
	16–31	—	90	85	150		5	
Nov.	1–15	35		90	95			
	16–30	10		50	55			
Dec.	1–15		180	95	140			Crop I
	16–31	55	90	100	150		35	
Jan.	1–15	150		105	150		45	
	16–31	10	90	110	140			
Feb.	1–15	190		110	150		70	
	16–28/29	120		100	150		20	
March	1–15	160		100	150		60	
	16–31	140		95	150		45	

total 430 mm

Generally, seep areas are only a fraction of the source areas. In this example the former area is only about 10% of the latter so that the required discharge capacity for the seep area would have to be of the order of 40 mm.day^{-1} (not counting evaporative drying). As seepage emergence is often highly non-uniform, still higher rates may be required locally.

Flows paths along which the seepage water moves from the source to the seep area are often difficult to trace, requiring in any event considerable investigations. Correct identification of the main water transporting layers (often just a vein or crack) is pre-conditional to effective interception. Costly investigation and remedial drainage measures may not be worthwhile. Seep areas may sometimes be used for rice growing or for grazing/hay-making without much drainage being required.

References and further reading

BEAR, J 1979: *Hydraulics of Groundwater*, McGraw-Hill, New York

BEAR, J, D ZASLAVSKY and S IRMAY 1968: 'Physical principles of water percolation and seepage', Arid Zone Research XXIX, Unesco, Paris

BOUWER, H 1955: 'Tile Drainage of Sloping Land', *Agr. Engineering 36* (6)

BUREAU OF RECLAMATION, 1977: *Groundwater Manual*, United States Department of the Interior, Washington DC

CEDERGREN, H R 1967: *Seepage, Drainage and Flow Nets*, Wiley, New York

EDELMAN, J H 1972: 'Groundwater hydraulics of extensive aquifers', Technical Bulletin 13, International Institute for Land Reclamation and Improvement, Wageningen

ILRI 1972: *Fieldbook for Land and Water Management Experts*, International Institute for Land Reclamation and Improvement, Wageningen

LUTHIN, J N and J C GUITJENS 1955: 'Transient solutions for drainage of sloping lands', *Journal of Irrigation and Drainage Div., Proc. of the ASCE 93* (3)

POLUBARINOVA-KOCHINA, P Ya 1962: *Theory of Groundwater Movement*, Princeton University Press, Princeton

VERRUIJT, A 1970: *Theory of Groundwater Flow*, MacMillan, London

Chapter 14 Reclamation and drainage of unripened soils

Unripened soils are encountered in the reclamation of sea, lake beds, marshes and swamps, floodplains, coastal forelands, etc. The overall appearance of these soils after emergence from under water is that of soft mud. Upon drainage certain changes take place in the mud which together are referred to as '*ripening*' of the soil. Ripening precedes the common soil formation processes leading to full development of the soil profiles.

Much of the current knowledge about the reclamation of unripened soils has been gained during the impoldering of the former Zuiderzee in the Netherlands which started in 1927 and continues to the present day. Impoldering and reclamation of fresh subaqueous sediments are of course often interdependant, the first activity typically creating the necessary hydraulic control to proceed with the latter. The technology on impoldering and reclamation gained in the Netherlands has been applied in many other parts of the world, adding further knowledge but also showing that no universal solutions exist.

14.1 THE SOIL RIPENING PROCESSES

Soil ripening of fresh subaqueous sediments starts with the loosely packed, oversaturated mud soil losing (permanently) part of its excess water, initially mostly by evaporative drying, later also by drainage. The watertable falls and the soil above the watertable becomes exposed to capillary forces that pull the soil particles into a closer packing which is enhanced by the fact that they no longer 'float', supported by the buoyant forces, but are subjected to increase particle pressures. In this process the soil particles are also spatially re-arranged, all in all resulting in a permanent reduction of the soil volume, being partly shrinkage (caused by the capillary forces), and partly settlement (caused by the increase in particle pressures). This induces subsidence, cracking and in the end, structural development of the soil, together comprising the physical ripening of the soil. The higher the colloid content of the soil the more intensive the ripening, and in fact sandy sediments, lacking colloids, hardly ripen at all.

14.1.1 Physical ripening

A prominent feature of the physical ripening process is the change in phase composition (see figure 14.1). The pore volume (V_{pores}) reduces markedly and as a consequence the bulk density (BD) of the soil increases as the soil ripens. As long as the soil remains saturated, this decrease in pore space shows by a decrease in the water content of the soil as measured by the *A-value* (= water content of the saturated unripened soil in weight %, ie gm water per 100 gm dry soil). Under these conditions, it also holds that:

$$BD = \frac{100}{100/\gamma_s + A} \text{ gm.cm}^{-3} \qquad \text{......(eq. 14.1)}$$

$$\% \text{ pores} = BD \times A \qquad \text{......(eq. 14.2)}$$

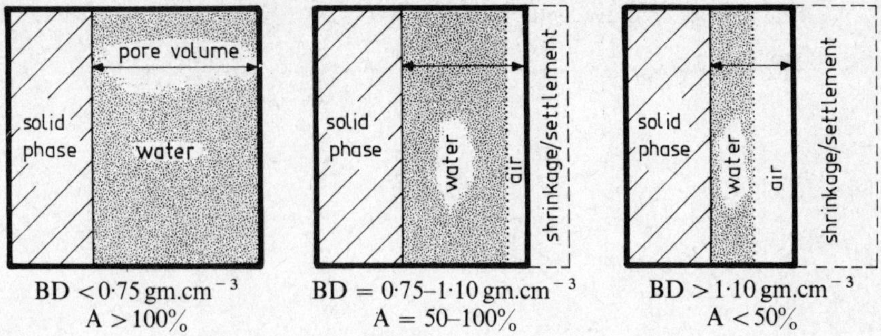

FIGURE 14.1 *Changes in the phase composition of the soil during the ripening of the soil (data applies to a clay soil)*

where γ_s = specific density of the solid phase. Taking for the average specific density of the soil mineral and the organic material (% OM) respectively 2·65 and 1·45 gm.cm^{-3}, $\gamma_s = 2{\cdot}65 - (\%\,\mathrm{OM}/100)\,1{\cdot}20$.

The consistency of the soil changes as the soil ripens, from nearly liquid to firm/hard. For field assessment, consistency grade scales have been developed enabling the experienced soil surveyor to arrive at fair estimates of the current stage of ripening of the soil.

Progress of the ripening with depth and time

The ripening of a mud soil normally starts at the surface, initiated by evaportive drying, from where it slowly extends to the deeper soil layers. The rate of ripening at all depths decreases as the ripening of the soil progresses while the rate of extension into the soil also decreases as the ripening reaches deeper in the soil. Both trends are clearly shown in figure 14.2. The first trend is an inherent characteristic of the ripening process while the second trend is mostly due to the slower rate of drying deeper in the soil. Full ripening of a soil profile to 1·0–1·5 m depth may take centuries and typically subsoils remain unripened long after reclamation.

Since much of the water-loss during the initial ripening period is due to evaporative drying, it generally holds that the ripening progresses more during dry years than during wet years. Good drainage, however, is most essential for ripening.

14.1.2 Other ripening processes

Parallel with the physical ripening, various chemical and biological changes occur in the soil (respectively referred to as the chemical and biological ripening of the soil). The *chemical ripening* comprises the oxidation processes and the adjustments in the cation composition of the adsorption complex of the soil. The oxidation of the sulphides, commonly found in marine sediments and which may lead to the formation of acid-sulphate soils, is of particular importance (see section 14.3). Many marine soils will also be salty and will have to be leached before they can be used to grow crops. The *biological ripening* of the soil involves the development of aerobic micro-biological life within the soil. This may take time, especially the development of an adequate micro-biological nitrification capacity in the soil. Inocculation of the soil may be required to establish an effective nodule type of bacteria population for the fixation of atmospheric nitrogen.

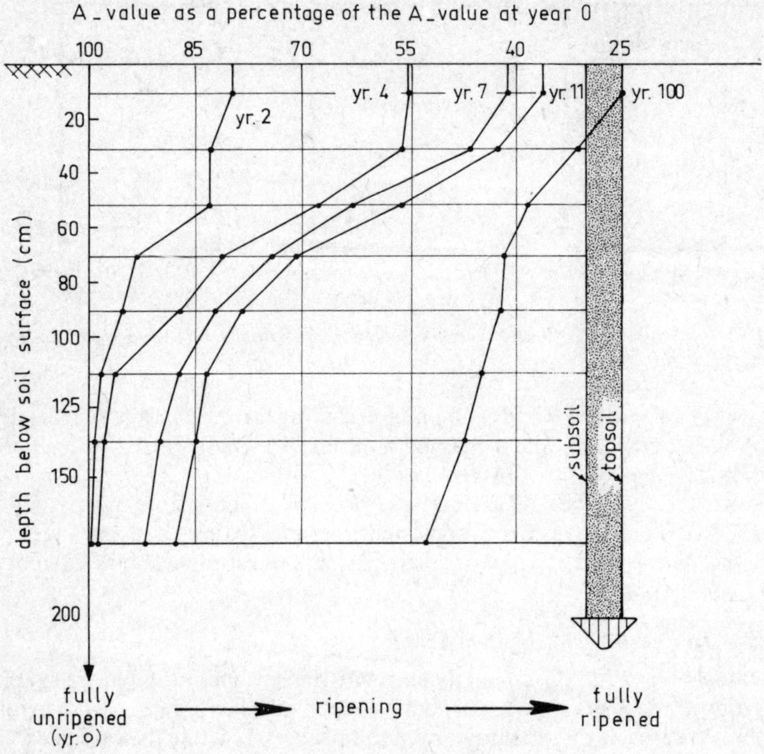

FIGURE 14.2 *The progress of the ripening in the soil profile with time and with depth (light clay soil in the Zuiderzee polder, the Netherlands)*

14.2 RECLAMATION AND DRAINAGE METHODS

Reclamation and drainage of fresh subaqueous sediments are of course closely related. During the initial ripening period they are in fact largely synonymous.

Early reclamation stage

During this period the hydraulic conductivity of the still unripened soil is usually very low and drainage only entails maintaining the surface clear of ponded water. As such, however, it performs a very useful function in that it maximises the vaporisation of water from in the soil rather than on the surface, thus promoting the ripening. This surface drainage is most conveniently undertaken using parallel field ditches (see the system described in section 4.2.1(2). Spacings of the order of 10 m are generally satisfactory. The ditches may be made with a rotary ditcher although special equipment is required to be able to work on the soft soils (low wheel pressures). Amphibious equipment is sometimes used to gain access at a very early stage (just after emergence from under water before wheel/track equipment can enter) to press small drainage pathways into the mud in order to lead ponded water/rain water off the land. The field ditches should be deepened gradually, in parallel with the progress of the ripening. Ditches could of course be made deep at the outset but these would be liable to cave in, in the unripened subsoil, and anyway would not draw water from the soil due to its low hydraulic conductivity and the immobility of the soil water. Starting with ditches of 30–40 cm depth, a ditch depth of 60–70 cm is reached in a period of some five years. Gradually the ditches start to perform profile drainage, mostly shallow drainage through the already partially ripened top layer.

Development of the hydraulic conductivity (K-value)

The low hydraulic conductivity of the unripened clayey sediment, in spite of the large pore volume, is due to the fact that all pores are micro-pores which offer a very high resistance to water movement. Upon ripening, cracks form through which excess water can move rapidly and the hydraulic conductivity improves dramatically, especially in the case of clay soils which crack most. The overall K-value of heavy clay soils often increases from only a few mm per day in the unripened sediment to a few m per day in the cracked, ripening soil (the conductivity of the soil mass between the cracks remains low). Where the crack volume represents irreversible shrinkage, the cracks remain intact when the soil is re-wetted. This is common with the ripening of fresh marine sediment in The Netherlands. Ripening cracks in some soils, however, are known to close again by swelling in the rainy season, probably due to differences in the clay mineralogy between soils.

In tropical coastal swamps, roots of the cleared mangrove vegetation may leave well preserved, good sized channels which confer on the soil a good hydraulic conductivity. The ripening history also has an influence. Naturally pre-ripened sediments generally developed less cracking upon reclamation, resulting in lower conductivity values than fully unripened sediments. Future K-values, therefore, depend on a number of factors, the influence of which are not well understood, making predictions difficult. Much depends also on the soil structural development taking place during the later stages of the physical ripening, in parallel and integrated with the chemical and biological ripening of the soil. In general, very little improvement of the hydraulic conductivity may be expected due to ripening cracking when the existing BD values of the soil have attained values of $1{\cdot}5$–$1{\cdot}6\,gr.cm^{-3}$. Also, ripening cracking is generally to be expected rather more in humid climates than in semi-humid/semi-arid climates since in the latter some ripening has often already occurred during the prolonged dry periods.

Advanced reclamation stage

By the time the ripening front has reached well into the subsoil (say to some 60–70 cm depth) pipe drains may be intsalled if desired. In the Netherlands this is normally done after some five years ripening, using drain depths of 1·0–1·2 m. These pipe drains function initially very much like the drains described in section 4.4 (see figure 4.13). The excess rain water sinks in vertically through the cracks and then flows laterally over the 'impermeable' unripened subsoil to the trenches and then down to the pipes. The trenches should be backfilled with well-ripened soil and when this is done carefully and under dry conditions, clay soils generally do not need a pipe envelope. Earlier installation of pipes is generally not advisable as firstly their effectiveness is restricted by the limited ripened depth, secondly the pipe lines may become disturbed by the large subsidence still to occur, while thirdly, due to this subsidence, the drain depth would also in time become too shallow. Even after five years the pipes should be laid at some 10–20 cm over-depth to allow for subsequent subsidence of the overlying soil. The spacing to be used depends very much upon how the K-value develops upon ripening. When a good number of stable ripening cracks become established, spacings up to 30–50 m may well be used for clay soils. Where cracking of the subsoil will develop further in the years after installation the pipe drains may be extra widely spaced, keeping some of the field ditches for the time being to supplement the pipe system.

Drainage, evaporation and ripening

Good drainage enhances the ripening considerably. Drainage can however only remove the soil water above field capacity (FC). In the case of unripened soils most of the pore water (the pores being almost all micro-pores) is held below field capacity, and so drainage is mostly the removal of surface water and water from the cracks. The removal of pore water from unripened soil depends mostly on evaporative drying, which is of course most effective when all the non-pore water has already been removed by drainage. Evaporation losses from bare soil diminish in time due to the formation of a protecting mulch layer (see section 11.2.2), although in this case the effectiveness of these layers is partly defeated by the cracks which form. Through the latter the subsoil is able to continue to lose water by evaporation. This occurs also, and to a much larger extent, when there is vegetation on the land.

A vigorous vegetation with a deep and extensive root system can thoroughly deplete the rootzone, promoting and extending the ripening to deeper layers. Mainly for this reason, the land to be reclaimed in the Netherlands is sown with reed just before emergence (broadcast by planes). This has the additional advantage that after emergence the land will not be infested by all kinds of weeds. This reed is followed by the first crops, usually colza (rape seed), followed by wheat and barley. Alfalfa is also useful during this early cropping period as it is deep rooting and adds nitrogen to the soil. The ripening is considerably retarded by seepage inflow into the area, often the ripening stagnating at a shallow depth. The ripening in such areas can be accelerated by intercepting the seepage inflow.

Tropical lowland soils used for wet rice cultivation will generally remain unripened except for a shallow top layer which periodically dries out during the off-season. The application of irrigation will in general of course also retard the rate of soil ripening.

14.3 ACID SULPHATE SOILS

Sea water is normally rich in sulphates and as a consequence so are marine and brackish water sediments. In the presence of fresh organic matter, the sulphates in unripened/non-aerated sediments change into sulphides by microbiological reduction. These then combine with the Fe ions to FeS (ferrous sulphide), giving these sediments their typically dark blue appearance. In time the FeS will turn into pyrite (FeS_2) with soil colours becoming somewhat greyer. The pyrite may accumulate, sea bed soils usually having a higher concentration than tidal lands. Although river water is usually low in sulphur, low lying basin sediments may become enriched by accumulation. For all sediments it holds that they contain more sulphur the higher their clay content.

Acidification processes

Upon drainage and ripening of these sediments, air will enter the soil and the sulphides will become oxidized through one of the following reactions:

$$4\,FeS + 10\,H_2O + 9\,O_2 \rightarrow 4\,Fe(OH)_3 + 4\,H_2SO_4$$

$$4\,FeS_2 + 14\,H_2O + 15\,O_2 \rightarrow 4\,Fe(OH)_3 + 8\,H_2SO_4$$

$$12\,FeS_2 + 30\,H_2O + 45\,O_2 + 4\,K^+ \rightarrow \underset{\text{(jarosite)}}{4\,KFe_3(OH)_6(SO_4)_2} + 16\,H_2SO_4 + 4\,H^+$$

The oxidation of the FeS is a rapid process but some forms of aged FeS_2 are rather inert to oxidation, especially when the soil reaction is alkaline. The oxidation processes are promoted by the sulphur-oxidizing bacteria in the soil while the role of the Fe ions can be taken over by Al ions, in the latter case leading to the formation of $Al(OH)_3$. The sulphuric acid may react with the $Fe(OH)_3$ or the $Al(OH)_3$ to form salts which, because they are composed of weak bases and a strong acid, will readily hydrolise to form some Fe/Al hydroxide-sulphate compound and again sulphuric acid. In these circumstances the soil reaction becomes strongly acidic with the pH falling as low as 2–3, although a pH value around 4 is more common as some of the hydrogen ions formed will normally be leached from the soil. Under such conditions, crops suffer from lack of phosphorus (becomes unavailable at low pH) and from toxicity (Al toxicity for dry land crops; Fe toxicity for wetland rice). The generally poor physical conditions of the soils also limit crop growth. Where such soils develop, the land is often abandoned.

Neutralisation and reclamation

In calcareous soils the sulphuric acid is neutralised:

$$H_2SO_4 + CaCO_3 \rightarrow CaSO_4 + H_2O + CO_2$$

and the pH values will remain neutral when the oxidation processes described take place in the ripening soil. This neutralisation may of course also be achieved by adding lime to the soil, the liming requirements often being reduced by allowing some leaching before liming of the soil (eg waiting for one to two rainy seasons to pass or by leaching the soil by irrigation). For clay soils up to 50 tonnes $CaCO_3$ per ha may be required although lighter textured soils can be neutralised with much less. Deep ploughing may sometimes be used to bring up calcareous subsoil to mix with the acidic top layer. Flooding with base rich (Ca^{++}, Mg^{++}) water is also useful, to replace H^+ on the complex and subsequently leach it from the rootzone; this may even be done with sea water provided that the resulting salinisation/sodification of the soil can readily be corrected afterwards.

The acidification can of course be stopped/prevented by maintaining high watertables (keeping the soil under water, out of contact with air). In theory watertables may be lowered gradually at such a rate that leaching, alone or in combination with liming, maintains the balance between H^+ production and removal/neutralisation, but few examples of success on a practical scale exist. For crops like grass (temperate climates) and rice (tropical climates) shallow rootzones suffice and the subsoils may be left unreclaimed, under water. Of course there will be little leaching under such high watertable conditions and the alternative (low watertables maintained by deep drains) might be preferable when neutralisation can be achieved at low costs. Pipe drains should generally not be installed until the neutralisation has reached an advanced state as the pipes are liable to become clogged by ochre when installed too early.

Diagnosis

The chances of acid sulphate soil formation during the ripening/drainage of muddy sediments can roughly be assessed by comparing the lime and the pyrite content of the sediment. There is a clear hazard when the lime concentration of the sediment is equal to or less than the pyrite concentration (both expressed in equivalents, the latter as SO_4^{--}). The unripened sediment may also be left exposed to aeration and drying and the pH development be studied over a period of one to two months. To maintain the acidification processes, the sample should be kept moist and the temperature should not be too low. The changes in colour of the sample are often already indicative (yellow jarosite mottling).* A number of rapid evaluation tests have also been developed (ILRI, 1973). Pyrite accumulation is especially to be expected in tropical mangrove swamps due to the abundant presence of organic matter. Reclamation in such areas almost universally faces the problem of acid sulphate soil formation.

14.4 SUBSIDENCE PREDICTION

The shrinkage and settlement of unripened soil upon drying (and the underlying compression/re-arrangement of the soil skeleton) is not entirely irreversible. Some of the original volume is usually regained by swelling once the soils are rewetted. The irreversible part of it will show as a long term gradual decrease of the soil volume, superimposed on which shorter cycled reversible changes in soil volume may occur (see figure 14.3).

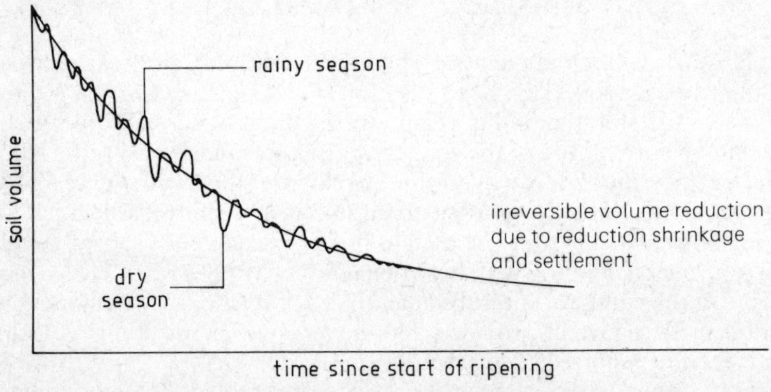

FIGURE 14.3 *Reversible and irreversible soil volume reduction during the ripening of clay soils*

The soil volume reduction exhibits itself in the field by subsidence of the soil surface (*vertical contraction* of the soil skeleton) and by the formation of cracks (*horizontal contraction*). In unripened soils, subsidence and crack formation will occupy most of the soil volume vacated by the water when the soil ripens, the air-filled pore space often increasing very little, see figure 14.4 (this contrasts with non-

* The popular name *cat clays* for acid sulphate soils stems from these mottlings which give the soils an appearance rather similar to cat-droppings.

shrinking, ripened soils which have rigid skeletons where the water lost from the pores will be wholly replaced by air).

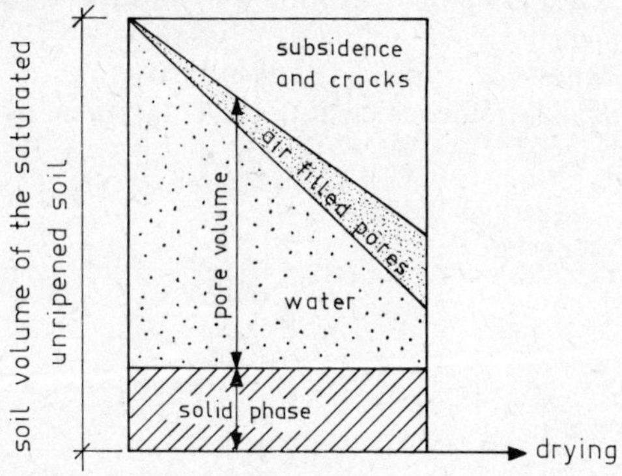

FIGURE 14.4 *Changes in phase composition of the unripened soil upon drying*

14.4.1 Ripening subsidence

Cracks fill in from the sides and/or from the surface and in the long run all ripening shrinkage and settlement translates itself into a lowering of the soil surface (subsidence). The total subsidence due to ripening may thus be estimated as:

$$D_t = \frac{BD_o}{BD_t} \cdot D_o \qquad \text{......(eq. 14.3)}$$

$$\Delta D = D_o - D_t = D_o \frac{BD_t - BD_o}{BD_t} \qquad \text{......(eq. 14.4)}$$

where D_o and D_t = thickness of the soil layer in respectively the unripened state (yr o) and after 't' years ripening (yr t)

where BD_o and BD_t = bulk density of the soil at respectively yr o and yr t

The values of BD_o and BD_t may be determined on (core) samples of respectively the unripened mud and of t years old reclaimed/ripened soil of the same type from older reclamations (see also table 14.1).

14.4.2 Settlement subsidence due to lowering of the watertable

In addition to the subsidence caused by the ripening of the soil above the watertable, there will be subsidence due to settlement of the soil caused by the lowering of the watertable. In general, the watertable before reclamation will be at or above the soil surface and a lowering of the watertable will increase particle pressures throughout the whole of the soil (see figure 14.5). This type of settlement subsidence in the soil may be estimated according to TERZAGHI (ILRI 1974) as:

$$\Delta z = \frac{D}{c} \ln \frac{\sigma_p + \Delta\sigma_p}{\sigma_p} \qquad \text{......(eq. 14.5)}$$

where: Δz = the settlement subsidence (cm)
σ_p = the average particle pressure in the soil layer considered (gm.cm^{-2})
$\Delta\sigma_p$ = increase in particle pressure due to the lowering of the watertable (gm.cm^{-2})
D = thickness of the soil layer considered (cm)
c = consolidation coefficient (to be determined by means of a laboratory test).

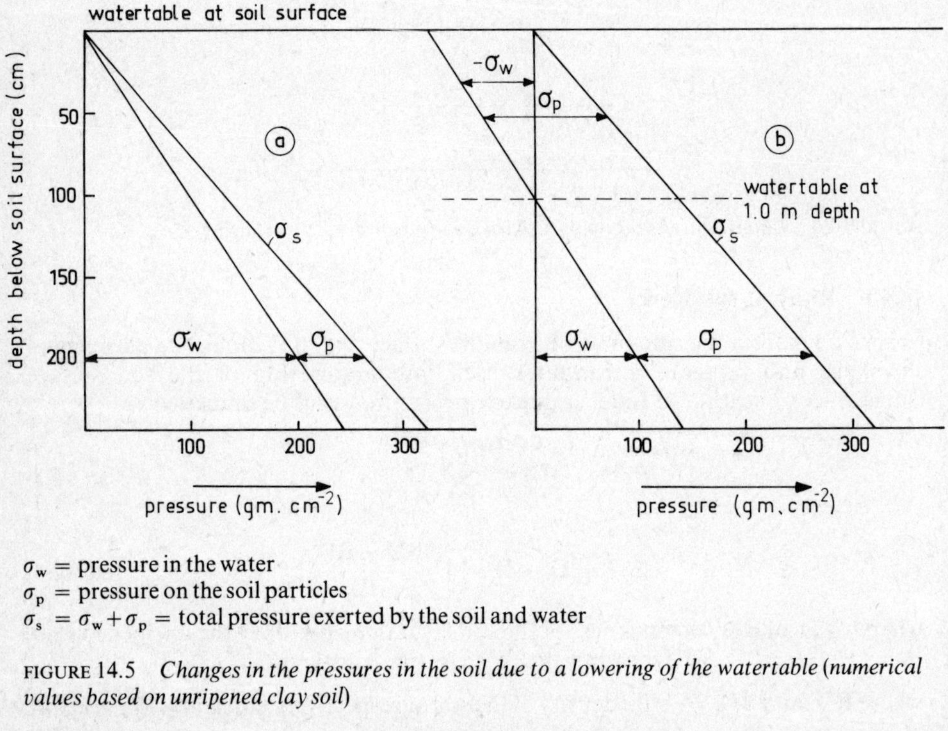

σ_w = pressure in the water
σ_p = pressure on the soil particles
$\sigma_s = \sigma_w + \sigma_p$ = total pressure exerted by the soil and water

FIGURE 14.5 *Changes in the pressures in the soil due to a lowering of the watertable (numerical values based on unripened clay soil)*

Under hydrostatic conditions σ_w is positive below and negative above the watertable, its value corresponding to the heights respectively below and above the watertable. The value of σ_s in a fully saturated soil at depth H below the soil surface is:

$$\sigma_s = H\left(\frac{\%\,\text{pores}}{100} + BD\right) \qquad \text{......(eq. 14.6)}$$

It may generally be assumed for unripened soil that as the watertable is lowered, the soil above the watertable remains (nearly) saturated. The value of the particle pressure σ_p follows from $\sigma_p = \sigma_s - \sigma_w$ which gives, again for a fully saturated soil:

– watertable at soil surface: (figure 14.5-A)

$$\sigma_p = H\left(\frac{\%\,\text{pores}}{100} + BD - 1\right)$$

– watertable at depth W below the soil surface (figure 14.5-B):

$$\sigma_p = H\left(\frac{\%\,\text{pores}}{100} + BD + \frac{W - H}{H}\right)$$

The value of σ_p in equation 14.5 may thus be readily determined given the values of the above soil constants. Ideally these should be determined for each case but for rough estimates the values given in table 14.1 may be used. The value of $\Delta\sigma_p$ (in $gm.cm^{-2}$) under these saturated conditions is numerically equal to the fall of the watertable in cm. For details on the applied calculation methods see ILRI 1973.

TABLE 14.1 *Average values of the physical soil constants for settlement subsidence prediction*

	Specific density solid phase ($gm.cm^{-3}$)	*% pores*	*BD* ($gm.cm^{-3}$)	*c Consolidation coefficient*
Sand: loose packing	2·65	45	1·45	100
dense packing	2·65	40	1·60	150
Loam, ripened	2·65	50	1·30	25
Clay: ripened	2·65	60	1·00	20
unripened	2·65	80	0·50	8–10
Peaty clay, half-ripened (50% OM)	2·00	60	0·80	6–8
Peat, drained	1·40–1·60	80–90	0·15–0·30	4–5

14.4.3 Oxidation of peat soils

Additional subsidence is to be expected when peat/peaty soils are reclaimed/drained, as the improved aeration of the soil will lead to accelerated oxidation of the organic material. Under temperate conditions peat soils may be expected to loose some 1–2 mm soil depth per annum under normal drainage conditions (watertable in summer at 50–60 cm depth). At the higher temperatures of the (sub)tropics, the rates of soil loss due to oxidation are always much higher, up to several cm per annum under well drained conditions. To minimise subsidence/soil loss by oxidation of peat soils, watertables should be maintained as high as is compatible with the (agricultural) use of the land. On swamp/marsh soils the typical covering of organic debris will also mostly disappear in a matter of a few years after reclamation/drainage, the organic matter content of the topsoil stabilising at a few percent in warm climates, slightly higher in temperate climates. The oxidation of organic matter deeper in the soil is much slower, becoming negligible in the permanently saturated zone below the lowest watertable level (summer/dry season level).

14.4.4 Experiences in the Netherlands

No great accuracy can be expected from the predictions of subsidence on the basis of equations 14.4 and 14.5, the accuracy in fact depending almost entirely upon the accuracy with which the final bulk density value of the ripened/half-ripened soil can be estimated. The Terzaghi formula (equation 14.5) was developed for instantaneous loading while in land reclamation the loading due to lowering of the watertable is rather gradual. Best results are obtained when the coefficient c is determined in the laboratory under loading conditions rather similar to those existing in the field. The Terzaghi formula also supposes the increase in load to be fairly uniform across the area considered and the pressures in the groundwater at all times to correspond to the prevailing hydrostatic pressures. This last condition will prevail in freely draining soils though in clay soils of low hydraulic conductivity excess soil water pressures may continue to exist for a considerable time after the lowering of the watertable.

Rates of subsidence

The ripening subsidence may be calculated with equation 14.4 at any time after the start of the reclamation. In contrast, equation 14.5 only enables the final settlement subsidence to be predicted. In peat soils some 90% of the final subsidence is often reached within two years of the initial reclamation. In total, peat layers may reduce to some 60–70% of their original thickness upon reclamation/drainage (ignoring the loss due to oxidation which continues *ad-infinitum*). Conditions, however, vary considerably depending upon the prevailing conditions during the formation of the peat. Peat may, for example, have an extremely low bulk density (and consequently subside considerably upon reclamation) or it may be rather firm (higher bulk density) either due to the nature of the peat itself or to preceding ripening and settlement during historically low rainfall/better drainage periods.

The settlement in clay soils is much slower than in peat soils. Much of the subsidence in fresh subaqueous mud is due to ripening, the rate of which depends to a great extent upon the prevailing drainage and weather conditions. Typically, ten years after reclamation, some 50% of the final subsidence will have occurred. The ripening subsidence is usually restricted to a depth of about 100–150 cm (= average watertable depth). The settlement subsidence affects a much deeper layer. In the Netherlands it is estimated that during the first 100 years after reclamation, the total subsidence of the subsoil below 150 cm (post-subsidence), amounted to some 10, $7\frac{1}{2}$, 5 and $2\frac{1}{2}$ cm per m soil layer for respectively peat, clay, sandy clay and sand. The subsidence continues beyond this period but at such a slow rate that it only needs to be taken into account in very long term planning (periods of 500 to 1 000 years).

Effects

The subsidence affects the planning of the drainage system for the reclaimed area. Starting from the present levels (pre-subsidence), the soil map and the subsidence calculations enable future levels and lie of the land to be predicted, on the basis of which canal alignments and locations of outlets may be planned. Medium term (10 to 20 years) and long term ($\sim$ 100 years) subsidence predictions should be made and be related to those engineering works with a life (technical and/or economic) of the same order of magnitude. Separate calculations, based on detailed site investigations should be made for all important structures.

The subsidence of peat soils is roughly proportional to the (original) thickness and so the post-reclamation topography of peat land often closely reflects the relief of the underlying substratum. Much depends also on the drainage conditions. In the Netherlands, for example, high watertable levels are often maintained in peat land during the summer to stop subsidence and oxidation.

References and further reading

ECAFE 1964, 1971 and 1978: Proceedings of the First, Second and Third Regional Symposia on the Development of Deltaic Areas, Economic and Social Commission for Asia and the Far East, Water Resources Series nos 25, 39 and 50, United Nations, New York

GLOPPER, RT de 1973: 'Subsidence after drainage of deposits in the former Zuyder Zee and in the brackish and marine forelands in the Netherlands', Series 'Van Zee tot Land', no. 50, Rijksdienst voor de IJsselmeerpolder, Staatsuitgeverij, The Hague

ILRI 1973: 'Acid sulphate soils', Proceedings of an International Symposium, Bulletin 18, International Institute for Land Reclamation and Improvement, Wageningen

ILRI 1982: *Proceedings of the Bangkok Symposium on Acid Sulphate Soils*, Publication 31, International Institute for Land Reclamation and Improvement, Wageningen

KAMERLING, GE 1974: *Bodemfysisch en agrohydrologisch onderzoek in de jonge kustvlakte van Suriname*, PUDOC, Wageningen (in Dutch)

MOORMANN, FR and N VAN BREEMEN 1978: *Rice: soil, water and land*, International Rice Research Institute, Los Banos, Philippines

PONS, LJ and IS ZONNEVELD 1965: *Soil ripening and soil classification*, Publication 13, International Institute for Land Reclamation and Improvement, Wageningen

SCHOTHORST, CJ 1982: 'Drainage and behaviour of peat soils'. In: *Proceedings of the International Symposium on Peat Soils Below Sea Level*, Publication 30, International Institute for Land Reclamation and Improvement, Wageningen

SEGEREN, WA and H SMITS 1974: 'Drainage of newly reclaimed marine clayey sediments, peat soils and acid sulphate soils'. In: *Drainage Principles and Applications*, volume IV, publication no 16, International Institute for Land Reclamation and Improvement, Wageningen

PART VI
INVESTIGATIONS

Chapter 15 General investigations

15.1 CLIMATE

Most drainage problems result from rainfall exceeding the evapotranspiration during short or long periods. Climatic analyses, therefore, are able to contribute a great deal to a better understanding and diagnosis of these problems. More specifically such analyses contribute to:

- an assessment of the scope for solving the problem as well as identification of the most appropriate drainage method;
- the formulation of the design discharges.

Useful types of climatic analyses in this respect are outlined below.

(a) *Soil moisture balance calculations*

An example of this type of calculation is presented in figure 15.1. The starting point of the calculations is the situation at the end of January when under the prevailing climatic conditions it may be safely assumed that the watertable is at its highest permitted level while the overlying soil profile is at field capacity. The soil moisture storage (S) thus is maximal ($S = S_{max}$).* The soil moisture storage remains at S_{max} as long as $P > E$. As no more water can be stored in the soil, all positive (P–E) values count as *excesses*, to be drained. From April onwards $P < E$ and the stored soil moisture starts to become consumed. Negative (P–E) values are deducted from the storage until $S = 0$. Further negative (P–E) values count as *deficits* (water shortages to plants). S_{max} thus represents the water that can be accommodated in the fully depleted soil profile and is available to the crop in the fully replenished profile. For the soils in which soil moisture balance calculations are relevant, values of S_{max} are typically of the order of 50–150 mm (in this example taken at 75 mm).

The deficits last through the summer, till September, when the precipitation again overtakes the evapotranspiration ($P > E$). The depleted storage replenishes, reaching S_{max} in November and remaining at this level until the end of March. All positive (P-E) values occurring during this period count as excesses.

Calculations with average monthly rainfall and evapotranspiration as in figure 15.1 show the main periods of excess (drainage required) and of deficit (irrigation required). A more refined and realistic picture of the occurrence of excesses and deficits can be obtained when these calculations are made for shorter periods (5–10 day periods instead of monthly periods), and are made for a number of years. Figure 15.2 shows the results of 11 years of 10-day period calculations for the same station to which the calculations in figure 15.1 refer. The situation varies from year to year, giving insight into the frequency of occurrence of excesses and deficits. Projected against calendars of cropping and farm operations, critical periods can be identified, growing conditions for crops and workable days for farm machinery can be assessed and drainage and irrigation requirements defined. Information as in figure 15.2 is

* In dry climates calculations may start at the end of the dry season when it may be safely assumed that the soil moisture storage is fully depleted ($S = 0$).

	P (*precipi-tation*) mm	*E* (*evapo-transpi-ration*) mm	*P–E* mm	*S* (*storage*) mm	*Excesses* mm	*Deficits* mm
End of January	→			$S = S_{max} = 75$		
February	54	—	+54	75	54	
March	49	24	+25	75	25	
April	69	70	−1	74		
May	90	102	−12	62		
June	101	116	−15	47		
July	85	136	−51	0		4
August	81	112	−31	0		31
September	80	65	+15	15		
October	67	30	+37	52		
November	102	10	+92	75	69	
December	79	5	+74	75	74	
January	60	—	+60	75	60	

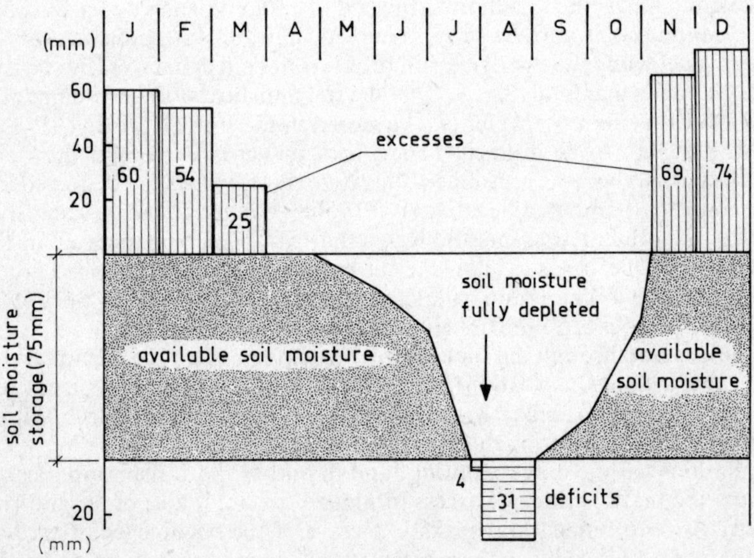

FIGURE 15.1 *Simple soil moisture balance calculations*

especially useful in establishing how drainage problems can be alleviated by adapting the farming of the land, eg changing the crop calendar to avoid a coincidence of planting or harvest with a wet period. Although primarily intended to analyse drainage problems in a qualitative sense, soil moisture balance calculations also provide useful quantitative information for drainage design.

The procedure used in the example is straightforward. All the rainfall is assumed to be totally effective and evaporation is assumed to be independent of the moisture

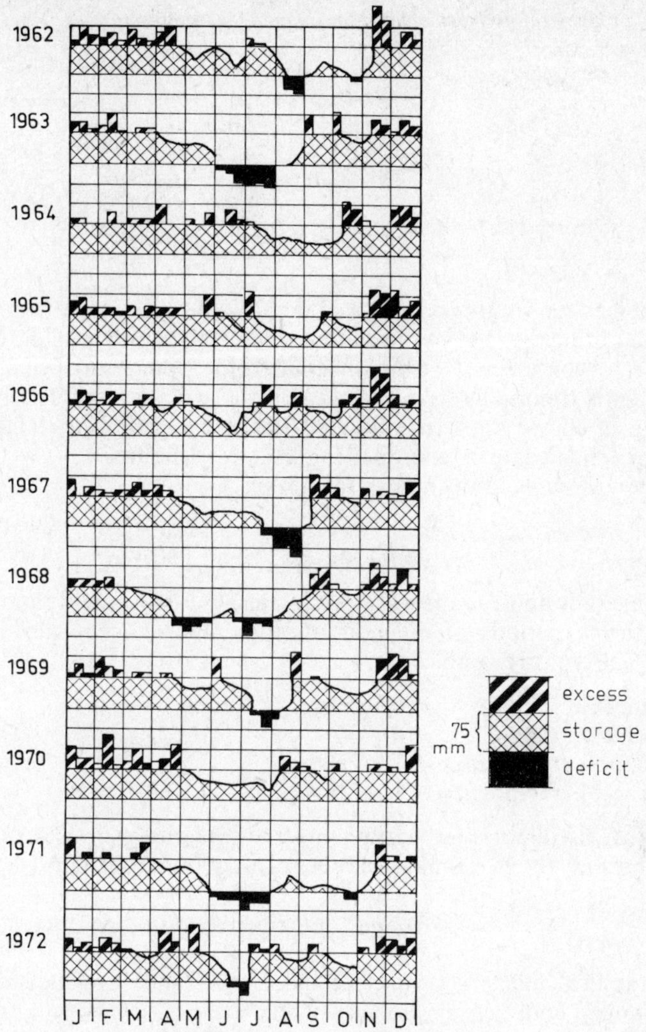

FIGURE 15.2 *Soil moisture balance calculations for* 10-*day periods* (*Zagreb, Yugoslavia*)

status within the soil. Other more sophisticated methods of analysis exist but are often too pretentious and the simpler approach presented in the example has much to recommend it.

(b) *Rainfall depth–duration–frequency studies*

These studies are used to derive design criteria, especially to derive values for q (drainage coefficient, see chapters 7 and 8). Depth-duration-frequency data as presented in table 15.1 are available for many rainfall stations or may be compiled from daily rainfall data (covering minimally some 20–30 yr).

TABLE 15.1 *Example of rainfall depth-duration-frequency relationships* (Tanzania)

Frequency	*1 day*	*2 days*	*3 days*	*5 days*	*10 days*
1 × 1 yr	27 mm	47 mm	56 mm	76 mm	90 mm
1 × 2 yr	42	65	74	97	111
1 × 3 yr	52	75	87	109	118
1 × 5 yr	64	84	102	127	143
1 × 10 yr	82	114	132	160	218

As mentioned earlier, rainy periods of some 3–5 days are normally most critical for groundwater drainage, the intermediate periods of $\frac{1}{2}$–2 days for shallow drainage and the short duration storms (<6 hr) for runoff from sloping land. For agricultural drainage, design is commonly based on the control of 1 × 5 yr rainfall events. The rainfall should of course apply to the critical calendar period (as established by considering the rainfall data in relation to the farming in the area).

Data vary considerably from one area to another:

	1 day	*3 days*	*10 days*
1 × 5 yr rainfall	40–150 mm	65–200 mm	110–350 mm

High values generally apply to the warmer climates. For many areas ratio's between the rainfall during periods of different duration and/or frequency have been established. Some typical examples are:

- 6 hr rainfall/24 hr rainfall = 0·5–0·7
- 12 hr rainfall/24 hr rainfall = 0·6–0·8
- 5 yr frequency/1 yr frequency = 1·5–2·0
- 10 yr frequency/1 yr frequency = 1·7–2·5

Such ratios are useful for interpolation and extrapolation purposes, especially in situations where only rough estimates of order of magnitude are required.

15.2 TOPOGRAPHY

Good topographical maps, showing the 'lie of the land', are indispensable in drainage planning and design. For feasibility studies, maps with a scale of 1:10 000/25 000 showing 0·5–1·0 m interval contour lines, will generally suffice for the planning of the main system; sample areas may be mapped in more detail, to be used to assess aspects of field drainage (including cost estimates).

For final planning and design, more detailed maps are required with map scales usually 1:5 000/10 000 and contour line intervals of 0·25–0·50 m. Contour lines at 0·50 m interval will generally suffice for sloping land. For regular, flat land an interval of not more than 0·25 m is normally required. Map scales should match the contour line interval so that the contour lines (measured on the map) are not more than 5–10 cm and not less than 0·5–1·0 cm apart (this implies the use of a smaller scale map for sloping land than for flat land).

Detailed topographic maps are especially required for the design of surface drainage systems for flat land. Small differences in elevation are important and contour lines should be based on an adequate number of points to give a good picture of the meso/micro-topography. For the design of groundwater drainage systems somewhat lower standards for the mapping of the in-field topographic situation apply. The mapping should, however, be adequate to establish alignments and grades.

For the planning and design of runoff control systems for sloping land, the maps should show such features as slope pattern, length and degree of slope, uniformity, etc.

The topographic maps should also show the main elements of any existing drainage system and all relevant infra-structural features such as roads, powerlines, settlements, etc. To assess whether existing drains can be used, longitudinal profiles (scale 1:5 000/10 000) with cross-sections (every 100–200 m, scale 1:100) will be needed as well as the characteristic dimensions and levels of all structures in these systems (bridges, culverts, etc). To assess outlet conditions it may be necessary to extend the topographic mapping to well outside the project area.

15.3 Soils and land

Information on the soil and land conditions in the area is used for many purposes in drainage planning and design, eg:

- to diagnose the drainage problem
- to suggest/evaluate possible solutions
- to formulate design criteria

The main soil and land characteristics to be considered are:

(a) *Soil/land surface*

Signs of wetness such as indicator plants (sedges, rushes, etc), poor decomposition of organic matter, water on the land after rain, water in the tracks, springs/spring-lines and seepage water (often occurs with emergence of rock strata onto surface, frequently associated with iron ochre deposits), condition of crops, smell, etc.

Slope features such as overall relief, length, degree and regularity of slope, etc (partly also to be derived from topographic maps), to assess overland flow conditions.

Retention/detention features such as micro-relief, surface roughness, type of cover, presence of terraces or bunds, etc.

(b) *Topsoil* (0–20/30 cm)

Erodibility, eg expressed by the erodibility index of the USDA universal soil loss equation or by some other integrated erodibility parameter, or to be deduced from related characteristics such as soil texture and structure, rainfall intensity, length and degree of slope, etc.

Infiltrability, eg expressed by measured infiltration rates/curves, or to be deduced from soil texture, structure, etc (see section 1.10); liability of the soil to capping; swelling and cracking characteristics, etc.

FC–LPL relation, to assess workability as influenced by soil moisture conditions (see section 1.9, figure 1.16).

Thickness and hydraulic conductivity of the top soil, to assess conditions for lateral topsoil drainage (usually interflow).

(c) *Main soil profile/rootzone* (0–75/150 cm)

General stratification and main properties of the different layers, to assess:

- root development potential
- available moisture, including extent and limitations of capillary rise
- limitations to earth movement
- most appropriate drainage base depth, including required measures for the installation of pipe drains (method, envelope, pipe material, etc), mole drains (soil stability, method, depth, etc) and ditches (side slopes, bank protection, etc)
- subsidence to bc expected; state of ripening
- degree of clogging of pipes by iron ochre to be expected.

K and/or KD value of the different layers, including any existing anisotropy, mainly to assess:

- conditions for vertical downward percolation, with emphasis on the identification of impeding layers; depth of occurrence, thickness, nature, genesis, etc, of these layers; these enable the feasibility of subsoiling or other measures to improve vertical flow to be assessed and can guide the planning and execution of such measures
- conditions for horizontal and radial flow towards pipe drains and ditches (to be assessed in conjunction with the substratum characteristics).

Vertical flow conditions in the soil may also partly be deduced from the hydromorphic characteristics in the profile (gley features, mottles, concretions, colours, etc) and from the rooting pattern.

Soil-water-air relations of the different layers:

study of pF curves to assess the harmful effects of excess water in the profile and the beneficial effects of its control (from pore distribution, field capacity, available moisture, drainable pore space, etc).

(d) *Substratum* (75/150 cm–5/20 m)

KD values of the substrata and the depth of an underlying impermeable base, to assess drainage flow conditions below the drainage base.

(e) *Soil salinity*

Visual signs (see section 10.3.2); EC and ESP data of the soil at different depths; EC and SAR values of the groundwater; salt composition; seasonal variations; critical capillary height and critical watertable depth (to be deduced from profile characteristics).

(f) *Watertables*

Real and perched watertables and piezometric gradients in the groundwater (see next section).

Standard soil surveys usually do not supply all of this information and additional investigations are required to make a satisfactory diagnosis and to establish and specify the required measures. These investigations may be integrated with a normal soil survey or be carried out separately. The first procedure is usually followed in drainage projects where no soil mapping has as yet been done, while the second procedure is applicable when plans for improved drainage are made for an area for which soil maps are available. The results of these drainage investigations may be incorporated in the soil maps but some aspects may also be conveniently compiled as 'single value' maps:

- soil hydraulic conductivity maps showing the conductivity conditions above and below the drainage base, the latter, eg in the form of the composite KD value (directly suitable for the calculation of the required drain spacing)
- mapping of the depth and thickness of impeding layers in the profile (basis for the planning of a subsoiling programme)
- mapping of the depth of soil ripening (basis for the calculation of the subsidence to be expected).

Many other characteristics may be 'single value' mapped such as waterable depth (including seasonal fluctuations), soil and groundwater salinity, infiltrability, slope features, etc.

15.4 Watertable and groundwater surveys

Watertable surveys provide valuable information on the (groundwater) drainage conditions in the area. Watertable levels reflect the prevailing balance between the different groundwater recharge/discharge components. As the balance changes, so does the watertable level. When the watertable is permanently or seasonally too close to the soil surface, control by groundwater drainage systems may be required.

Watertable observation wells

Watertable levels (phreatic levels) may be measured in bore holes reaching well into the groundwater (see figure 1.3). In an area wide survey, these observation wells may be placed in a grid or another regular pattern, although the (conceived) groundwaterflow pattern in the area should also be taken into account. Groundwaterflow will generally be down the slope, towards depressions and (natural) drains. A few judiciously located observation wells may give as representative a picture of the phreatic surface as that obtained with a dense grid system.

Open water levels in (natural or excavated) wells and drains are also closely related to the phreatic surface. During periods of discharge, those levels will tend to be lower than the watertables in the adjoining land. Only when there is static equilibrium (several days after rain) are these levels more or less the same. Well and canal levels may also be higher than the watertable in the adjoining land, eg when the well or canal walls have become sealed by sediments in the water.

In watertable surveys interest is primarily in long term, seasonal variations (short term fluctuations after rain or irrigation may be studied separately, usually in connection with the functioning of a drainage system). The seasonal trends in watertable levels are often closely related to the rainfall or irrigation regime in the area and may be identified adequately by observations at a few selected key sites on key dates (selections made on the basis of previous detailed studies).

The prevailing seasonal watertable regime can often also be deduced fairly accurately from the *hydromorphic characteristics* of the soil profile. The soil below the average low watertable (= average summer/dry season watertable level) typically has a uniform light grey to dark blue colour associated with permanent non-aerated soil. The soil above the average high watertable (= average rainy-/winter season level) is typically uniformly brown to grey-brown. In between a mottled *gley-zone* occurs, resulting from the intermittant oxidation and reduction of iron and other elements. The upper and lower gley levels can usually be identified in a soil profile (after some initial guidance on the basis of available watertable records) and these recorded in soil surveys provide valuable information on the current watertable regime in the soil. Hydromorphic characteristics in the soil only change slowly and it should be confirmed that the observed gley features reflect the present groundwater regime (*fossil gley* can occur in alluvial soils well below the present low watertable levels, dating from a previous sedimentation and water regime period).

Piezometric studies

Watertable wells provide no information on potential differences within the groundwater body. Where such differences are expected to occur, piezometers should be used instead. Piezometers measure the pressure in a water body at the point where its filter is placed (constituting the bottom 5–10 cm of the piezometer tube, see figure 1.4 and figure 15.3). A typical use of piezometers is shown in figure 15.4. A battery of piezometers, 1–2 m apart, is installed at different depths.* The readings in the piezometers ① and ② show that the pressure distribution with depth in the highly permeable layer III is hydrostatic, indicating that there is no flow in a vertical direction (or the flow is too small to generate any significant head loss in this highly permeable soil). Piezometers ② and ③, however, show that there is an upward flow from layer III to the rootzone (generally termed seepage flow or in this specific case also *leakage* from an underlying aquifer). The rate of this flow may be determined with eq. 7.1 $\left(\text{flow rate } v = K_{II}\frac{h}{D}\right)$. This leakage adds to the drainage load due to rainfall or irrigation.

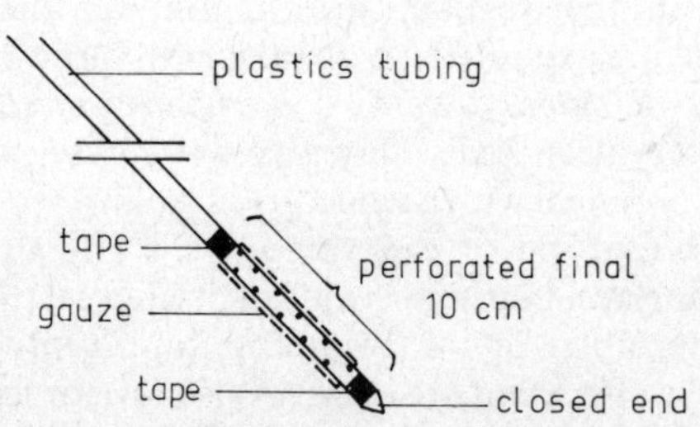

Made from common plastics electrical conduit with an inside diameter of about 2 cm, closed at the end by melting or by means of a stopper. The final 10 cm of the lower end is perforated with some ten holes of 2 mm diameter. This part is covered with durable gauze (nylon stocking will do), taped to the tube. The piezometer is installed in an 8 cm diameter augerhole. The initial backfilling is with sand (surrounding the whole of the perforated lower end), followed by bentonite powder which will swell upon becoming wet to form a sealing ring as shown in figure 1.4. Further backfilling may then be done with normal soil.

FIGURE 15.3 *Simple piezometer suitable for drainage studies*

* When there is also horizontal flow, the line of piezometers should be aligned at right angles to the direction of this flow to avoid horizontal head losses over the 1–2 m distance between the piezometers being falsely interpreted as vertical head losses.

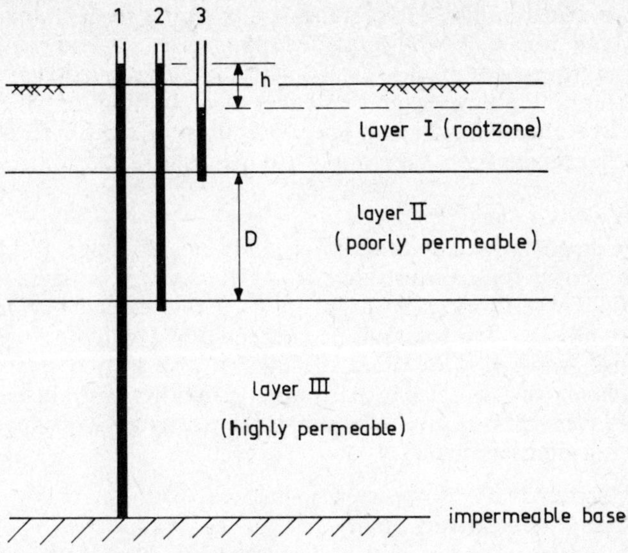

FIGURE 15.4 *Typical use of piezometers*

15.5 HYDROLOGY AND GEOHYDROLOGY

The following aspects should be investigated, both for diagnosis and for planning purposes:

(a) *Runoff and flooding*

The origin of floodwater, frequency and seasonal period of occurrence, affected area and resultant damage should be established, mostly on the basis of hydrological and topographical surveys. Together with information on climate soils, agriculture, etc, they form the basis for a flood control scheme.

(b) *Outlet conditions*

The duration, frequency and seasonal periods of occurrence of high and low water levels at the outlet point should be determined. If a gauging station exists nearby, available data should be compiled into stage-frequency-duration relationships for selected calendar periods. Where such a station does not exist, data from a more remote station with a similar character may be used. The stage heights recorded at such a station may be transposed to the outlet point either by:

- hydraulic calculations, using the Chézy-Manning formula for canal flow (with measured or estimated roughness values)
- carrying out a short measurement programme, correlating the corresponding stage heights at the two sites.

Where data from gauging stations are unavailable, a new station may be established at the selected outlet point. A short measuring prriod of 1–2 years may yield sufficient useful information while extreme conditions not covered by this period may be assessed by correlative analysis (eg correlating the stage height and/or discharges recorded in the 1–2 years measuring period to the rainfall data for this period, the latter hopefully forming part of a long series, covering 10–20 years) or by other methods (eg estimating discharges on the basis of rainfall-runoff relationships and then determining stage heights by either extrapolating the short period rating curve or by hydraulic calculations). The procedures described refer to outlets onto a river but rather similar procedures apply to outlets onto a lake or the sea.

(c) *Geohydrological conditions*

Satisfactory diagnosis and solution of groundwater drainage problems will often require information on the substrata beyond the depth investigated in normal soil survey work. These investigations do not as a rule extend beyond some 2–5 m whereas groundwater flow to drains may extend down to depths of $\frac{1}{6}$–$\frac{1}{4}$ L below the drainage base, where L is the drain spacing (see also section 16.5). Where widely spaced drains are used, but also in many other groundwater drainage situations, the normal soil investigations should be supplemented by geohydrological investigations, collecting information on:

- the overall groundwater flow pattern in the area
- the hydraulic characteristics of the substrata (eg to some 10 m below the drainage base for L = 50 m when no shallower impermeable layer occurs)
- piezometric levels and gradients at different depths and sites
- patterns and rates of natural drainage and seepage flow
- groundwater salinity at different depths.

Deep geohydrological investigations (beyond 5–10 m) require special equipment and experience and as such are often contracted to specialised firms.

15.6 AGRICULTURE

The agricultural investigations should:

(a) establish the land use alternatives for the area, taking into account the present drainage conditions and the technical/economic feasibility of improved drainage (see section 2.4)
(b) help to formulate the drainage requirements on the basis of these land use alternatives and the prevailing soil and climatic conditions
(c) help to assess the anticipated benefits from drainage (implicit in the objectives (a) and (b) above but needed explicitly to carry out economic evaluations as in chapter 17).

Investigations to be carried out would normally include:

- survey of present land use and farm practices
- survey of damage done by excess water
- various crop and farm management studies.

As pointed out in section 6.4, the relationships between drainage and farming are complicated and even experimental results are of limited value since the actual conditions may differ in almost any way from the experimental conditions. A deep understanding of the local agricultural conditions in combination with a sound analysis of the prevailing soil and climatic conditions, offers the best basis and hope for arriving at the correct drainage requirements and a sound drainage plan.

15.7 MISCELLANEOUS

Various infra-structural features may influence the drainage plan by posing specific drainage requirements. Outside interests may also be affected by the improved drainage. Administrative and legal aspects need to be studied, ecological consequences to be assessed and rights of way to be established. When planning the drainage of irrigated land, the irrigation practices should be studied.

Information on local construction conditions need to be collected (available machinery, materials and skills, labour situation, accessibility and workability conditions, unit costs, etc).

References and further reading

BUREAU OF RECLAMATION 1978: *Drainage Manual*, US Department of the Interior, Washington DC

BROWN, RH et al (eds) 1972: *Groundwater Studies*, UNESCO, Paris (loose leaf publication) latest supplement 1977

ECAFE 1968: 'The use and interpretation of hydrological data', Water Resources Series no 34, Economic and Social Commission for Asia and the Far East, United Nations, New York

FAO 1974: 'Soil survey in irrigation investigations', Soils Bulletin (in draft), Food and Agriculture Organisation of the United Nations, Rome

ILACO 1981: *Agricultural Compendium for Rural Development in the Tropics and Subtropics*, Elsevier, Amsterdam

ILRI 1972: *Fieldbook for Land and Water Management Experts*, International Institute for Land Reclamation and Improvement, Wageningen

VAN DER MEER, K and RH MESSEMAECKERS VAN DE GRAAFF 1974: 'Hydropedological Surveys'. In: *Drainage Principles and Applications*, volume III, publication 16, International Institute for Land Reclamation and Improvement, Wageningen

DE RIDDER, NA 1974: 'Groundwater Survey'. In: *Drainage Principles and Applications*, volume III, publication 16, International Institute for Land Reclamation and Improvement, Wageningen

REICH, BM 1963: Short duration rainfall-intensity estimates and other design aids for regions of sparse data, *J of Hydrology*, *3*: 3–28

SOIL CONSERVATION SERVICE 1971: *Drainage of Agricultural Land*, National Engineering Handbook, section 16, US Department of Agriculture, Washington DC

USDA 1962: 'Field Manual for Research in Agricultural Hydrology', Agricultural Handbook no 224, US Department of Agriculture, Washington DC

WMO 1974: 'Guide to hydrological practices', Technical Publication no 168, 3rd edn, World Meteorological Organisation, Geneva

Chapter 16 Determination of the hydraulic conductivity

The hydraulic conductivity value K of the soil is obviously a very important characteristic in relation to drainage (especially in relation to groundwater drainage). The natural drainage of the soil and the scope for and costs of improved drainage all depend greatly upon it. The determination of the K value of the soil is thus an important aspect of almost any drainage investigation.

16.1 Laboratory measurement

With these methods the K-value is measured in the laboratory on a soil sample taken in the field.

The quality of the results depends very much upon the quality of the sample, which should be representative of the site under investigation. Owing to soil variability, a large number of locations may have to be sampled and at each location 2–3 replicate samples may be required. It is also vitally important to minimise soil disturbance since the K-values depend to a large extent upon the soil's pore geometry. *Undisturbed samples* (also referred to as *core samples*) may be taken by pressing an open cylinder ring into the soil (see figure 16.1). Rings in common use have a depth and inside diameter of about 5 cm (volume exactly 100 cm^3, a convenient volume for bulk density determination). Samples may be taken vertically or horizontally, depending on whether K is to be measured in the vertical (K_h) or in the horizontal (K_v) direction.

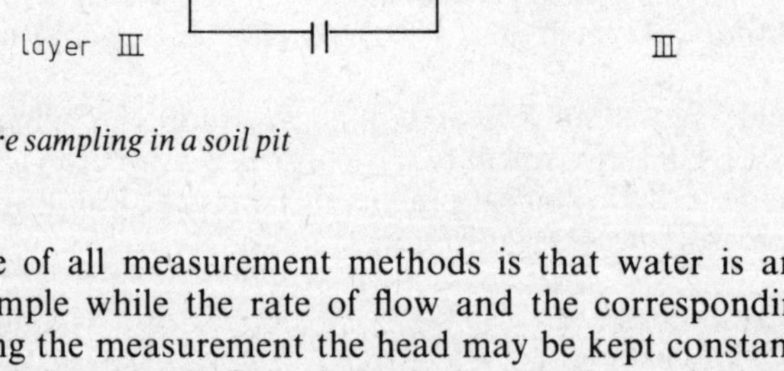

FIGURE 16.1 *Core sampling in a soil pit*

The principle of all measurement methods is that water is arranged to flow through the sample while the rate of flow and the corresponding headloss are recorded. During the measurement the head may be kept constant (*constant head method*), resulting in a steady flow through the sample, or the head may decrease (*falling head method*), resulting in decreasing rates of flow during the measurement. The hydraulic conductivity may then be determined from the flow and head record, using Darcy's Law (see for example the arrangement in figure 1.10). Of course, the sample must be fully saturated during the measurement, especially making sure that no entrapped air remains.

Evaluation

(a) The main drawback of laboratory measurements is that the K-value only relates to a small part of the soil (sample of only 100 cm^3). Many samples could be taken but this becomes costly. Equipment has been developed to take large, relatively undisturbed, samples down to great depths although for routine drainage investigations this type of equipment is unlikely to be available.

(b) Inconsistent, highly variable values are obtained in clay soils in which K depends almost wholly upon the development of cracks. A sample taken in between the cracks may have zero conductivity whereas an adjacent sample containing a crack might have a very high value. Clays are also difficult to saturate and are liable to swell, when wetted, rather more in the unconfined situation in the laboratory than in the field situation.

(c) In general determination of the K-value in the field is preferable to laboratory measurements, the latter only being used to supplement field methods to:

- determine the K above the watertable; field measurements for this purpose are almost equally as unsatisfactory as laboratory methods;
- separate K_h and K_v;
- help to predict how salt affects the hydraulic conductivity and other physical properties, especially whether deflocculation will occur (see also section 10.2.3).

16.2 FIELD MEASUREMENTS BELOW THE WATERTABLE

A distinction is made between determination of hydraulic conductivity *below* the watertable and *above* the watertable. Reliable methods have been developed for measurement below the watertable whereas methods employed above the watertable are generally less satisfactory. The latter are discussed in section 16.3.

16.2.1 Augerhole method

Hooghoudt (1936) developed an empirical equation relating hydraulic conductivity to the rate of rise of the water level in a bailed out augerhole. Hooghoudt's analysis was subsequently refined by KIRKHAM and VAN BAVEL (1948) and by ERNST (1950).

Figure 16.2 illustrates the basic steps involved in carrying out the augerhole method. An augerhole is first made to a depth well below the watertable and deep into the layer to be measured. The water level should then be allowed to rise in the hole until it establishes itself at an equilibrium coinciding with the watertable level. In highly permeable soils this will take 10 to 15 minutes whereas in poorly permeable soils it may take days. The water level in the hole is then lowered, by bailing or pumping water out of the hole. Flow immediately commences from the soil into the augerhole under the hydraulic gradient created by the difference in level between the watertable and the water level in the hole. A detailed record of the (rising) water level in the hole is then maintained over an appropriate time period.

The equations used are based upon the assumptions that the watertable level remains constant and horizontal throughout the text and that flow enters the augerhole over its entire area below the watertable (situation A in figure 16.2). This situation exists shortly after lowering of the water-level although a cone of

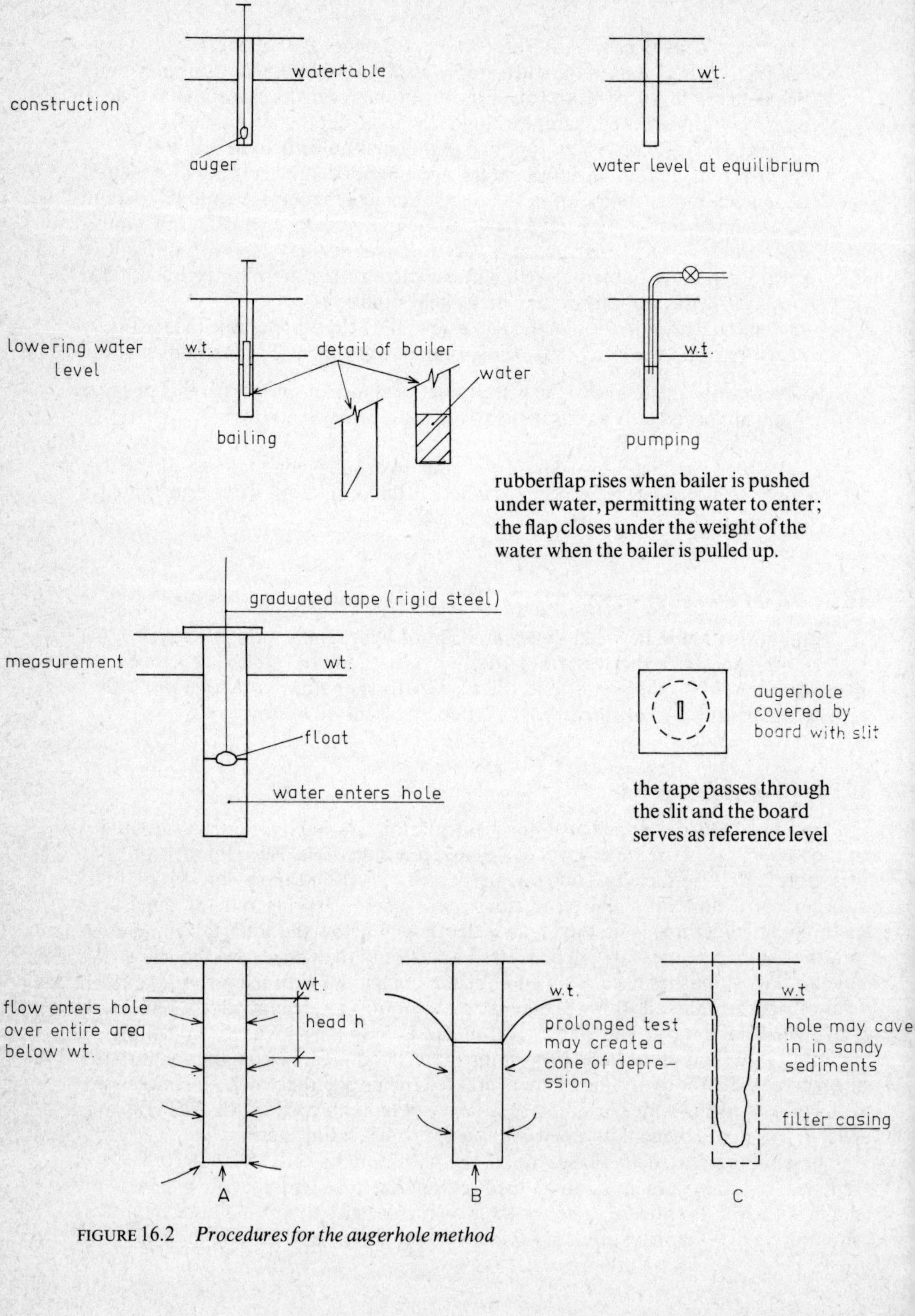

FIGURE 16.2 *Procedures for the augerhole method*

depression (B) will eventually develop invalidating the assumptions. No significant errors are made as long as the measurements are completed before 25% of the water initially removed from the hole, has been replaced. This condition will be complied with provided the test is completed before $h \leqslant \frac{3}{4}h_o$ (see figure 16.3).

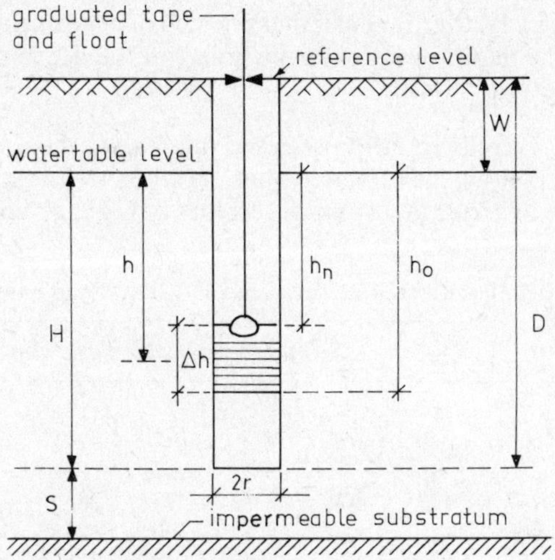

H = depth of hole below watertable
S = depth to the impermeable substratum below the bottom of the hole
r = radius of the hole
W = depth of the watertable below the reference level
h_o = distance between watertable level and elevation of the water level in the hole at the time of the first reading
h = the same at any time t after the first reading
h_n = the same at the end of the period of the measurement
Δh = the rise of water level in the hole over a period of measurement Δt

FIGURE 16.3 *Standard nomenclature for the augerhole method*

The value of h_o must also be considered in relation to soil stability and hydraulic conductivity. In highly permeable soils, take $h_o \sim 40$ cm for a larger value may result in caving-in of the hole. Caving-in, always a problem in sandy soils, should be prevented by use of a filter casing (situation C in figure 16.2). For soils of low K value take h_o up to 80 cm to create sufficient head for a good rate of inflow.

In clay soils smearing of the wall of the hole may occur hindering the inflow. This can sometimes be overcome by prior installation in dry soil (during the summer or before irrigation). The adverse effects of smearing may also be partly overcome by bailing out the hole several times to flush out clogged pores prior to the actual test.

Calculations

KIRKHAM and VAN BAVEL (1948) and ERNST (1950) separately analysed the flow to auger holes and developed the following equation:

$$K = C\frac{\Delta h}{\Delta t} \qquad \text{......(eq. 16.1)}$$

where:

$$K = \text{hydraulic conductivity, m.day}^{-1}$$

$$\frac{\Delta h}{\Delta t} = \text{rate of rise of the water level in the hole, cm.sec}^{-1}$$

C = geometry factor (dimensionless) which depends upon the radius of the hole, the depth of water in the hole and the depth to an impermeable substratum*

Kirkham and van Bavel's equation is based upon a solution of the fundamental differential equation of flow, the Laplace equation, whereas ERNST used the so-called relaxation technique to arrive at a solution. Values of C are given in table 16.1.

TABLE 16.1 *Values of C for use with the augerhole method* (BOAST and KIRKHAM 1971)

H/r	$\bar{h}$/H**	*S/H for impermeable substratum*								
		0	*0·05*	*0·1*	*0·2*	*0·5*	*1*	*2*	*5*	∞
1	1	447	423	404	375	323	286	264	255	254
	0·75	469	450	434	408	360	324	303	292	291
	0·5	555	537	522	497	449	411	386	380	379
2	1	186	176	167	154	134	123	118	116	115
	0·75	196	187	180	168	149	138	133	131	131
	0·5	234	225	218	207	188	175	169	167	167
5	1	51·9	48·6	46·2	42·8	38·7	36·9	36·1		35·8
	0·75	54·8	52·0	49·9	46·8	42·8	41·0	40·2		40·0
	0·5	66·1	63·4	61·3	58·1	53·9	51·9	51·0		50·7
10	1	18·1	16·9	16·1	15·1	14·1	13·6	13·4		13·4
	0·75	19·1	18·1	17·4	16·5	15·5	15·0	14·8		14·8
	0·5	23·3	22·3	21·5	20·6	19·5	19·0	18·8		18·7
20	1	5·91	5·53	5·30	5·06	4·81	4·70	4·66		4·64
	0·75	6·27	5·94	5·73	5·50	5·25	5·15	5·10		5·08
	0·5	7·67	7·34	7·12	6·88	6·60	6·48	6·43		6·41
50	1	1·25	1·18	1·14	1·11	1·07	1·05			1·04
	0·75	1·33	1·27	1·23	1·20	1·16	1·14			1·13
	0·5	1·64	1·57	1·54	1·50	1·46	1·44			1·43
100	1	0·37	0·35	0·34	0·34	0·33	0·32			0·32
	0·75	0·40	0·38	0·37	0·36	0·35	0·35			0·35
	0·5	0·49	0·47	0·46	0·45	0·44	0·44			0·44

A clear-cut impermeable substratum might not be present in which case a layer may be considered to be 'impermeable' if it has a substantial thickness (min. 1–2 m)

* Boast and Kirkham (1971) also give values of C for the case where the layer of soil in which the augerhole has been formed is underlain by a layer of much greater hydraulic conductivity.
**$\bar{h}$ is the value of the mean head during the measuring period.

and a hydraulic conductivity value less than one tenth of the value estimated to apply in the more permeable overlying layer.

Example: W + H = 154 cm
W = 74 cm
S = 40 cm
r = 4 cm

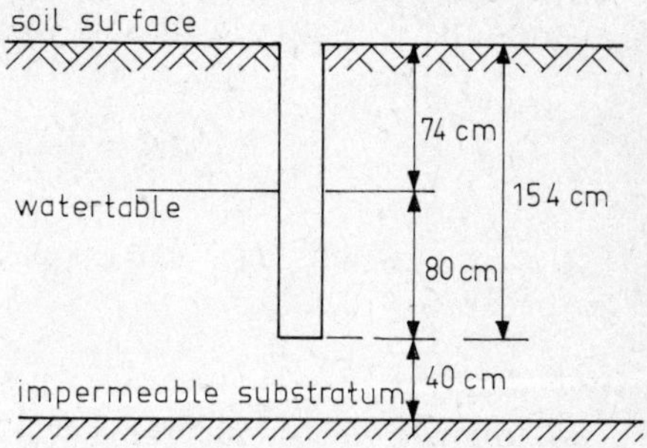

Recorded values:

Time	*Water level*	*Head*	*Change in head*	*Time interval*
t	W+h	h	Δh	Δt
(*sec*)	(*cm*)	(*cm*)	(*cm*)	(*sec*)
0	116·8	42·8	—	—
20	115·6	41·6	1·2	20
40	114·4	40·4	1·2	20
60	113·3	39·3	1·1	20
80	112·2	38·2	1·1	20
100	111·2	37·2	1·0	20

Calculations (for the complete time period)

$\Delta h = 42{\cdot}8 - 37{\cdot}2 = 5{\cdot}6$ cm
$\Delta t = 100 - 0 = 100$ secs.

$$\bar{h} = \frac{42{\cdot}8 + 37{\cdot}2}{2} = 40 \text{ cm (mean head)}$$

$H = 154 - 74 = 80$ cm
$H/r = 80/4 = 20$; $S/H = 40/80 = 0{\cdot}5$,
$\bar{h}/H = 40/80 = 0{\cdot}5$, so $C = 6{\cdot}60$ (table 16.1)

$$K = C\frac{\Delta h}{\Delta t} = 6{\cdot}60\frac{5{\cdot}6}{100} = 0{\cdot}37 \text{ m.day}^{-1}$$

Use of the augerhole method in layered profiles

The augerhole method can also be used to estimate the K – value of two layers separately (when both layers occur below the watertable). In this case two augerholes are made, a shallow one wholly within the upper layer and a deeper hole reaching into the underlying layer (see figure 16.4). A measurement is first carried out

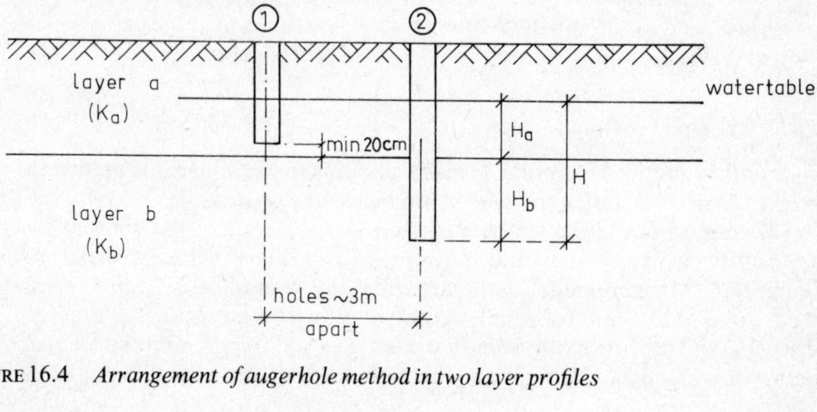

FIGURE 16.4 *Arrangement of augerhole method in two layer profiles*

in the deeper hole (2) yielding a hydraulic conductivity $K_{\overline{ab}}$. A second measurement is then made in the shallower hole (1) yielding $K = K_a$. The unknown K_b may then be calculated using the known values of $K_{\overline{ab}}$, K_a, H_a and H_b in the formula (see equation 16.5 in section 16.4):

$$K_b = \frac{K_{\overline{ab}}(H_a + H_b) - K_a H_a}{H_b}$$

Of course when $K_a \gg K_b$, values of K_b thus obtained should not be expected to be very accurate.

Evaluation

(a) Inflow into the augerhole will be mostly due to horizontal flow passing through the vertical walls, and the calculated K-value will as a result be mainly an estimate of the K_h of the soil.

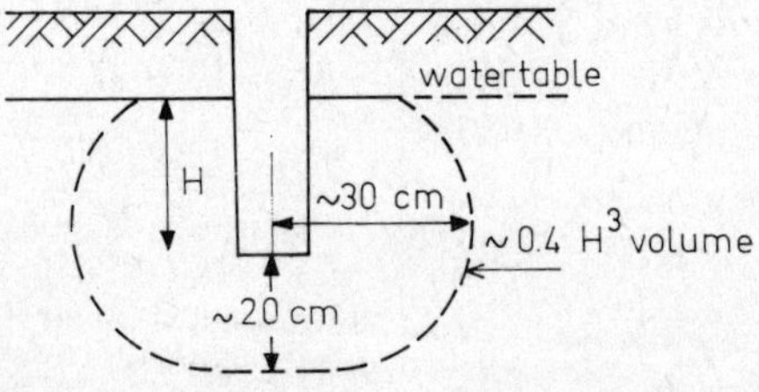

FIGURE 16.5 *Volume of soil normally included in the augerhole method*

(b) A much larger volume of soil is included in the measurement compared for example to the 100 cc of a normal sample for laboratory measurement (see figure 16.5). This averages out much of the micro-variability in soil structure. The remaining variability should be covered in the selection of sites and number of measurements. Considerable variability will, however, always remain due to:

– measurement errors (~10–20% error) – soil variability (~10–50% error in silty/sandy soils, much greater in clays)	in total approximately 10–30% variation between averages of 5 replicates

16.2.2 Piezometer method

The piezometer method is similar in many respects to the augerhole method. A tube (ID = 3 to 8 cm) is installed to some depth below the watertable. A cavity is formed at the lower end of the tube which may simply consist of the flat circular soil base under the tube or it may be cylindrical in shape and extend below the tube as shown in figure 16.6. Measurements can start once the waterlevel within the tube has reached an equilibrium (normally coinciding with the watertable level). The waterlevel in the tube is first lowered by bailing or pumping and its subsequent rate of rise, as the water flows from the soil through the cavity into the tube, is recorded.

The tube may be installed by sliding it into a previously prepared augerhole or the formation of the hole and the positioning of the tube may proceed simultaneously, using an auger worked within and beyond the tube (latter to form the cavity). This allows a series of tests to be carried out at increasing depths by alternately augering beyond and then lowering the tube. In some soils the piezometer tube may also be installed by jetting (suitable for deep installations).

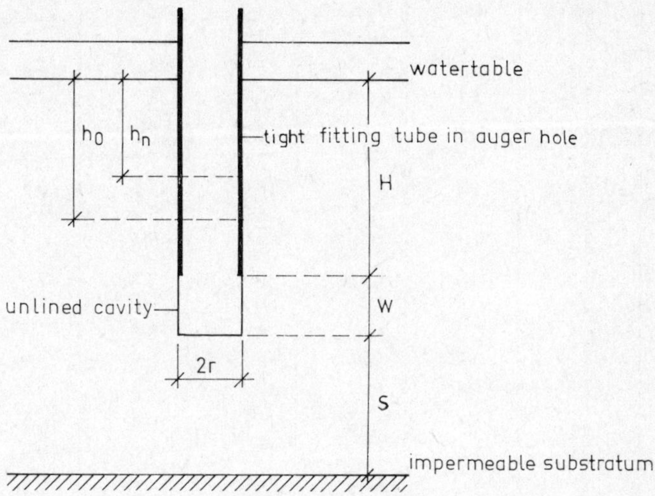

FIGURE 16.6 *Nomenclature for the piezometer method*

Kirkham (1945) solved the Laplace equation also for this case by adopting similar assumptions to those used in the case of the augerhole method (see section 16.2.1). The following equation was derived (nomenclature as in figure 16.6):

$$K = \frac{\pi r^2}{C} \frac{\ln \frac{h_o}{h_n}}{(t_n - t_o)} \qquad \text{......(eq. 16.2)}$$

where

r = the radius of the cavity (cm)
h_o, h_n = the head at times t_o, t_n etc (cm)
t_o, t_n = the time intervals (secs, minutes, etc)
C = geometry factor (dimensions of length, cm)

The geometry factor C functionally depends upon H, r, W and S. Values are presented in table 16.2 as the ratio $\frac{C}{r}$ (dimensionless). This table may be extended beyond the given range by graphical extrapolation.

TABLE 16.2 *Shape factors $\frac{C}{r}$ for use with the piezometer method* (YOUNGS 1968)*

W/r	H/r	S/r for impermeable substratum						
		∞	8·0	4·0	2·0	1·0	0·5	0
0	20	5·6	5·5	5·3	5·0	4·4	3·6	0
	16	5·6	5·5	5·3	5·0	4·4	3·6	0
	12	5·6	5·5	5·4	5·1	4·5	3·7	0
	8	5·7	5·6	5·5	5·2	4·6	3·8	0
	4	5·8	5·7	5·6	5·4	4·8	3·9	0
0·5	20	8·7	8·6	8·3	7·7	7·0	6·2	4·8
	16	8·8	8·7	8·4	7·8	7·0	6·2	4·8
	12	8·9	8·8	8·5	8·0	7·1	6·3	4·8
	8	9·0	9·0	8·7	8·2	7·2	6·4	4·9
	4	9·5	9·4	9·0	8·6	7·5	6·5	5·0
1·0	20	10·6	10·4	10·0	9·3	8·4	7·6	6·3
	16	10·7	10·5	10·1	9·4	8·5	7·7	6·4
	12	10·8	10·6	10·2	9·5	8·6	7·8	6·5
	8	11·0	10·9	10·5	9·8	8·9	8·0	6·7
	4	11·5	11·4	11·2	10·5	9·7	8·8	7·3
2·0	20	13·8	13·5	12·8	11·9	10·9	10·1	9·1
	16	13·9	13·6	13·0	12·1	11·0	10·2	9·2
	12	14·0	13·7	13·2	12·3	11·2	10·4	9·4
	8	14·3	14·1	13·6	12·7	11·5	10·7	9·8
	4	15·0	14·9	14·5	13·7	12·6	11·7	10·5
4·0	20	18·6	18·0	17·3	16·3	15·3	14·6	13·6
	16	19·0	18·4	17·6	16·6	15·6	14·8	13·8
	12	19·4	18·8	18·0	17·1	16·0	15·1	14·1
	8	19·8	19·4	18·7	17·6	16·4	15.5	14.5
	4	21·0	20·5	20·0	19·1	17·8	17·0	15·8
8·0	20	26·9	26·3	25·5	24·0	23·0	22·2	21·4
	16	27·4	26·6	25·8	24·4	23·4	22·7	21·9
	12	28·3	27·2	26·4	25·1	24·1	23·4	22·6
	8	29·1	28·2	27·4	26·1	25·1	24·4	23·4
	4	30·8	30·2	29·6	28·0	26·9	25·7	24·5

Evaluation

1 This method is particularly suited to determine the hydraulic conductivity of rather thin discrete layers, up to great depths (for which the augerhole method is less suited).
2 Errors may normally be expected to be of the same order of magnitude as for the augerhole method.
3 The shape of the cavity may be selected to mainly reflect the horizontal hydraulic conductivity K_h (by using a long narrow cavity W/r = 8) or the vertical K_v (by using a flat circular base W/r = 0) enabling meaningful estimates of the anisotropy ratio to be obtained.

* Youngs 1968 also gives $\frac{C}{r}$-values for the case described in the footnote to table 16.1

16.2.3 Drain outflow method

By this method, the hydraulic conductivity at a site is determined by analysing a simultaneous record of drain discharge and watertable head on installed (pilot) drains. The analysis is based upon the assumption that a particular drain spacing equation is valid for the situation under investigation. In this section Hooghoudt's equation is used to illustrate the principles involved.

The watertable level should be measured in open wells, ideally constructed for maximum sensitivity to changes in watertable level. Where conditions permit, a series of such wells should be installed between the drains to observe the watertable profile in order to verify the validity of the analysis. The layout of a suitable trial, the location and construction of suitable wells is shown in figure 16.7.

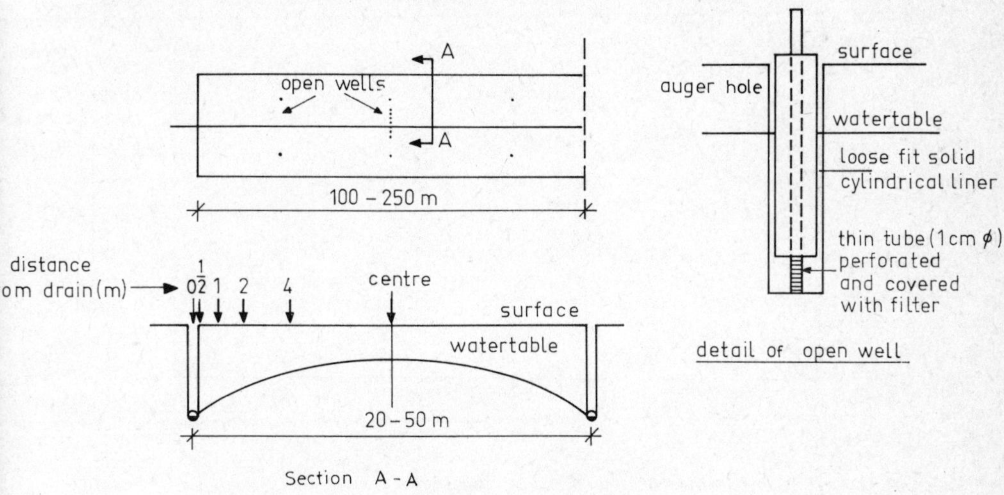

FIGURE 16.7 *Pilot trial layout and watertable observation wells*

The watertable level should ideally be recorded under controlled steady state conditions. However, this is rarely possible in practise and so a record has to be compiled from a series of continuous or even occasional single measurements (over say a few months) provided they are obtained during periods of recession (ie when infiltration of rain or irrigation water has ceased and watertable levels recede causing drain discharge to decline). Records obtained during periods of recharge to the watertable are often erratic, and should not be used.

It is also important to ensure that the watertable profile is approximately uniform along the whole drain length so that the measured discharge can be assumed to have resulted from a uniform inflow.

The discharge may be measured most simply by means of a bucket and stopwatch, though this is inaccurate for high discharges. A continuous record may also be obtained by using a drain flow meter (V notch weir, freely discharging orifice or purpose designed instrument), which can later be linked with a continuous record of the watertable elevation. The degree of sophistication of the measurements depends upon many factors although manual observation on a regular basis and using simple equipment has much to recommend it.

Analysis

The data given applies to a trial with drains at 2 m depth and a spacing of 50 m. Investigations have indicated that the soil is a highly uniform silty clay loam, underlain by a thick layer of impervious clay some 4·5 m from the surface.

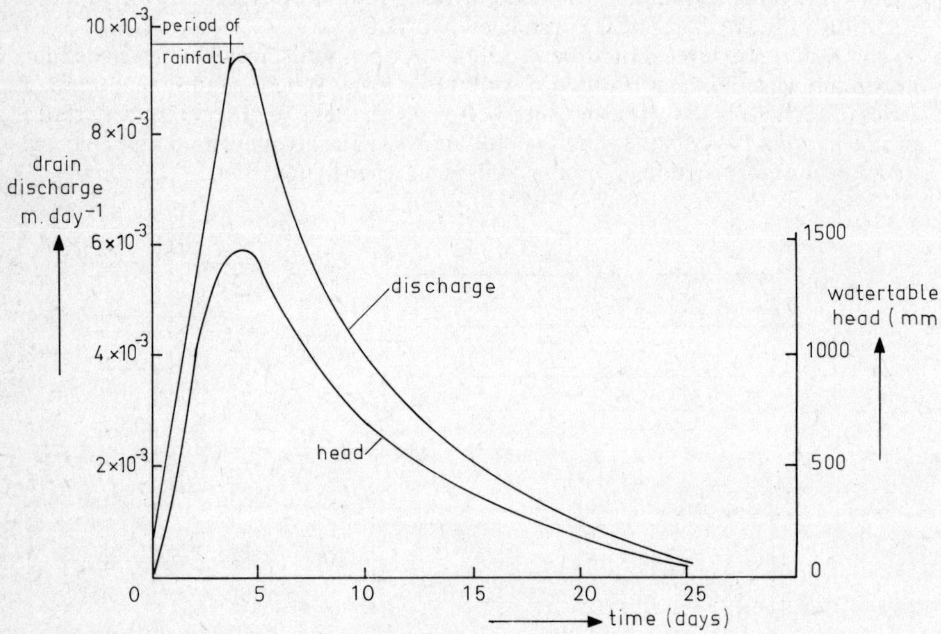

FIGURE 16.8 *Observed watertable head and discharge on a pilot trial*

Figure 16.8 illustrates typical features of the drainage response. Rainfall in the first five day period creates a discharge hydrograph which rises to a distinct peak value. The watertable also rises to a maximum value though the measured response is usually damped and lags behind the rising discharge hydrograph. Analysis of the recession period after the soil has been completely wetted and significant recharge to the watertable has ceased (1–3 days after rainfall) yields the relationship between drain discharge (q) and watertable head (h) as illustrated in figure 16.9.

It is assumed that in this case the relation between q and h obeys Hooghoudt's equation for drainspacing (eq. 7.6, section 7.3).

$$q = \frac{8K\,dh}{L^2} + \frac{4K\,h^2}{L^2}$$

The first function $\frac{8K\,dh}{2}$ describes the flow to the drain below drain depth. The second function $\frac{4K\,h^2}{L^2}$ describes the flow to the drain above drain depth. This

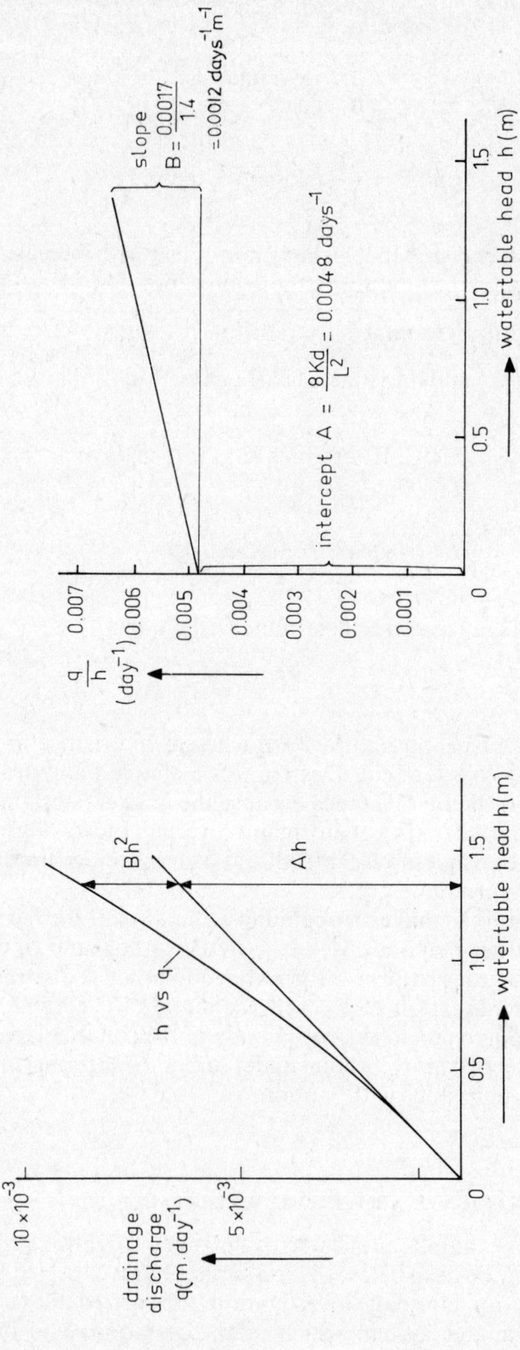

FIGURE 16.9 *Relationship between watertable head and drain discharge*

equation may be reformulated as:

$$q = Ah + Bh^2 \qquad \text{......(eq. 16.3A)}$$

$$\frac{q}{h} = A + Bh \qquad \text{......(eq. 16.3B)}$$

where $A = \frac{8Kd}{L^2}$ (intercept) and $B = \frac{4K}{L^2}$ (slope), see figure 16.9.

The plotting of q vs h becomes increasingly non-linear with increasing h due to the increasing contribution of the flow above the drain depth. The value of the unknown K may be more readily determined by plotting h against $\frac{q}{h}$. In figure 16.9 the intercept is 0·0048 days^{-1} and the slope is 0·0012 days^{-1}.m^{-1}. Thus:

$$\frac{4K}{L^2} = 0{\cdot}0012, \text{ from which } K = 0{\cdot}75 \text{ m.day}^{-1}$$

$$\frac{8Kd}{L^2} = 0{\cdot}0048, \text{ from which } d = 2 \text{ m}$$

(the value of d could of course also be determined from table 7.1).

Evaluation

(a) This example is rather straightforward whereas in practice the relationships are seldom quite so clear cut. Despite this a simple analytical procedure is recommended in the first instance because the answers obtained indicate the presence and characteristics of any major layering. Increasingly sophisticated analysis may be progressively introduced as a more accurate picture of the nature of the soil profile emerges.

(b) The K value obtained makes Hooghoudt's equation fit the recorded q, h data set. Its exact nature remains obscure, even when the real flow pattern exactly matches Hooghoudt's pattern as the K in Hooghoudt's equation is not unique but rather a combination of K_h and K_v, mainly the former. However, the estimated K value refers to the entire body of the soil involved in the flow to the drains. Use of this K value under other conditions and with other equations, should be done with caution.

16.3 FIELD MEASUREMENTS ABOVE THE WATERTABLE

These methods are, for example, applicable to a planned irrigation project where the present watertable may be deep below the soil surface but is expected to rise after the introduction of irrigation. Drainage investigations are carried out to assess the cost of a groundwater drainage system which might be required in future (*deferred drainage*).

16.3.1 Infiltrometer-method

An infiltration measurement is carried out, using either a single or a double ring infiltrometer. Measurements can be made at the surface or at different depths below the soil surface (on 'steps' in a profile pit see figure 16.1). The infiltration rate (I) of water into the soil is governed by Darcy's Law:

$$I = K_\theta \frac{h+z-P}{z} \qquad \text{......(eq. 16.4)}$$

where:
- I = infiltration rate ($m.day^{-1}$)
- K_θ = hydraulic conductivity of the soil at moisture content θ ($m.day^{-1}$)
- h = water depth on the soil surface (m)
- z = depth to the wetting front (m)
- P = soil water pressure at the wetting front, inside the transmission zone (m)

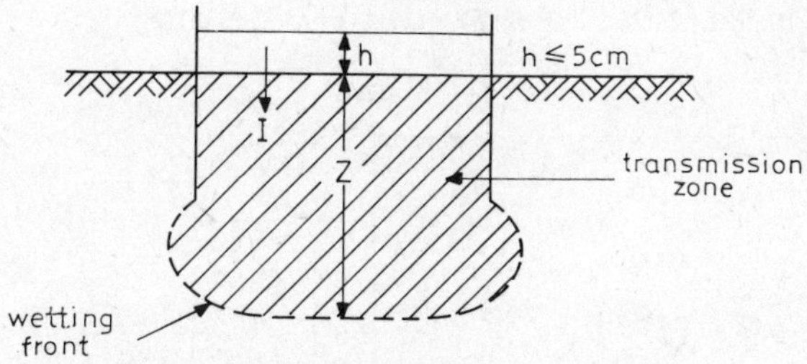

FIGURE 16.10 *Single ring infiltration measurement*

The moisture content in the transmission zone becomes virtually saturated so that $\theta \rightarrow \theta_{sat}$ and $P \rightarrow 0$ (this applies to most medium/heavy textured soils, but not for coarse textured soils). When $\theta \rightarrow \theta_{sat}$, also $K_\theta \rightarrow K_{sat}$. After prolonged infiltration, z becomes relatively large compared to $(h-P)$ so that the hydraulic gradient approaches unity $\left(\frac{h+z-P}{z} \rightarrow \frac{z}{z} \rightarrow 1\right)$.

So:

$$I_{final} = K_\theta \frac{h+z-P}{z} \sim K_{sat}$$

Evaluation

(a) This method measures K_v (vertical hydraulic conductivity).
(b) The method is simple but is not very accurate due to:

- I_{final} only approximates K_{sat} (see above)
- soil variability (small volume of soil involved; at least 3 replicates should be made to arrive at a reasonably reliable value)
- disturbance of soil when driving the infiltrometer ring into the soil.

(c) Large errors are to be expected in coarse textured (highly permeable) soils.
(d) The method may give erroneous results in swelling soils especially if water of low salt content is used in saline or sodic soils. Water having a salt content at least as high as the saturation extract of the soil should be used in the test.

16.3.2 Inverted augerhole method

An augerhole is made well into the layer in which the K-value is to be measured. The hole is then filled up to a certain level with water and its subsequent rate of fall is recorded as water flows from the hole into the surrounding soil.

The same type of equipment can be used as for the conventional augerhole method (see figure 16.2).

The readings should only be made after sufficient water has seeped into the surrounding soil to create a fairly thick nearly saturated zone. Using the same reasoning to that used in connection with the infiltrometer method, it is assumed that the rate of seepage from the hole into the surrounding soil approximately equals K_{sat} (again only really valid in medium to heavy textured soils). Then using the notation detailed in figure 16.11 one may derive for time t:

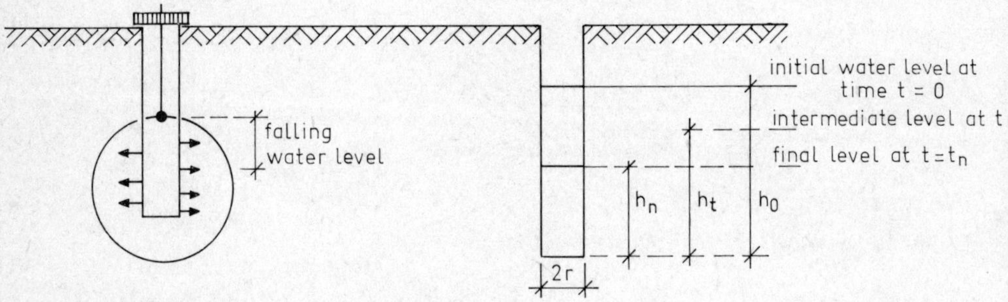

FIGURE 16.11 *The inverted augerhole method*

- outflow from hole:

$$Q = K(2\pi rh + \pi r^2) = 2K\pi r\left(h + \frac{r}{2}\right)$$

- from fall in waterlevel in hole:

$$Q = -\pi r^2 . \frac{dh}{dt}$$

Equating these two expressions for Q yields:

$$-\frac{2K}{r}dt = \frac{dh}{h + \frac{r}{2}}$$

Integration between the limits:

$$t = o, \ h = h_o$$
$$t = t, \ h = h_t$$

Gives:

$$\frac{2K}{r} t = \ln\left(h_o + \frac{r}{2}\right) - \ln\left(h_t + \frac{r}{2}\right)$$

or
$$K = \frac{1{\cdot}15\, r \log\left(h_o + \frac{r}{2}\right) - \log\left(h_t + \frac{r}{2}\right)}{t}$$

Recorded values of t and h_t may be plotted on log normal paper where they should form a straight line. The value of K may then be determined from the slope of this line (comparable to the determination of the reaction factor α in figure 7.11).

Evaluation

This method is essentially inhibited by the same drawbacks as the infiltrometer method (section 16.3.1). In fact, since the flow into the hole is more horizontal (through the walls) than vertical (through the bottom of the hole), there is less validity to the claim that the gradient ~ 1 than for the infiltrometer method (this claim only applies to vertical flow). The nature of the measured K-value, consequently, is also rather obscure.

16.4 Composed K-values

Field soils are often composed of layers of different hydraulic conductivity while individual layers are frequently highly anisotropic, especially below the root zone. Typical situations where these conditions occur are discussed in the following section. Composed K-values have been derived for flow through layered and anisotropic soils.

Horizontal flow through layered soil (see figure 16.12)

Darcy's Law:
$$Q_1 = K_1 D_1 i$$
$$Q_2 = K_2 D_2 i$$
$$Q = Q_1 + Q_2 = K_1 D_1 i + K_2 D_2 i$$

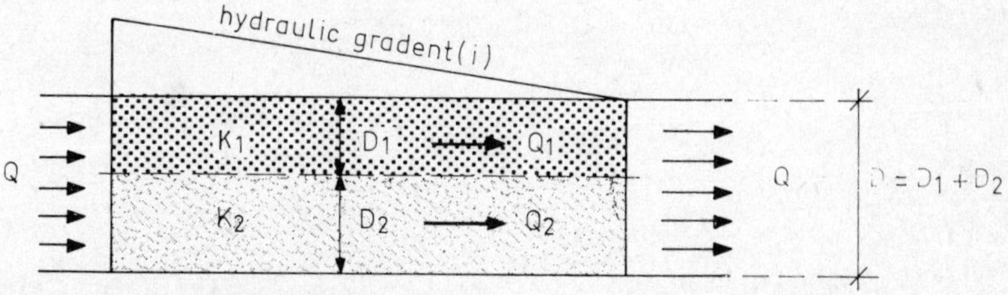

FIGURE 16.12 *Horizontal flow through layered soil*

Composed K for the two layers ($\bar{K}$):

$$Q = \overline{K}(D_1 + D_2)i$$

therefore

$$\overline{K} = \frac{K_1 D_1 + K_2 D_2}{D_1 + D_2}$$

or

$$\overline{KD} = K_1 D_1 + K_2 D_2 \qquad \text{...... (eq. 16.5)}$$

Vertical flow through layered soil (see figure 16.13)

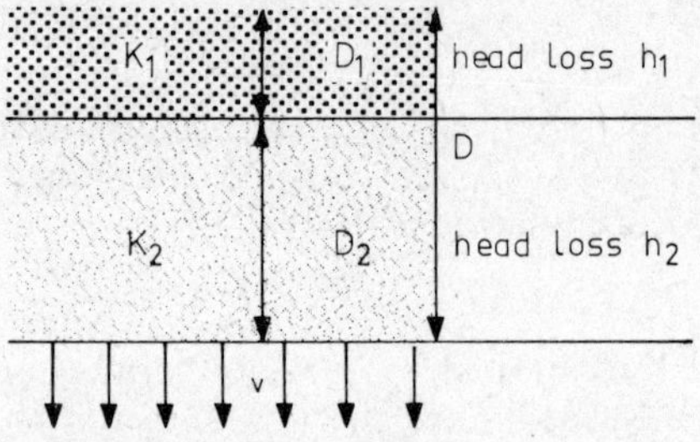

FIGURE 16.13 *Vertical flow through layered soil*

Darcy's Law:

$$v = K_1 \frac{h_1}{D_1}, \quad h_1 = \frac{vD_1}{K_1}$$

$$v = K_2 \frac{h_2}{D_2}, \quad h_2 = \frac{vD_2}{K_2}$$

thus

$$h = h_1 + h_2 = v\left(\frac{D_1}{K_1} + \frac{D_2}{K_2}\right)$$

also

$$v = \bar{K} \frac{h}{D_1 + D_2}, \quad h = v \frac{D_1 + D_2}{\bar{K}}$$

therefore

$$\bar{K} = \frac{D_1 + D_2}{\dfrac{D_1}{K_1} + \dfrac{D_2}{K_2}} = \frac{K_1 K_2 D}{D_1 K_2 + D_2 K_1} \qquad \text{...... (eq. 16.6)}$$

Radial flow and flow through anisotropic soil

In anisotropic soil quite distinct values of hydraulic conductivity exist in the horizontal direction (K_h) as compared to the vertical direction (K_v).

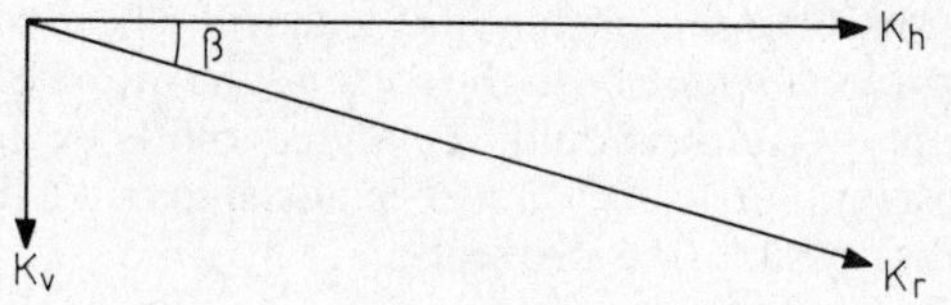

FIGURE 16.14 *Composed K-value for radial flow through anisotropic soil*

Where radial flow is under consideration (eg for converging flow to a drain) an appropriate hydraulic conductivity value may be calculated as:

$$K = (K_h K_v)^{\frac{1}{2}} \qquad \ldots\ldots\text{(eq. 16.7)}$$

Strictly this value of hydraulic conductivity will only apply to flow with a particular orientation β to the horizontal where $\tan\beta = \frac{K_v}{K_h}$. Transformations of scale may also be carried out (see section 16.7) which permit the soil to be treated as being isotropic.

16.5 SELECTION AND DENSITY OF TEST SITES

The test programme should preferably be conducted as part of a soil survey or against the background of an available soil map. Through detailed studies in sample areas, correlations may be established between measured K-values and those readily observable soil characteristics such as soil texture and structure. In general, coarse textured soils are more permeable than fine textured soils and useful correlations between K-value and soil texture may be established for the area under investigation (see table 16.3). Soils with identical texture may, however, have quite different K-values due to differences in structure. This applies especially to fine textured soils, some heavy clay soils for example having well developed structures and much higher K-values than those indicated in the table. Hydromorphic features such as colour, mottling, etc, are also often helpful is assessing K. All these soil characteristics can of course be observed most clearly in a pit although fair estimates of K may also be derived from examination of the soil obtained with the aid of an auger.

TABLE 16.3 *Typical ranges of hydraulic conductivities*

Textural class	*Typical range of K-values* *($m.day^{-1}$)*
very coarse (cS)	5 –10
coarse (fS, LS)	2 – 6
medium coarse (fLS, SL)	0·5 – 2
medium (fSL, SiL, L)	0·25– 0·8
medium fine (SCL, CL, SiCL)	0·10– 0·40
fine (SC, SiC, C)	<0·05– 0·10

These deduced K-values are used to supplement the network of measured K-values. The ultimate aim should, however, be to establish how the main soil characteristics influence the hydraulic conductivity of each soil type. Although not all soil characteristics used in soil mapping are relevant to drainage, many of them are, and so boundaries between drainability classes will often coincide with soil boundaries. In new projects where both the soil and the drainage survey still have to be done, the soil mapping criteria should be planned jointly by the soil surveyors and the drainage specialist (and possibly other specialists) to make the resulting maps as relevant to drainage as possible (see also section 15.3).

Test sites and depth

Initially, a random selection, for example, according to a grid pattern and using the site densities recommended in table 16.4 may be adopted. Once correlations with soil conditions have been established, a stratified selection on the basis of the known variation in soils within the area, should be followed. Even where delineation of drainability classes can largely be based on soil boundaries, a low density area wide measurement programme should be completed so as to check on the validity of the established correlation.

TABLE 16.4 *Densities of K measurements* (FAO 1980)

Soil condition	*Areas of land per test site*		
	Drainspacing about 30 m		*Drainspacing about 75 m*
Heterogeneous	<5 ha	→	10 to 15 ha
↓	↓		↓
Homogeneous	10 to 25 ha	→	40 to 75 ha

Separate tests should be carried out at each test site in at least two augerholes spaced a few metres apart. The hydraulic conductivities should be compared and, if similar in magnitude, accepted as valid. A third test should be carried out when there is no apparent explanation of the dissimilarity between results.

The required depth of investigation depends on the drain spacing and on the soil conditions. When the soil is homogeneous and permeable to great depth, the main drainage flow extends to depths of $\frac{1}{6}$L to $\frac{1}{4}$L beneath drain depth. In such soils the investigations should ideally extend down to eg 5 to 7·5 m below drain depth when L = 30 m. In most soils, however, the flow below drain depth is limited by poorly permeable layers and/or by the marked anisotropy ($K_V \ll K_h$) of the substrata (common in most alluvial soils). The general guidelines of table 16.5 are suggested which also take some account of the cost of deep investigations.

It should be emphasised that there is little point in collecting information on shallow soil layers in a situation where most of the drainage flow passes beneath the level of the drains. In these circumstances it is much wiser to reduce the number of tests whilst increasing the proportion of tests carried out at greater depth. Guidelines should also remain flexible, permitting increase/decrease of the density and/or the depth in the course of the survey when the results indicate this to be warranted. Guidelines may for example often be relaxed once reliable correlations with soil conditions have been established.

TABLE 16.5 *Depth of site investigations for drainage*

Soil conditions	*Depth of investigation in situations where there is no impermeable layer at shallower depth (m below drain depth)*		
	on all sites	*on 20% of sites*	*on 10% of sites*
	Depth shown in relation to spacing L		
Deep, reasonably permeable soil	$\frac{1}{10}$L	$\frac{1}{6}$L	
Stratified substrata			
– $K_V < K_h$	$\frac{1}{20}$L	$\frac{1}{10}$L	
– $K_V \ll K_h$	$\frac{1}{20}$L		$\frac{1}{10}$L
Poorly permeable substrata	$\frac{1}{20}$L		$\frac{1}{10}$L

16.6 PROCESSING OF HYDRAULIC CONDUCTIVITY SURVEY RESULTS

The preceding sections have emphasised the practical difficulty of evaluating hydraulic conductivity on a field scale. The methods which may be theoretically accurate in an ideal environment are much less accurate in practice. The number of tests is generally small with limited or no statistical significance. The values of hydraulic conductivity are, however, the foundation for rational drainage design and it is therefore absolutely essential that they should be critically appraised.

The values should be arranged in order of magnitude per soil type and the correlation should be analysed in the first instance to remove or adjust extreme values (ie a very high value of 5 m.day^{-1} in a clay known to have poor drainage, should be ignored).

Maps should be prepared at a suitable scale (1:5 000/10 000 for final design work) showing the K-values differentiated for K above and K below drain depth and the depths to the impermeable substratum. This permits areas which are uneconomic to drain to be identified. Areas having similar hydrophysical properties (similar K and depth to impermeable substratum) should then be delineated and drainspacings should be calculated per test site. The resulting spacings may later be rationalised into a limited range of spacings, 15, 20, 25, 30, 40 m etc. Generally on a large scheme single units of land (area ~ 60 Ha) surrounded by collector drains should be drained by one or at most two spacings.

16.7 DRAIN SPACING DETERMINATION IN ANISOTROPIC SOILS

As mentioned earlier, Hooghoudt's drain spacing equation can be interpreted to show that the total flow to the drain is composed of a component of flow occurring above the drainage base and a component of flow below the base (section 7.3). The flow above the drainage base is predominantly vertical so that a value of $K = K_v$ may be used in this part of the formula. Below the base, flow is largely horizontal and so a value of $K = K_h$ may be used for this part. The K value determined by means of the augerhole method is suited for this flow as this value refers to a comparable flow orientation (essentially horizontal with, possibly, a slight radial flow influence).

Alluvial soils in large river plains (Tigris, Indus, etc) are nearly always highly anistropic with K_h exceeding K_v by a factor of 20 to 30 (Boumans 1979). Anisotropy has a considerable influence on drain spacing. It may be dealt with by transforming scales in such a way that the soil may be considered to be isotropic. This transformation process can be done either by multiplying the vertical dimensions by the factor $R^{\frac{1}{2}}$, where $R = K_h/K_v$, or by dividing the horizontal dimentions by $R^{\frac{1}{2}}$. The 'isotropic' hydraulic conductivity value is calculated as $K' = (K_hK_v)^{\frac{1}{2}}$. The usual drain spacing equations may then be used to estimate the spacing. Transformation of the vertical scale is generally the most straightforward and will be illustrated for the case of figure 16.15.

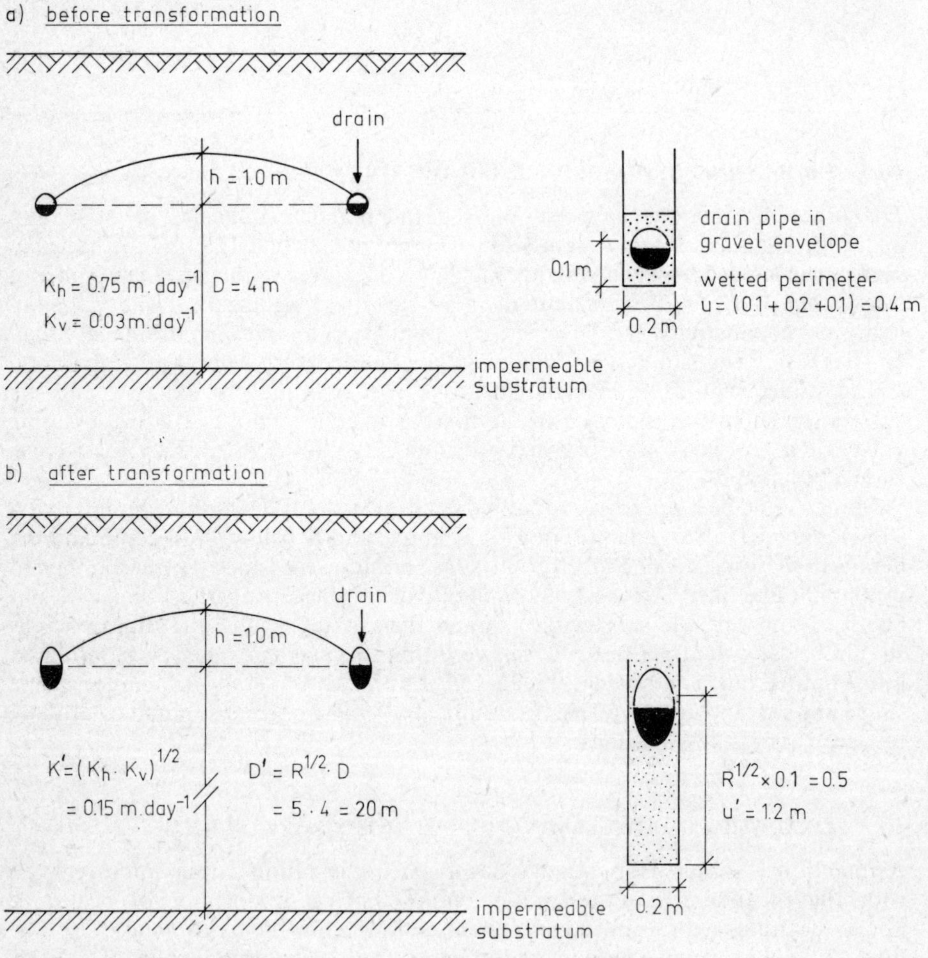

FIGURE 16.15 *Principal dimensions before and after transformation*

Example: basic design criteria $q = 2{\cdot}5\,mm.day^{-1}$
$h = 1{\cdot}0\,m$

The drainspacing is calculated using Hooghoudt's equation (eq. 7.9).

Steps: 1 Calculate the ratio $R = \frac{K_h}{K_v} = 25$, and $R^{\frac{1}{2}} = 5$.

2 Adjust the vertical scales by multiplying by $R^{\frac{1}{2}}$. Note that this transformation does not apply to the head h. Plot the transformed dimensions (figure 16.15).

3 Calculate the 'isotropic' K-value and apply Hooghoudt's formula:

$K' = (K_h K_V)^{\frac{1}{2}} = (0{\cdot}75 \times 0{\cdot}03)^{\frac{1}{2}} = 0{\cdot}15\,m.d^{-1}$

$$L^2 = \frac{8K'd'h}{q} \quad \text{where} \quad d' = \frac{D'}{\frac{8D'}{\pi L}\ln\frac{D'}{u'} + 1}$$

......(eq. 7.8)

$L^2 = \frac{8(0{\cdot}15)d'(1)}{0{\cdot}0025} = 480\,d'$; by trial and error $L = 50$; $d' = 5{\cdot}17\,m$

References and further reading

BEERS, WFJ VAN 1958: 'The augerhole method', Bulletin no 1, International Institute for Land Reclamation and Improvement, Wageningen

BOAST, CW and D KIRKHAM 1971: 'Augerhole seepage theory', *Proceedings of the Soil Science Society of America, 35*: 365–374

BOUMANS, J 1979: 'Drainage calculations in stratified soils using the anisotropic soil model to simulate conductivity conditions'. In: *Proceedings of the International Drainage Workshop*, J Wesseling (Ed.), publication 25 International Institute for Land Reclamation and Improvement, Wageningen

BUREAU OF RECLAMATION 1978: *Drainage Manual*, US Department of the Interior, Washington DC

ERNST, LF 1950: 'Een nieuwe formule voor het berekenen van de doorlaatfactor met de boorgatenmethode', Rapport Landbouwproefstation en Bodemkundig Instituut TNO, Groningen (mimeo in Dutch)

FAO 1976: 'Drainage Testing', Irrigation and Drainage Paper no 28, Food and Agriculture Organisation of the United Nations, Rome

FAO 1980: 'Drainage Design Factors', Irrigation and Drainage Paper 38, Food and Agriculture Organisation of the United Nations, Rome

HOOGHOUDT, SB 1936: *Bijdragen tot de kennis van enige natuurkundige grootheden van de grond*, Verslagen Landbouwkundig Onderzoek 42 (13) B: 449–541, The Hague (in Dutch)

ILRI 1974: *Drainage Principles and Applications*, volume III, publication 16, International Institute for Land Reclamation and Improvement, Wageningen

KIRKHAM, D 1945: 'Proposed method for field measurement of permeability of soil below the watertable', *Proceedings of the Soil Science Society of America, 11*: 93–99

KIRKHAM, D and CHM VAN BAVEL 1948: 'Theory of seepage into augerholes', *Proceedings of the Soil Science Society of America, 13*: 75–81

YOUNGS, EG 1968: 'Shape factors for Kirkham's piezometer method for determining the hydraulic conductivity of soil *in situ* for soils overlying an impermeable floor or infinity permeable stratum', *Soil Science, 106*: 235–237

Chapter 17 Cost evaluation

Land drainage is generally undertaken either to bring land into production or to increase the productivity of existing cultivated land. It represents a capital investment intended to result in future benefits and the viability of the drainage project may and should, be assessed like any other investment on the basis of sound economic analysis.

The scope of this chapter is restricted to the evaluation of the costs and benefits directly related to the improved drainage as would for example be relevant to the farmer who is considering improving the drainage of his land. In larger public drainage projects, other objectives (besides improved drainage) are often served by the project, while the costs and the benefits are often less straightforward economic or financial quantities. In the evaluation of these projects the contribution of the project to the national economy and welfare needs to be considered while for farmer's projects a more narrow view can be taken.

17.1 Basic information

In some cases the benefits of land drainage are so clearly in excess of the costs as to make detailed analysis unnecessary. In many cases, however, calculations are required so as to enable a decision to be reached about its worthwhileness. The information which must be established firstly is:

Costs

The construction costs of the project may be estimated on the basis of the recommended design or from contractors' estimates. In the case of most small projects, the funds will be required at the outset but larger projects are often phased over a period of years. In the case of phased development the amount and timing of the capital inputs must be established.

Operation costs arise on most projects for maintenance, re-moling, reconstruction of field ditches, and in the case of pumped drainage, for power. In some circumstances, eg shallow drainage with a bedding system, increased farm costs also arise because the farm operations take extra time and care.

In each case the amount and the timing of the future costs needs to be established.

The time period

For the purposes of cost evaluation a number of different time periods may be identified. Firstly, the drainage project will have an *actual life* which for a well maintained pipe drainage scheme may be anywhere between 50 and 100 years. Secondly, the project has an *economic life* corresponding to a notional lifetime at the end of which the project will be renewed. The inherent value of the project is at a maximum immediately after construction declining steadily thereafter until, at the end of the economic life, it has a zero terminal value. The economic lifetime could coincide with the actual lifetime but in economic terms future costs occurring after say, 30 years, are insignificant. For this reason the economic life of the project is generally taken to be between 20 and 30 years. Thirdly, the project has a *financial life*, often imposed by the lending agency. For example, a farmer having obtained a loan from the bank to finance the project may be required to pay back the loan over a ten year period. In this case, the financial life is very much shorter than either the economic or the actual life also the project has a terminal value at the end of its financial life which should be considered. The farmer's main concern is that the cash flow benefits equal or preferably exceed the costs arising during the financial life, the existance of a terminal value in this case being of little significance.

Financing

The cost of borrowed capital can be readily estimated knowing the period over which the loan has to be repayed and the interest rate imposed upon the loan. The cost of using available capital may be determined from the return likely to have been obtained from the alternative use to which the capital might have been put.

Benefits

The principal benefit of land drainage should of course be reflected in a yield increase but at the present time the response of a crop to drainage conditions is not fully understood. Also, the benefits of drainage vary in response to climatic factors, particularly rainfall which is also highly variable. In the UK, for example, the reduction of the crop yield due to poor drainage varies from zero in dry years to perhaps 50% in very wet years, a reduction which does not of course occur on well drained land. In fact the real timing of the benefits has a major effect on the overall cash flow but since these cannot be predicted in advance the financial evaluation has to be based upon an average benefit due to an average increase in crop yield (see section 17.4, example 1).

Important benefits may also arise from the improved workability and trafficability of drained land resulting, for example, in timely planting in spring (important for obtaining good yields) and more workable days (reducing the machinery costs). The same problems apply to evaluating these benefits as mentioned for the benefits due to yield increases.

With good drainage, the farmer is at all times assured of a reliable yield. This assurance can provide a firm foundation for initiating a whole train of consequential investments leading to benefits due to their mutual dependancy. The farmer may also obtain tax advantages by investing in improved drainage of his land, or, as happens in the UK, may obtain a direct grant from the Government.

17.2 DISCOUNTING

The costs and the benefits of a land drainage project occur at different times during the project period (the main costs occurring mostly during the construction phase, the benefits after the project has reached maturity). Economically, the timing of the costs and benefits is important as different interest periods apply. The *discounted cash flow* (DCF) method of analysis takes this into account. It is based upon the simple observation that a sum of money available now is worth more than the same sum of money in ten year's time. This is because the sum available now could be invested in, for example, a bank and increase in value due to accrued interest over the ten year period.

Numerically, discounting is the inverse of charging compound interest. For example, the sum of £100 now will with an interest rate of 10% increase to £110 in one year. Conversely, £110 a year later is equivalent to £100 today (the *present value*) the discount factor being 0·909. Discount factors for different interest rates and varying periods of time such as are presented in table 17.1, are readily available to enable any future sum to be converted to a present value. Thus a benefit of £90/ha arising in year 15 of the project has a present value at an interest of 10% $90 \times 0{\cdot}239 = £21{\cdot}51$. That is the sum of £21·51 invested now at a compound interest of 10% will increase in value to £90 in 15 years' time.

TABLE 17.1 *Discount factors for calculating present values of future sums*

Year	*Rate of interest*				
	3%	5%	10%	15%	20%
1	0·971	0·952	0·909	0·870	0·833
2	0·943	0·907	0·826	0·756	0·694
3	0·915	0·864	0·751	0·658	0·579
4	0·888	0·823	0·683	0·572	0·482
5	0·863	0·784	0·621	0·497	0·402
6	0·837	0·746	0·564	0·432	0·335
7	0·813	0·711	0·513	0·376	0·279
8	0·789	0·677	0·467	0·327	0·233
9	0·766	0·645	0·424	0·284	0·194
10	0·744	0·614	0·386	0·247	0·162
11	0·722	0·585	0·350	0·215	0·135
12	0·701	0·557	0·319	0·187	0·112
13	0·681	0·530	0·290	0·163	0·093
14	0·661	0·505	0·263	0·141	0·078
15	0·642	0·481	0·239	0·123	0·065
20	0·554	0·377	0·149	0·061	0·026

Interest rates in a free economy are closely linked to the rate of inflation and the prevailing interest rate tends to be 3–4% above the general level of inflation. Two strategies may be adopted in relation to the choice of interest rate in an inflationary economy. In the first, inflation may be neglected in which case an appropriate but rather low interest rate should be adopted. Current thinking indicates that this will be between 3% and 8%, although the final value chosen should be the subject of consultation with an economist. In the second strategy, the future costs and benefits are increased in value to take account of an assumed rate of inflation and the interest rate is then based upon the prevailing interest rates on opportunity capital. This latter procedure has been adopted for the first of two examples of cost analysis which are presented.

17.3 EVALUATION INDICES

The costs and the benefits may be evaluated on the basis of the following indices:

1 Net Present Value (NPV) being the difference between the present value of benefits and costs; a positive value clearly being desirable.

2 Benefit/Cost ratio (B/C) being the present value of benefits divided by the present value of costs; a value in excess of unity being desirable.

In many situations the benefits of land drainage are very difficult to quantify. In these cases, it is more meaningful to approach the problem from the opposite end to establish what the benefit would have to be in order to cover exactly the costs. The farmer or engineer can then consider whether this *break even benefit*, is likely to be obtained or even exceeded.

For large public projects, the *internal rate of return* (IRR) is often used as the evaluation index. The IRR reflects the interest made by the project on the investments in the project. This index is not relevant to most projects initiated by farmers who are mainly interested in financial cash flow considerations in order to meet predetermined interest rates on loans.

17.4 EXAMPLES OF COST EVALUATION OF DRAINAGE PROJECTS

Example 1

A farmer in the UK plans to drain an area of rough grazing land to enable it to be used as arable land.

Drainage system: pipe drainage + moling, construction cost £520/ha
remoling every seven years.............................. £ 20/ha
maintenance every five years............................ £ 15/ha
Financing: bank loan over ten years at 10% interest
Inflation: 5% over the life of the project
Benefits: expected additional profit on arable land........................ £ 90/ha

Calculations

The results of the calculations are embodied in table 17.2.

1 The future costs of moling and maintenance (column 3) and benefit (column 4) are increased by standard factors (column 5) to take account of the inflation rate of 5% (columns 7 and 8).
2 The annual repayment on the initial loan of £520/ha at 10% interest rate and over a ten year period is £84·60 (column 1)
3 The terminal value of the project at the end of the 20 year period considered is judged to be 20% of its installation cost or 0·2 × £520 = £104.
4 The procedure for evaluating costs/benefits in any one year is explained in relation to year 6. The annual repayment is £84.60. The benefit of £90/Ha (column 4) obtained at the outset has increased in value due to inflation to £120·60 (column 8). The excess of benefit over cost of £36·00 (column 9) has a present value of 0·56 × £36 = £20·20 (column 11).

Discussion

The NPV is £148·10/ha at the end of the ten year financing period and £566.30/ha at the end of the considered 20 year economic life time period (ignoring the terminal value). The present value of all costs arising over the first ten years is calculated as Σ (costs in columns 6 + 7 × discount factor) = 84·60 × 0·91 + 84·60 × 0·83 + ... (84·60 + 19·20) × 0·62 + ... = £555·44/ha. The present value of the costs after 20 years may

TABLE 17.2 *Cost and benefit analysis during the life of a drainage project*

Year	*Costs and benefits at actual prices (£/ha)*			*Actual future costs and benefits with 5% inflation (£/ha)*					*Present values of future sums (£/ha)*		
	loan	*moling main-tenance*	*benefits*	*inflation factor 5%*	*loan*	*moling main-tenance*	*benefits*	*benefits −costs*	*discount factor 10%*	*benefits −costs*	*NPV (column 11 summated)*
(1)	(2)	(3)	(4)	(5)	(6)	(7)	(8)	(9)	(10)	(11)	(12)
1	84·60		90·0	1·05	84·60		94·50	9·90	0·91	9·00	9·00
2	84·60		90·0	1·10	84·60		99·00	14·40	0·83	12·00	21·00
3	84·60		90·0	1·16	84·60		104·40	19·80	0·75	14·90	35·90
4	84·60		90·0	1·22	84·60		109·80	25·20	0·68	17·10	53·00
5	84·60	15	90·0	1·28	84·60	19·20	115·20	11·40	0·62	7·10	60·10
6	84·60		90·0	1·34	84·60		120·60	36·00	0·56	20·20	80·30
7	84·60	20	90·0	1·41	84·60	28·20	126·90	14·10	0·51	7·20	87·50
8	84·60		90·0	1·48	84·60		133·20	48·60	0·47	22·80	110·30
9	84·60		90·0	1·55	84·60		139·50	54·90	0·42	23·10	133·40
10	84·60	15	90·0	1·63	84·60	24·50	146·70	37·60	0·39	14·70	148·10
11			90·0	1·71			153·90	153·90	0·35	53·90	202·00
12			90·0	1·80			162·00	162·00	0·32	51·90	253·90
13			90·0	1·89			170·10	170·10	0·29	49·30	303·20
14		20	90·0	2·00		40·00	180·00	140·00	0·26	34·40	335·60
15		15	90·0	2·08		31·20	187·20	154·50	0·24	37·10	376·70
16			90·0	2·18			196·20	196·20	0·22	43·20	419·90
17			90·0	2·29			206·10	206·10	0·20	41·20	461·10
18			90·0	2·41			216·90	216·90	0·18	39·00	500·10
19			90·0	2·53			227·70	227·70	0·16	36·40	536·50
20		15	90·0	2·65		39·80	238·50	198·70	0·15	29·80	566·30
						terminal value	104·00	104·00	discounted terminal value	15·60	581·90

be similarly calculated at £579·30. The benefit/costs ratio after 10 years is (555.44 + 148.10)/555.44 = 1.27 while after 20 years and including the terminal value it is (579.30 + 581.90)/579.30 = 2.00. From these indices it appears that the project is financially viable in the short as well as in the long term.

The average annual *break even benefit*, ie the minimal benefit to be made to meet the costs, may be calculated as $\frac{\text{present value costs}}{\Sigma\ \text{discount factors}}$. Over the first ten years this gives £555.44/(0.91 + 0.83 + 0.75 + ... 0.39) = £555.44/6.14 = £90.46 which is clearly less than the real benefits which average £119.00 over this 10 year period (column 8).

The weakness of these analyses lies clearly in the assumed constant benefit in each of the project years. This figure itself must usually be based on inadequate basic data, while obviously it will vary from year to year in line with the sequence of occurrence of wet and dry years. When the first years after construction are wet the results of these analyses will be more favourable than when this early period is rather dry.

Example 2

In this example, the costs of two drainage systems for heavy land in Yugoslavia are compared. The first system is a regular pipe drainage scheme whilst the second is a parallel passable ditch system as described in section 4.2.1.

Calculations

The results of the calculations are embodied in table 17.3

1 Only those costs that differ between the two systems have been taken into account. This is justified in this case since the objective is simply to evaluate the additional benefits needed from the more expensive system.
2 Inflation may be ignored since it is likely to have a similar effect on both streams of costs considered.
3 The discount rate of 6% reflects the prevailing opportunity cost of capital in the country.
4 The costs are evaluated over a 30 year period since the present values for periods in excess of 30 years are relatively insignificant.

Basic cost differentials

System A (regular pipe drainage)

		Costs Dinar/ha
(Ai)	Cost of construction (plastics pipe spacing 15 m)...............	16,200
(Aii)	Pipe flushing 2 years after installation thereafter every 5 years at Dinar 0·5/m	650
(Aiii)	Annual replacement of some end pipes.........................	20

System B (parallel passable ditch system)

		Costs Dinar/ha
(Bi)	Cost of construction first year – (approx 50%)	3,000
	second year – (approx 25%)	1,500

third year –
(approx 25%) 1,500

(Bii) Field ditches
twice a year rotary ditching 100
part reconstruction every 5 years 1,000

(Biii) Extra farm costs
exra time and care in tillage, smoothing, transport, etc.. 400

TABLE 17.3 *Comparison of costs of two drainage systems*

Year	*Undiscounted costs (Dinar/ha)*		*Discounted costs (Dinar/ha)*	
	System A	*System B*	*System A*	*System B*
1	16,200	31,000	15,066	2,790
2	20	1,500	18	1,335
3	670	1,500	563	1,260
4	20	500	16	395
5	20	500	15	375
6	20	500	14	350
7	20	500	13	330
8	20	1,500	13	945
9	20	500	12	295
10	670	500	375	280
11	20	500	11	265
12	20	500	10	250
13	20	1,500	9	705
14	20	500	9	220
15	20	500	8	210
16	20	500	8	200
17	670	500	248	185
18	20	1,500	7	525
19	20	500	7	165
20	20	500	6	155
21	20	500	6	145
22	20	500	6	140
23	20	1,500	5	390
24	670	500	168	125
25	20	500	5	115
26	20	500	4	110
27	20	500	4	105
28	20	1,500	4	300
29	20	500	4	92
30	20	500	4	88
			16,638	12,845

Discussion

System A costs Dinar 3793/ha more than System B over thirty years which amounts to an annual average undiscounted cost of Dinar 276/ha. The net profit on wheat or maize is approximately Dinar 2/Kg so that System A would have to produce 138 Kg/ha more yield than System B in order to be as financially attractive.

References and further reading

DASPUTA, AK and DW PEARCE 1978: *Cost–Benefit Analysis*, MacMillan, London

GRANT, EL, WG IRESON and RS LEAVENWORTH 1976: *Principles of Engineering Economy*, Ronald, New York

JAMES, LD and RR LEE 1971: *Economics of Water Resource Planning*, McGraw-Hill, New York

MAFF 1977: 'Drainage and the Economy of the Farm', Technical Management Note no 23. Ministry of Agriculture, Fisheries and Food, London

MISHAN, EJ 1972: *Cost-Benefit Analysis*, George Allen and Unwin, London

Index

NOTES